Communication and Cybernetics 12

Editors: K.S.Fu W.D.Keidel H.Wolter

J. D. Markel A. H. Gray, Jr.

Linear Prediction of Speech

With 129 Figures

Springer-Verlag
Berlin Heidelberg New York 1976

John D. Markel
Speech Communications Research Laboratory, Inc.,
Santa Barbara, California 93109, USA

Augustine H. Gray, Jr.
Department of Electrical Engineering and Computer Science,
University of California Santa Barbara, California 93106, USA
and Speech Communications Research Laboratory, Inc.
Santa Barbara, California 93109, USA

ISBN 3-540-07563-1 Springer-Verlag Berlin Heidelberg New York
ISBN 0-387-07563-1 Springer-Verlag New York Heidelberg Berlin

Library of Congress Cataloging in Publication Data. Markel, John D. 1943- Linear prediction of speech. (Communication and cybernetics; 12). Bibliography: p. Includes index. 1. Speech processing systems. 2. Speech synthesis. I. Gray, Augustine H., 1936- joint author. II. Title. TK7882.S65M37. 621.38. 75–40003

Printed in Germany.
Typesetting, printing, and binding: Oscar Brandstetter Druckerei KG, Wiesbaden.

To Virginia and Averill

Preface

During the past ten years a new area in speech processing, generally referred to as linear prediction, has evolved. As with all scientific research, results did not always get published in a logical order and terminology was not always consistent.

In mid-1974, we decided to begin an extra hours and weekends project of organizing the literature in linear prediction of speech and developing it into a unified presentation in terms of content and terminology. This effort was completed in November, 1975, with the contents presented herein.

If there are two words which describe our goals in this book, they are unification and depth. Considerable effort has been spent on showing the interrelationships among various linear prediction formulations and solutions, and in developing extensions such as acoustic tube models and synthesis filter structures in a unified manner with consistent terminology. Topics are presented in such a manner that derivations and theoretical details are covered, along with Fortran subroutines and practical considerations. Using this approach we hope to have made the material useful for a wide range of backgrounds and interests.

The organization of material reflects our particular approach to presenting many important topics. The material itself reflects considerable research performed primarily by a rather small number of colleagues: Drs. Bishnu Atal, Fumitada Itakura, John Makhoul, Shuzo Saito, and Hisashi Wakita, as can be seen from the references. It is very gratifying to all of us who have been involved in developing this area to see the significant impact that linear prediction techniques have made in speech processing, and to see the rather large continuing research interest in this area.

Valuable technical comments and criticisms of certain portions of the manuscript were obtained from Drs. B. S. Atal, J. Makhoul, and H. Wakita. We are also indebted to Dr. Wakita for his assistance in the preparation of Chapter 4 and for providing us with several illustrations and examples.

We are grateful to the following individuals who provided us with information and illustrations for use in the book: B.S. Atal, R.G. Crichton, W.B. Kendall, G.E. Kopec, S.S. (McCandless) Seneff, J. Makhoul, L.R. Morris, A.V. Oppenheim, L.L. Pfeifer, M.R. Sambur, L.R. Rabiner, C.J. Weinstein, and J. Welch. We would also like to thank D. Cohen for his assistance in preparing the high-contrast computer program listings.

For the countless hours of careful typing, and willingness to work within our rather strict deadlines, we are indebted to Marilyn Berner. The excellent layout work and drafting by David Sangster for the large number of original figures in this book is also greatly appreciated. We would also like to thank Dr. Beatrice Oshika for her careful proofreading of the final manuscript and page proofs.

A special thanks is given to Dr. David Broad who read the complete manuscript for both content and grammar. His many valuable suggestions have greatly improved the logic and readability of the book.

Finally, the authors are indebted to Dr. June E. Shoup-Hummel, under whose guidance SCRL has provided a stimulating environment for studying the myriad problems that focus upon the basic nature of speech.

Santa Barbara, California
January 1, 1976

J. D. Markel
A. H. Gray, Jr.

Table of Contents

1. Introduction . . . 1
   1.1 Basic Physical Principles . . . 1
   1.2 Acoustical Waveform Examples . . . 3
   1.3 Speech Analysis and Synthesis Models . . . 5
   1.4 The Linear Prediction Model . . . 10
   1.5 Organization of Book . . . 16

2. Formulations . . . 18
   2.1 Historical Perspective . . . 18
   2.2 Maximum Likelihood . . . 20
   2.3 Minimum Variance . . . 23
   2.4 Prony's Method . . . 25
   2.5 Correlation Matching . . . 31
   2.6 PARCOR (Partial Correlation) . . . 32
      2.6.1 Inner Products and an Orthogonality Principle . . . 35
      2.6.2 The PARCOR Lattice Structure . . . 38

3. Solutions and Properties . . . 42
   3.1 Introduction . . . 42
   3.2 Vector Spaces and Inner Products . . . 44
      3.2.1 Filter or Polynomial Norms . . . 46
      3.2.2 Properties of Inner Products . . . 47
      3.2.3 Orthogonality Relations . . . 48
   3.3 Solution Algorithms . . . 50
      3.3.1 Correlation Matrix . . . 51
      3.3.2 Initialization . . . 53
      3.3.3 Gram-Schmidt Orthogonalization . . . 54
      3.3.4 Levinson Recursion . . . 55
      3.3.5 Updating $A_m(z)$ . . . 56
      3.3.6 A Test Example . . . 57
   3.4 Matrix Forms . . . 58

4. Acoustic Tube Modeling . . . 60
   4.1 Introduction . . . 60
   4.2 Acoustic Tube Derivation . . . 61
      4.2.1 Single Section Derivation . . . 63
      4.2.2 Continuity Conditions . . . 65

4.2.3 Boundary Conditions 68
4.3 Relationship between Acoustic Tube and Linear Prediction 71
4.4 An Algorithm, Examples, and Evaluation 77
4.4.1 An Algorithm 78
4.4.2 Examples 80
4.4.3 Evaluation of the Procedure 82
4.5 Estimation of Lip Impedance 84
4.5.1 Lip Impedance Derivation 84
4.6 Further Topics 88
4.6.1 Losses in the Acoustic Tube Model 88
4.6.2 Acoustic Tube Stability 90

5. Speech Synthesis Structures 92
5.1 Introduction 92
5.2 Stability 93
5.2.1 Step-up Procedure 94
5.2.2 Step-down Procedure 95
5.2.3 Polynomial Properties 98
5.2.4 A Bound on $|F_m(z)|$ 99
5.2.5 Necessary and Sufficient Stability Conditions 101
5.2.6 Application of Results 102
5.3 Recursive Parameter Evaluation 103
5.3.1 Inner Product Properties 103
5.3.2 Equation Summary with Program 110
5.4 A General Synthesis Structure 113
5.5 Specific Speech Synthesis Structures 117
5.5.1 The Direct Form 118
5.5.2 Two-Multiplier Lattice Model 118
5.5.3 Kelly-Lochbaum Model 121
5.5.4 One-Multiplier Models 123
5.5.5 Normalized Filter Model 123
5.5.6 A Test Example 126

6. Spectral Analysis 129
6.1 Introduction 129
6.2 Spectral Properties 130
6.2.1 Zero Mean All-Pole Model 130
6.2.2 Gain Factor for Spectral Matching 130
6.2.3 Limiting Spectral Match 132
6.2.4 Non-uniform Spectral Weighting 134
6.2.5 Minimax Spectral Matching 136
6.3 A Spectral Flatness Model 139
6.3.1 A Spectral Flatness Measure 139
6.3.2 Spectral Flatness Transformations 141
6.3.3 Numerical Evaluation 142
6.3.4 Experimental Results 143
6.3.5 Driving Function Models 144

6.4 Selective Linear Prediction 146
6.4.1 Selective Linear Prediction (SLP) Algorithm 148
6.4.2 A Selective Linear Prediction Program 149
6.4.3 Computational Considerations 150
6.5 Considerations in Choice of Analysis Conditions 151
6.5.1 Choice of Method 151
6.5.2 Sampling Rates 153
6.5.3 Order of Filter 154
6.5.4 Choice of Analysis Interval 156
6.5.5 Windowing 157
6.5.6 Pre-emphasis 158
6.6 Spectral Evaluation Techniques 159
6.7 Pole Enhancement 161

7. Automatic Formant Trajectory Estimation 164
7.1 Introduction 164
7.2 Formant Trajectory Estimation Procedure 165
7.2.1 Introduction 165
7.2.2 Raw Data from $A(z)$ 167
7.2.3 Examples of Raw Data 169
7.3 Comparison of Raw Data from Linear Prediction and Cepstral Smoothing 172
7.4 Algorithm 1 176
7.5 Algorithm 2 180
7.5.1 Definition of Anchor Points 181
7.5.2 Processing of Each Voiced Segment 181
7.5.3 Final Smoothing 183
7.5.4 Results and Discussion 184
7.6 Formant Estimation Accuracy 185
7.6.1 An Example of Synthetic Speech Analysis 185
7.6.2 An Example of Real Speech Analysis 187
7.6.3 Influence of Voice Periodicity 188

8. Fundamental Frequency Estimation 190
8.1 Introduction 190
8.2 Preprocessing by Spectral Flattening 191
8.2.1 Analysis of Voiced Speech with Spectral Regularity 191
8.2.2 Analysis of Voiced Speech with Spectral Irregularities 193
8.2.3 The STREAK Algorithm 197
8.3 Correlation Techniques 199
8.3.1 Autocorrelation Analysis 200
8.3.2 Modified Autocorrelation Analysis 201
8.3.3 Filtered Error Signal Autocorrelation Analysis 203
8.3.4 Practical Considerations 206
8.3.5 The SIFT Algorithm 206

9. Computational Considerations in Analysis . . . 212
9.1 Introduction . . . 212
9.2 Ill-Conditioning . . . 213
9.2.1 A Measure of Ill-Conditioning . . . 214
9.2.2 Pre-emphasis of Speech Data . . . 216
9.2.3 Prefiltering before Sampling . . . 216
9.3 Implementing Linear Prediction Analysis . . . 217
9.3.1 Autocorrelation Method . . . 217
9.3.2 Covariance Method . . . 219
9.3.3 Computational Comparison . . . 220
9.4 Finite Word Length Considerations . . . 222
9.4.1 Finite Word Length Coefficient Computation . . . 223
9.4.2 Finite Word Length Solution of Equations . . . 224
9.4.3 Overall Finite Word Length Implementation . . . 225

10. Vocoders . . . 227
10.1 Introduction . . . 227
10.2 Techniques . . . 229
10.2.1 Coefficient Transformations . . . 229
10.2.2 Encoding and Decoding . . . 233
10.2.3 Variable Frame Rate Transmission . . . 235
10.2.4 Excitation and Synthesis Gain Matching . . . 239
10.2.5 A Linear Prediction Synthesizer Program . . . 242
10.3 Low Bit Rate Pitch Excited Vocoders . . . 245
10.3.1 Maximum Likelihood and PARCOR Vocoders . . . 246
10.3.2 Autocorrelation Method Vocoders . . . 249
10.3.3 Covariance Method Vocoders . . . 255
10.4 Base-Band Excited Vocoders . . . 260

11. Further Topics . . . 263
11.1 Speaker Identification and Verification . . . 263
11.2 Isolated Word Recognition . . . 265
11.3 Acoustical Detection of Laryngeal Pathology . . . 267
11.4 Pole-Zero Estimation . . . 271
11.5 Summary and Future Directions . . . 275

References . . . 278

Subject Index . . . 285

1. Introduction

Many different models have been postulated for quantitatively describing certain factors involved in the speech process. It can be stated with certainty that no single model has been developed which can account for all of the observed characteristics in human speech (nor would one probably desire such a model because of its inevitable complexity). A basic criterion of modeling is to find mathematical relations which can be used to represent a limited physical situation with a minimum of complexity and a maximum of accuracy. One of the most successful models of acoustical speech behavior is the *linear speech production model* developed by Fant [1960]. This model will be referred to throughout as the speech production model.

In recent years the mathematical technique of *linear prediction* has been applied to the problem of modeling speech behavior. The linear prediction model can be related to the speech production model, with the significant feature that the parameters of the speech production model are easily obtained using linear mathematics.

In this chapter, the linear prediction model is developed and the relationship to the speech production model is shown. A starting point which can be used in developing the speech production model is speech physiology. Speech physiology is the springboard for many different areas which are relevant to a better understanding of speech. The discussions here, however, will consider only briefly the physical principles of speech. The main focus in this book will be the acoustical properties of speech. Detailed discussions of both the physiological and acoustical characteristics of speech are contained in Fant [1960] and Flanagan [1972], and in Peterson and Shoup [1966a, 1966b].

1.1 Basic Physical Principles

The acoustical speech waveform is an *acoustic pressure wave* which originates from voluntary physiological movements of the structures shown in Fig. 1.1. Air is expelled from the lungs into the trachea and then forced between the *vocal folds*. During the generation of *voiced sounds* such as /i/* in *e*ve, the air pushed

---

* The symbol /·/ is used to denote the phoneme, a basic linguistic unit. Definitions of the phonemes of the General American dialect are given in Flanagan [1972, pp. 15–22].

toward the lips from the lungs causes the vocal folds to open and close at a rate dependent upon the air pressure in the trachea and the physiological adjustment of the vocal folds. This adjustment includes changes in the length, thickness, and tension of the vocal folds. The greater the tension, the higher the perceived *pitch* or acoustically measured *fundamental frequency* of the voice. The opening between the vocal folds is defined as the *glottis*. The subglottic air pressure and the time variations in glottal area determine the volume velocity of glottal air flow *(glottal volume velocity waveform)* expelled into the *vocal tract*. The rate at which

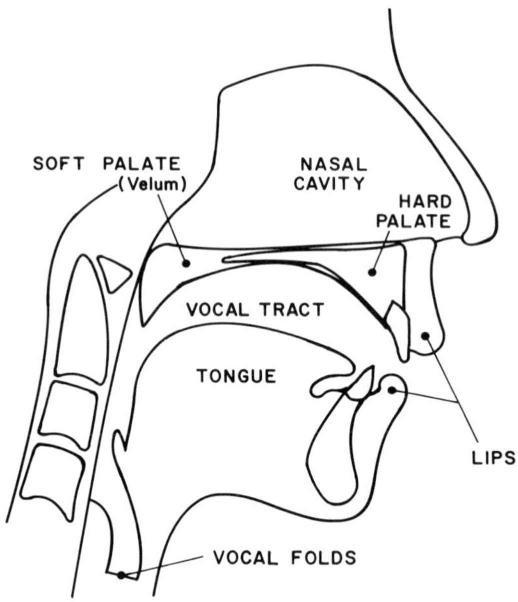

Fig. 1.1 Cross-sectional view of the vocal mechanism showing some of the major anatomical structures involved in speech production.

the glottis opens and closes can be approximately measured acoustically as the inverse of the time interval between observed periods of the acoustic pressure wave. It is the glottal volume velocity waveform that defines the acoustic energy input or driving function to the vocal tract.

The vocal tract is a non-uniform acoustic tube which extends from the glottis to the lips and varies in shape as a function of time. The major anatomical components causing this time-varying change are the *lips*, *jaw*, *tongue*, and *velum*. For example, the cross-sectional area of the lip opening can be varied over the considerable range of 0 cm^2 with the lips closed to about 20 cm^2 with the jaw and lips open. During the generation of the non-nasal sounds, the velum closes off the vocal tract from the *nasal cavity*. The nasal cavity constitutes an additional acoustic tube for sound transmission used in the generation of the nasal sounds /n/, /m/, and /ŋ/ as in ru*n*, ru*m*, and ru*ng*, respectively.

Unvoiced sounds such as /f/ in *fish* are generated by voluntarily holding the vocal folds open, forcing air past them, and then using the articulators to create a constriction (such as setting the upper teeth on the lower lip for *fish*). With both a constriction and vocal fold vibration, *voiced fricatives* such as /v/ in *van* are generated. *Plosive sounds* such as /p/ in *pop* are generated by building up air pressure in the mouth and then suddenly releasing the air.

1.2 Acoustical Waveform Examples

To illustrate the acoustical implications of speech production in both the time domain and frequency domain, the phrase "linear prediction" was spoken into a microphone, recorded on audio tape, and then analyzed. Figure 1.2A shows the acoustic waveform of the utterance on a scale of amplitude versus time. This waveform was obtained by low-pass filtering the tape recorder output to a 5 kHz bandwidth, performing analog-to-digital conversion into a computer system at a sampling rate of 10 kHz, and displaying the speech samples with linear interpolation, giving the appearance of a continuous display. By displaying a long segment of speech, the general envelope characteristics of the time-varying waveform can be seen.

In Figs. 1.2B and C, 25.6 ms intervals from a voiced and unvoiced portion of the utterance are shown. The waveform in Fig. 1.2B is nearly periodic. The distance between major peaks shows the pitch period P of the glottal vibrations. The waveform of Fig. 1.2C exhibits no discernible pitch period since the sound /ʃ/ (from predic*ti*on) is produced by turbulence noise generated by directing the air stream past a constriction formed by the tongue and teeth.

Figure 1.2D shows one pitch period of the waveform in Fig. 1.2B. The frequency of these decaying oscillations (the reciprocal of the oscillation period) in this example determines the approximate location of the major resonance of the vocal tract in the frequency domain, while the rate of decay approximates the bandwidth of the resonance.

Frequency domain representations of this utterance from *sonograms* (also referred to as *voiceprints* or *spectrograms*) using a Kay Sonograph are shown in Figs. 1.2E and F, using the wideband and narrowband filters, respectively. The sonogram shows speech energy as a parameter on a scale of continuous frequency versus time. The time scale in Fig. 1.2A has been adjusted to match that of the sonogram. During voiced regions, the dark bars indicate the locations of the resonances as functions of time. The voiced regions show vertical striations, corresponding to the epoch or beginning of each pitch period. During unvoiced intervals, the dark areas indicate major concentrations of energy. With the wideband sonograph filter, pitch-period resolution is obtained in the time domain, but a larger amount of averaging or smearing in the frequency domain occurs. With the narrowband sonograph filter, frequency resolution is obtained at the expense of time resolution.

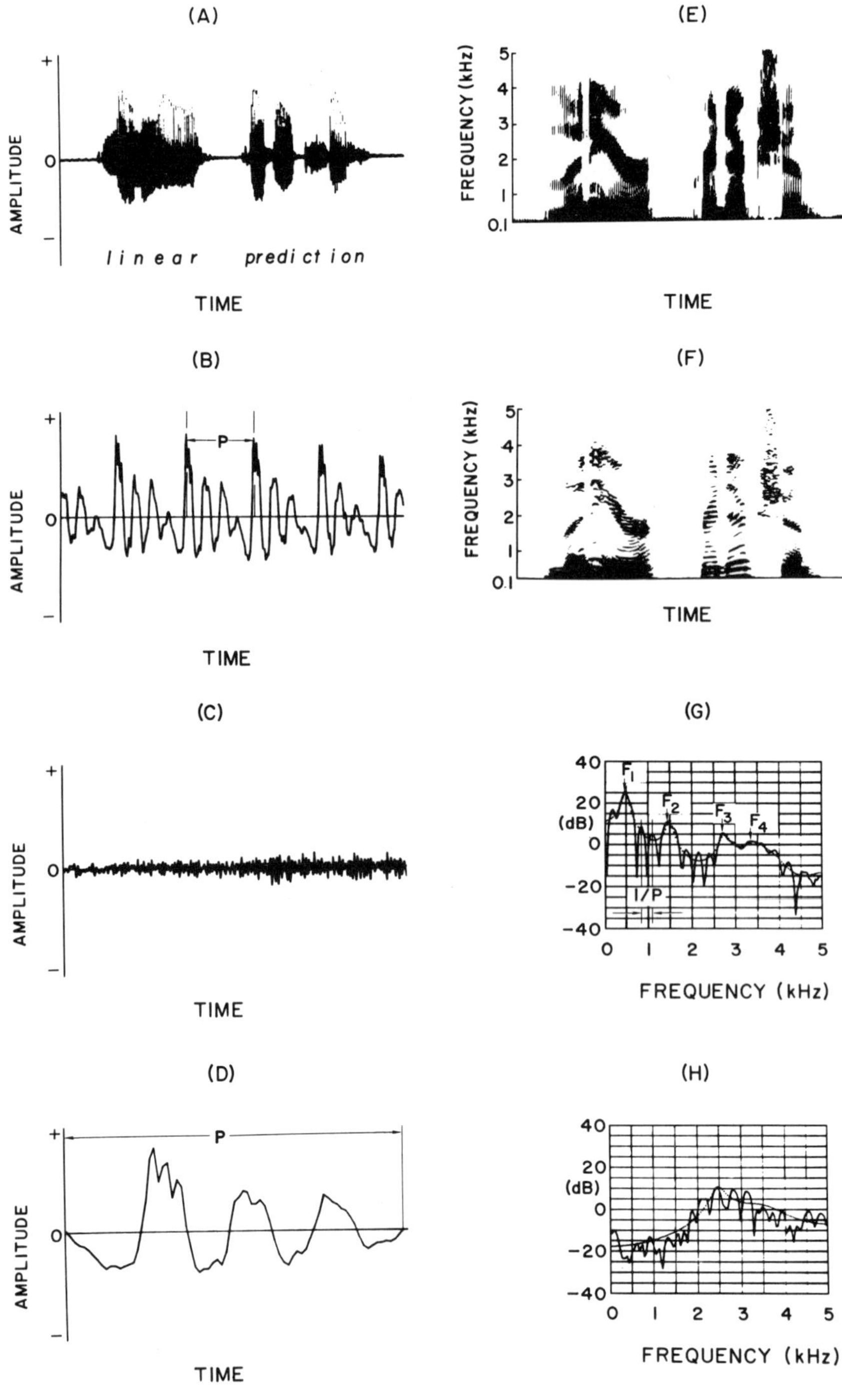

Fig. 1.2 Time and frequency domain representations of the utterance "linear prediction".

With the narrowband filter, the harmonic structure of voiced speech is obtained; that is, the narrow horizontal lines indicate the harmonics of the fundamental frequency. During unvoiced sounds no discernible harmonic structure is evident.

A final important display is based upon the log spectrum. It shows log amplitude (in units of dB) versus frequency. The log spectra for the voiced and unvoiced intervals for central portions of the /ɪ/ and /ʃ/ are shown in Figs. 1.2G and H, respectively. In addition, a smooth envelope is superimposed upon each log magnitude spectrum. The locations of the major peaks of the envelope in Fig. 1.2G, labeled as F_1, F_2, F_3, and F_4, define estimates of the *formant frequency* locations. (For now, it is sufficient to consider formants as resonances of the vocal tract.)

It should be clear from these figures that the acoustic speech waveform has a very complex structure. In an attempt to understand more about the speech process, models can be postulated, and then tested for various conditions. Ideally, it is desirable to have models that are both linear and time-invariant. Unfortunately, the human speech mechanism does not precisely satisfy either of these properties. Speech is a continually time-varying process. In addition, the glottis is not uncoupled from the vocal tract, which results in non-linear characteristics [Flanagan, 1968]. However, by making reasonable assumptions, it is possible to develop linear time-invariant models over short intervals of time for describing important speech events.

Reasonably short time intervals of the order of tens of milliseconds were illustrated in Figs. 1.2B and C. The voiced sound is nearly periodic, with each period being quite similar to the previous one. The unvoiced sound shows no discernible pitch periodic behavior. The frequency domain representation for the voiced sound shows cyclic behavior every $1/P$ units of frequency and that for the unvoiced sound shows random behavior with major energy at around 3 kHz.

The model of speech production to be introduced, separates the smoothed envelope structure from the actual spectrum, and attaches a physiological significance to each of the components of the model. Later it will be seen that the smooth envelope structure shown in Figs. 1.2G and H is easily obtainable by linear prediction of speech (in fact, the smoothed envelopes shown were automatically computed using linear prediction analysis of the time-domain waveforms shown in Figs. 1.1 B and C).

1.3 Speech Analysis and Synthesis Models

A linear model of speech production was developed by Fant in the late 1950's [Fant, 1960]. The speech production model is shown in Fig. 1.3. The assumptions behind this model are covered in detail by Fant [1960] and Flanagan [1972, pp. 214–246].

The glottal volume velocity waveform $u_G(t)$ is modeled as the output of a two-pole low-pass filter with an estimated cutoff at about 100 Hz. The filter input $e(t)$ is an impulse train with period P for voiced sounds, and random noise having

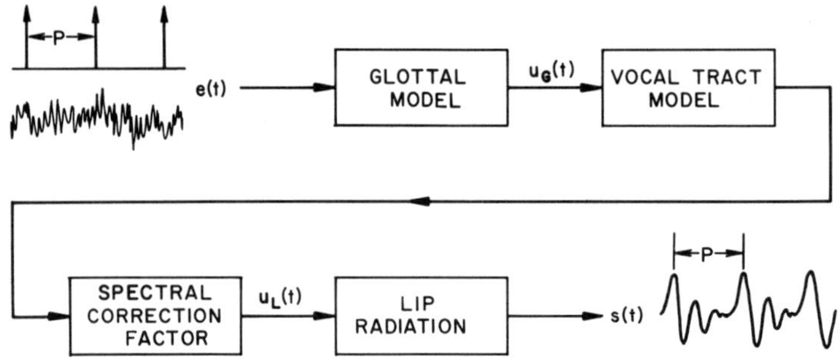

Fig. 1.3 A linear speech production model.

a flat spectrum for unvoiced sounds. Note that this model already represents a rather special case in that no provisions are made for mixing the inputs to simulate *voiced frication* or to couple in a filter branch for modeling nasal sounds. The vocal tract model which results is an all-pole model consisting of a cascade of a small number of two-pole resonators. Each resonance is defined as a *formant* with a corresponding center frequency and bandwidth.

A more accurate model would consist of an additional infinite number of resonances whose main effect at the lower frequencies is to raise the spectral level. Therefore, when only the lower-frequency behavior of the system, e.g., the important part of the audio range from 20 Hz to several kHz, is to be accurately modeled, this shaping can be accounted for by a *higher pole correction factor* which represents the lower-frequency effects of all the higher poles nearly independently of the detailed behavior of the higher poles. The volume velocity waveform at the lips $u_L(t)$ is transformed into an *acoustic pressure waveform* some distance away from the lips (representing the speech waveform $s(t)$), by a lip radiation model. The mathematical justification and detailed derivation of this model are presented in Fant [1960] and Flanagan [1972], while Flanagan presents the results of some carefully conducted experiments on acoustic radiation that corroborate this model.

The above model can be described in z-transform notation [Jury, 1964] for computer implementation by the following equation:

$$S(z)=E(z)G(z)V(z)L(z) \quad \text{[linear speech production model]} \tag{1.1}$$

where

$$S(z)\leftrightarrow s(nT)=s(t)\Big|_{t=nT} \tag{1.2}$$

defines the correspondence between the continuous waveform $s(t)$, the sampled data or discrete signal $s(nT)$ obtained by sampling $s(t)$ every T units of time, and the z-transform $S(z)$. As a shorthand notation, it is customary to assume a normalized sampling interval, $T=1$, so that $s(n)$ describes the sampled version of $s(t)$. A similar correspondence is used for the other variables. The driving function to the glottal shaping model is $E(z) \leftrightarrow e(n)$, a train of scaled unit samples, spaced by the pitch period $P=IT$ where I is a positive integer, i.e.,

$$E(z)=\sigma \sum_{n=0}^{\infty} (z^{-I})^n$$
$$= \frac{\sigma}{1-z^{-I}} \tag{1.3}$$

for $|z|>1$. The glottal shaping model $G(z)$ is of the form

$$G(z)=\frac{1}{(1-e^{-cT}z^{-1})^2} \tag{1.4}$$

and the lip radiation model $L(z)$ is of the form

$$L(z)=1-z^{-1}. \tag{1.5}$$

These are simplifying assumptions that do not necessarily predict the actual behavior of a particular speech sample.

The all-pole vocal tract model $V(z)$, consisting of K formants, is described by

$$V(z)=\frac{1}{\prod_{i=1}^{K}[1-2e^{-c_iT}\cos(b_iT)z^{-1}+e^{-2c_iT}z^{-2}]} \tag{1.6}$$

where the i*th* formant frequency and bandwidth are computed from $F_i=b_i/2\pi$ and $B_i=c_i/2\pi$, respectively. It was noted by Rabiner [1968] that in a digital representation, the higherpole correction term can be eliminated.

It should be noted that zeros at $z=0$ do not affect the definition of all-pole or all-zero filters. For example, $G(z)$ can be written in two equivalent forms as

$$G(z)=\frac{1}{(1-e^{-cT}z^{-1})^2}$$
$$=\frac{z^2}{(z-e^{-cT})^2}.$$

In other words, poles and zeros at $z=0$ are customarily not included in counting the number of poles and zeros. The model, as described, allows for defining only a *periodic pulse train* or *noise*, and a number of fixed *formant frequencies* and *bandwidths*. Therefore, only *steady state* vowel or fricative sounds are defined. However, it is easy to implement an arbitrary input $e(n)$, and coefficients for $V(z)$ that are changed or updated at desired intervals of time to represent the time-varying nature of speech. Common procedures used in speech synthesis are, to update the coefficients either at a constant rate such as 50 to 100 times per second, or at the initiation of every pitch period sample in the driving function or sequence, $e(n)$ (referred to as *pitch synchronous synthesis*).

The combination of the glottal term $G(z)$, the vocal tract model $V(z)$, and the lip radiation term $L(z)$ is of the form

$$G(z)V(z)R(z)=\frac{(1-z^{-1})}{(1-e^{\;cT}z^{\;1})^2\left\{\prod_{i=1}^{K}\left[1-2e^{\;c_iT}\cos(b_iT)z^{-1}+e^{-2c_iT}z^{-2}\right]\right\}} \tag{1.7}$$

where K formants are defined in the model. There is only one numerator term $(1-z^{-1})$ and it is nearly canceled by one of the denominator terms $[1-\exp(-cT)z^{-1}]$ since cT is generally much less than unity. A further simplification of the discrete model can then be made in an all-pole synthesis model

$$\boxed{S(z)=E(z)\frac{1}{A(z)}\quad \text{[synthesis model]}}\;, \tag{1.8}$$

by defining

$$\boxed{\begin{aligned} A(z)&=\sum_{i=0}^{M}a_iz^{-i}\quad (a_0=1)\\ &\simeq\frac{1}{G(z)V(z)L(z)}\end{aligned}}\;, \tag{1.9}$$

with $M\geq 2K+1$.

The filter $A(z)$ is an *all-zero filter* and will be referred to as an *inverse filter*. The filter $1/A(z)$ is an *all-pole filter* which represents the smooth spectral behavior of the speech model as shown in Figs. 1.2G and H to within a constant. Eq. (1.8) is referred to as the *synthesis model* since if $E(z)$ is applied to the all-pole filter $1/A(z)$, the output is $S(z)$, the z-transform of the speech signal model.

Multiplication of both sides of (1.8) by $A(z)$ results in the analysis model

$$\boxed{E(z)=S(z)A(z)\quad \text{[analysis model]}}\;. \tag{1.10}$$

This equation is referred to as the *analysis model* since if the speech signal $S(z)$ is input to the inverse filter $A(z)$ (whose coefficients are obtained by analyzing the speech waveform), the output is then $E(z)$, the driving function to the synthesis model. The analysis and synthesis model is illustrated in Fig. 1.4 for a

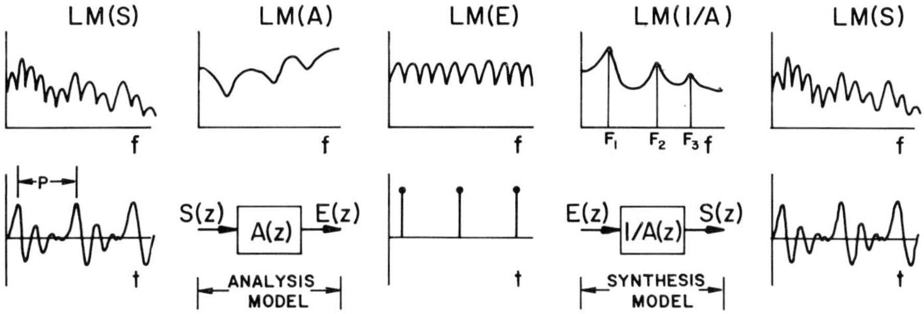

Fig. 1.4 Illustration of analysis and synthesis models with idealized waveforms for a voiced sound.

voiced sound. Using z-transform representation, the speech signal $S(z)$ is analyzed to obtain $A(z)$. The output obtained by passing $S(z)$ through the inverse filter results in $E(z)$ of (1.10), ideally, a series of unit samples or impulses having the same period as $S(z)$. In other words, the inverse filter acts as a pre-whitening filter. If the output samples are applied to the synthesis model, then from (1.8) the result is $S(z)$, the speech signal. The frequency domain representations of the various signals are also shown in the figure. The shorthand notation $LM[\cdot]$ is used to denote the log magnitude spectrum obtained from the corresponding discrete signal. For example,

$$LM(S) = 10\log_{10}|S(e^{j\theta})|^2$$

where

$$S(e^{j\theta}) = S(z)\Big|_{z=e^{j\theta}} ,$$

and θ is the normalized frequency, $\theta = 2\pi f/f_s, f_s = 1/T$ is the sampling frequency, and f is the continuous frequency variable.

In the synthesis model, $LM(E)$ has a flat or constant trend since $e(n)$ is assumed periodic with period P and constant amplitude. The closely-spaced ripples in the frequency spectrum are separated by $F_0 = 1/P$, the *fundamental frequency*. The combined effect of the smooth glottal shaping spectrum, vocal tract spectrum, and lip radiation spectrum is shown as $LM(1/A)$. The location of the first three major peaks in $LM(1/A)$ defines the three formant frequencies, F_1, F_2, and F_3. The output $s(n)$ is obtained by discrete convolution of the unit sample response from $1/A(z)$ with $e(n)$. In the frequency domain from (1.8), $|S[\exp(j\theta)]|^2 = |E[\exp(j\theta)]/A[\exp(j\theta)]|^2$, or $LM(S) = LM(E) + LM(1/A) = LM(E) - LM(A)$.

In the analysis model, an attempt is made to separate the speech log spectrum $LM(S)$ into the smoothed log spectrum $LM(1/A) = -LM(A)$ (containing the vocal tract model plus shaping due to the glottal and lip radiation spectra) and a log spectrum $LM(E)$ having a flat trend characteristic.

The parameters that define the speech production or synthesis model are the coefficients of $A(z)$, $a_i, i=1,2,\ldots,M$, and the parameters of $E(z)$, the pitch period P, and gain σ. Attention will now be focused upon the application of linear prediction for directly determining the coefficients of $A(z)$.

1.4 The Linear Prediction Model

Although linear least squares estimation (or prediction) dates from Gauss in 1795 [Sorenson, 1970], it appears that the first specific use of the term *linear prediction* was in Wiener's 1949 book in Chapter 2, "The Linear Predictor for a Single Time Series" [Wiener, 1966]. Since that time, under various other names and formulations, the same or similar linear prediction mathematics have been widely applied in many fields. The first researchers to directly apply linear prediction techniques (or equivalences) to speech analysis and synthesis were Saito and Itakura [1966] and Atal and Schroeder [1967].

In the sampled data domain, (1.10) can be equivalently written as

$$\begin{aligned} e(n) &= \sum_{i=0}^{M} a_i\, s(n-i) \\ &= s(n) + \sum_{i=1}^{M} a_i\, s(n-i) \end{aligned} \tag{1.11}$$

where $A(z)$ is defined from (1.9). An M*th* order linear predictor of the sample $s(n)$ would require a linear combination of the previous M samples. Defining $\hat{s}(n)$ as the predicted sample, then

$$e(n) = s(n) - \hat{s}(n), \tag{1.12}$$

where

$$\hat{s}(n) = -\sum_{i=1}^{M} a_i s(n-i). \tag{1.13}$$

The driving function term $e(n)$ can now be interpreted as the *prediction error* between the actual data sample $s(n)$ and the predicted sample $\hat{s}(n)$. The terms $-a_i, i=1,2,\ldots,M$, define the *predictor coefficients* to be found. The minus sign is chosen so that the error is based upon a difference of two variables. The choice of sign is completely arbitrary as a different variable $b_i = -a_i$ could just as easily be used.

Figure 1.5 shows a portion of a spoken vowel sound and its samples obtained by performing analog-to-digital conversion of the continuous waveform $s(t)$ every T ms. At each sample index n, $s(n)$ is predicted by a linear combination of the previous M samples $s(n-1), s(n-2), \ldots, s(n-M)$. The z-transform of (1.13) is

$$\hat{S}(z) = F(z) S(z), \tag{1.14}$$

where

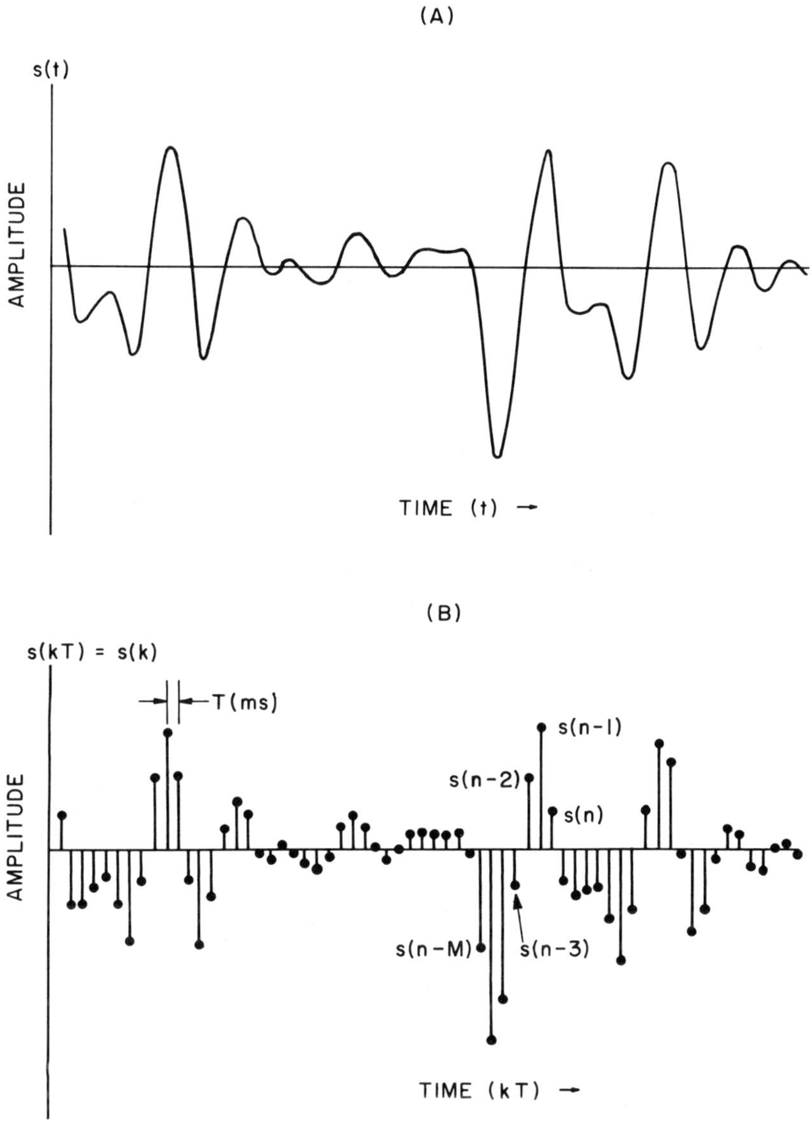

Fig. 1.5 A portion of a spoken vowel sound.
A) continuous waveform representation.
B) sampled data representation.

$$F(z) = -\sum_{i=1}^{M} a_i z^{-i} \tag{1.15}$$

defines the *linear predictor filter*, and $\hat{S}(z) \leftrightarrow \hat{s}(n)$ and $S(z) \leftrightarrow s(n)$ define the relationship between the sampled data and z-transform pairs. From (1.12) and (1.13), the linear prediction model in the z-transform domain is described by

$$
\boxed{\begin{array}{l} E(z)=S(z)[1-F(z)] \\ \text{or} \\ E(z)=S(z)A(z) \end{array}}\;, \tag{1.16}
$$

where

$$
\boxed{\begin{aligned} A(z)&=1+\sum_{i=1}^{M} a_i z^{-1} \\ &=1-F(z) \end{aligned}} \tag{1.17}
$$

and $E(z) \leftrightarrow e(n)$. Equivalent block diagrams of the analysis model in terms of the predictor filter $F(z)$ and the inverse filter $A(z)$ are shown in Fig. 1.6.

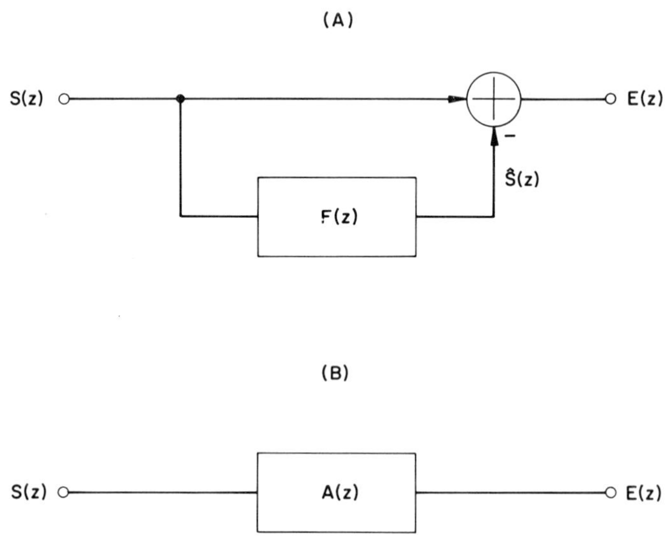

Fig. 1.6 Linear prediction model. A) explicit representation of predictor $F(z)$. B) equivalent representation of an inverse filter $A(z)$ where $A(z)=1-F(z)$.

The formulation in terms of a linear predictor filter was presented by Atal [1970a, d] and in terms of an inverse filter by Markel [1971b]. The conclusion is that *linear prediction of the speech samples results in an equivalent linear model for speech production.* The importance of linear prediction is that the parameters of the model $A(z)$ can be determined directly from the speech waveform by applying a least squares criterion to (1.12).

Since $e(n)$ defines the error between the sample $s(n)$ and the predicted value $\hat{s}(n)$, it would seem reasonable to choose the coefficients $a_1, i=1,2,\ldots,M$ so that $e(n)$ is minimized in some sense. Historically, the optimization criterion has been minimization of the sum of the squares of some specified number of error samples with respect to the coefficients. The main reasons for this choice of optimization criterion are simply that the resulting equations are linear, tractable, and they produce excellent results in the analysis of speech.

The *total squared error* α is defined by

$$\boxed{\alpha=\sum_{n=n_0}^{n_1} e^2(n)} \tag{1.18}$$

$$=\sum_{n=n_0}^{n_1}\left[\sum_{i=0}^{M} a_i s(n-i)\right]^2$$

$$=\sum_{n=n_0}^{n_1}\sum_{i=0}^{M}\sum_{j=0}^{M} a_i s(n-i)s(n-j)a_j,$$

where n_0 and n_1 define the index limits over which error minimization occurs. Defining

$$\boxed{c_{ij}=\sum_{n=n_0}^{n_1} s(n-i)s(n-j)}\quad, \tag{1.19}$$

the total squared error α can then be equivalently written as

$$\boxed{\alpha=\sum_{i=0}^{M}\sum_{j=0}^{M} a_i c_{ij} a_j}\quad. \tag{1.20}$$

Eq. (1.20) shows that the total squared error α is a quadratic form, i.e., at most, powers of two in the coefficients are obtained. Minimization of α is obtained by setting the partial derivation of α with respect to a_k, $k=1,2,\dots,M$ to zero and solving. Therefore, from (1.20),

$$\frac{\partial\alpha}{\partial a_k}=0=2\sum_{i=0}^{M} a_i c_{ik},$$

or since $a_0=1$,

$$\boxed{\sum_{i=1}^{M} a_i c_{ik}=-c_{0k}\quad k=1,2,\dots,M}\quad. \tag{1.21}$$

The M unknown predictor coefficients $\{a_i\}$ are obtained by solving this set of M linear simultaneous equations. The known parameters c_{ik}, $i=0, 1, \ldots, M$, $k=1, 2, \ldots, M$ are defined from the data being predicted by (1.19), which shows that the samples $s(n)$ from n_0-M to n_1 are required.

For the general linear predictor model (1.16) and its solution ((1.19) and (1.21)), two specific cases have been investigated in detail. These are referred to as the *covariance method* and the *autocorrelation method* [Makhoul and Wolf, 1972].

Assuming that a sequence of N speech samples $\{s(n)\}=\{s(0), s(1), \ldots, s(N-1)\}$ is available, the *covariance method* is defined by setting $n_0=M$ and $n_1=N-1$ so that the error is minimized only over the interval $[M, N-1]$, and all N speech samples are used in calculating the covariance matrix elements c_{ij}.

The *autocorrelation method* is defined by setting $n_0=-\infty$ and $n_1=\infty$ and defining $s(n)=0$, for $n<0$ and $n \geq N$. These limits allow c_{ij} to be simplified as

$$\begin{aligned} c_{ij} &= \sum_{n=-\infty}^{\infty} s(n-i)s(n-j) \\ &= \sum_{n=-\infty}^{\infty} s(n)s(n+|i-j|) \\ &= \sum_{n=0}^{N-1-|i-j|} s(n)s(n+|i-j|) \\ &= r(|i-j|). \end{aligned} \tag{1.22}$$

Thus, although the error $e(n)$ is minimized over an infinite interval, equivalent results are obtained by minimizing only over $[0, N+M-1]$ as seen from (1.11), since $s(n)$ is truncated to zero for $n \geq N$ and $n<0$. In summary,

Covariance Method

Solve

$$\sum_{i=1}^{M} a_i c_{ij} = -c_{0j}, \tag{1.23a}$$

for $j=1, 2, \ldots, M$,

where

$$c_{ij} = \sum_{n=M}^{N-1} s(n-i)s(n-j) \tag{1.23b}$$

with error

$$e(n) = \sum_{i=0}^{M} a_i s(n-i) \quad (a_0=1) \tag{1.23c}$$

$n=M, M+1, \ldots, N-1.$

Autocorrelation Method

Solve

$$\sum_{i=1}^{M} a_i r(|i-j|) = -r(j), \quad (1.24\text{a})$$

for $j=1, 2, \ldots, M$,

where

$$r(l) = \sum_{n=0}^{N-1-l} s(n)s(n+l) \quad (l \geq 0) \quad (1.24\text{b})$$

with error

$$e(n) = \sum_{i=0}^{M} a_i s(n-i) \quad (a_0=1) \quad (1.24\text{c})$$

for $n=0, 1, \ldots, N+M-1$.

The covariance method draws its name from control theory literature [Sage, 1968] where for a zero mean signal, c_{ik} defines the row i, column k element of a covariance matrix. The autocorrelation method draws its name from the fact that for the conditions stated, c_{ik} reduces to the definition of the short-term autocorrelation $r(l)$ at the delay $l=|i-k|$. It should be emphasized that the terms *covariance method* and *autocorrelation method* are *not* based upon the standard usage of the covariance function as the correlation function with the means removed.

From (1.23a) and (1.24a), it is seen that the solution from either method defines the all-zero inverse filter $A(z)$, or equivalently the all-pole filter $1/A(z)$. But this result is precisely equal to (1.9). The conclusion is that, based upon the linear prediction model, a linear solution for the parameters of the speech synthesis model $1/A(z)$ has been derived. Depending on the particular analysis method, analysis conditions, and computational techniques, rather different resultant values for $A(z)$ or $1/A(z)$ are possible. It is therefore important to consider the cause of these variations for the analysis of speech. The fact that these differences exist and are important for speech analysis use defines one of the major differentiations from linear prediction studies in other fields. In an area such as geophysics, linear prediction is widely studied under the name of predictive convolution [Peacock and Treitel, 1969], but with rather different goals. For example, in a geophysical study the stability of $1/A(z)$ may not be of significance and numerical computation is generally performed using large-scale, long word-length computers without dominant concern for computational speed. In the area of speech processing, the opposite is true. For linear prediction speech transmission systems, the stability of $1/A(z)$ is a significant problem, computation is often done with small word-length mini-computers or special-purpose computers, and the computations must often be done in real time or near real time.

Different parameters of particular significance to the field of speech can be obtained by application of linear prediction principles. Three important sets of parameters are formant frequencies (and possibly bandwidths), fundamental frequency, and vocal tract area functions.

All of the above topics have been discussed in the literature during the past few years. Unfortunately, as in most new areas, differing notation is used, logical interrelationships to other formulations are not always presented, and sufficient information is not always presented to be of practical value to the reader. This book is therefore an attempt to present the theory of linear prediction and its particular application to speech in a unified manner, including both important theoretical considerations and practical final results by way of Fortran programs and specific speech analysis examples.

1.5 Organization of Book

The present chapter has introduced the speech production model and has shown that linear prediction of speech can be used to derive the parameters of the model. Two different methods of linear prediction, the covariance method and the autocorrelation method, were defined on the basis of the interval over which error minimization occurs.

The next five chapters (Chapters 2–6) contain a detailed presentation of these two methods and their properties. Chapter 2 presents a number of formulations which are shown to result in the same set of linear equations to be solved as (1.21). Since the formulations start from widely different assumptions, considerable insight into the properties of the two methods can be gained.

In this introductory chapter, it was illustrated that by setting the partial derivative of the total squared error α with respect to the filter coefficients to zero, a set of linear simultaneous equations was obtained that defines the model $1/A(z)$. Implicit was the assumption that α was minimized instead of maximized, and that a solution did indeed exist. In Chapter 3, these assumptions are considered in detail, and a unified development of techniques for solving the linear simultaneous equations obtained from the various formulations is presented. Although it would be possible to simply apply a general linear simultaneous equation solver that would exist in most computer program libraries, the detailed study of the special properties of the covariance equations (1.23) and the autocorrelation equations (1.24) leads to at least two significant and practical results. First, faster computational solutions are possible because of certain properties of each method. Second, recursive solutions are developed which contain an important set of intermediate variables. These parameters in the autocorrelation method will be referred to as *reflection coefficients*. They define the stability properties of the synthesizer model. It is shown that a set of *generalized reflection coefficients* can be obtained from a unified solution of both the covariance and autocorrelation methods.

In Chapter 4, the reflection coefficients of the autocorrelation method are shown to be equivalent to the acoustic reflection coefficients of an acoustic tube model of the vocal tract $V(z)$ when the speech is sampled at an appropriate rate and pre-emphasized before analysis. The discrete area functions of an acoustic tube model are shown to be easily obtained from the bilinear transformation of the reflection coefficients. The acoustic tube can be viewed as a synthesis model implemented by using the reflection coefficients of the analysis model.

In Chapter 5, various formulations of synthesizer models or structures for implementing the acoustic tube model are developed in detail. Although they are mathematically equivalent for the time-invariant analysis-synthesis model (Fig. 1.4), they have slightly different characteristics under time-varying analysis-synthesis conditions and significantly different characteristics under small word-length computer implementation conditions. Techniques for efficiently determining the stability of the synthesis structure $1/A(z)$ from either the covariance or autocorrelation method are presented. In addition, various theoretical stability properties of the synthesis structures are developed.

In Chapter 6, properties of the spectral model $\sigma/A(z)$ are investigated, where σ is a gain term for matching the energy of the signal to that of the model. Considerations in the choice of analysis parameters are discussed and techniques for evaluating the frequency response on the unit circle, $\exp(j\theta)$, within the unit circle, $\exp(-\alpha+j\theta)$, $\alpha>0$, or on separate portions of the unit circle, are presented.

The following two chapters (Chapters 7 and 8) consider the difficult problem of automatically estimating the formant parameters and the fundamental frequency of the speech production model from real speech under the conditions of time-varying parameters. The properties that make linear prediction desirable in the estimation of those parameters are emphasized and illustrated with a number of examples. In Chapter 7, it is shown that the raw data obtained by simple peak picking of the spectrum $1/|A[\exp(j\theta)]|^2$ or $LM(1/A)$ for each analysis interval define the locations of the formant frequencies (for speech with a fundamental frequency less than about 300 Hz) about 85–90 percent of the time. For the remaining percentage of cases, two automatic formant tracking algorithms of differing complexity and accuracy are presented. In Chapter 8, the application of linear prediction in estimating the fundamental frequency F_0 is presented. A computationally efficient algorithm which takes advantage of known speech properties is discussed in detail with a number of examples.

In performing linear prediction analysis of speech, digital computation is necessary. Chapter 9 considers efficient implementation techniques and important computational problems that can occur, together with techniques for solving them. For example, even though the autocorrelation method is theoretically stable (that is, the synthesis filter $1/A(z)$ is stable), implementation of the analysis using integer arithmetic may result in an unstable filter unless corrective steps are taken.

Chapter 10 discusses linear prediction techniques as they apply to voice-coding (vocoding). A number of linear prediction vocoder simulation and hardware systems are discussed.

In Chapter 11, a number of other applications of linear prediction are briefly discussed along with the authors' view of future research directions.

2. Formulations

2.1 Historical Perspective

The first application of linear prediction to speech from a maximum likelihood formulation was in the work of Saito and Itakura in 1966. In 1967 and 1968, Atal and Schroeder [1967, 1968 b, 1968 c] published results on predictive coding of speech signals. Itakura and Saito also published a condensed English version of the maximum likelihood estimation procedure in 1968 and a detailed presentation in 1970. In 1970, Atal [1970 a] presented the first use of the term *linear prediction* for speech analysis. Since that time, the term has become universally accepted as referring to those methods of speech analysis that result in the solution of a predictor filter or inverse filter based upon solving a set of M linear simultaneous equations (as discussed in Chapter 1). In 1970, Markel [1971 a] discussed the application of Prony's method to speech analysis and noted the relationships to the linear prediction method developed by Atal and the maximum likelihood method developed by Itakura and Saito.

In 1971, Markel [1971 b] presented a deterministic formulation for designing an inverse filter $A(z)$ with a final result equivalent to the non-deterministic maximum likelihood formulation. In addition, a number of existing equivalent non-speech formulations such as Wiener filtering [Levinson, 1947; Robinson and Treitel, 1967], prediction deconvolution [Peacock and Treitel, 1969], and spiking filters [Robinson, 1967] were discussed. Atal and Hanauer [1971 b] then published the details of their linear prediction formulation.

In 1972, Makhoul and Wolf published a report containing a unified presentation of both the covariance and autocorrelation methods and also presented several additional autocorrelation method formulations, such as correlation matching, and spectral matching, which is the equivalent to the inverse filter formulation using Parseval's theorem in the frequency domain. Also in 1972, Itakura and Saito [1972 b, 1972 c] presented an important new PARCOR formulation of linear prediction using partial correlation techniques. (In Japan, their work on the PARCOR formulation had been presented in 1969.) Wakita [1972] then showed that the PARCOR formulation could be interpreted as the detailed structure of the inverse filter $A(z)$.

Since then, several other equivalent linear prediction formulations have been developed, such as a spectral flatness formulation [Gray and Markel, 1974 b], and a geometrical Hilbert space formulation [Narasimha, et al., 1974].

What has become clear during the past few years is that there is an essentially limitless possibility for formulations that finally result in the same general solution form as that presented in Chapter 1. A comprehensive history and bibliography of linear filtering theory which includes topics related to linear prediction can be found in Kailath [1974]. Other sources of linear prediction bibliography outside of strictly speech applications are contained in Flinn, et al. [1967] and Makhoul [1975a].

The advantage to studying a number of different formulations which lead to the same result is the insight into the nature of the problem that is gained. The formulations mentioned above are not based upon identical assumptions, and yet they result in identical relations, falling into either the covariance or the autocorrelation category. It is for this reason that assumptions which may not be intuitively satisfying can still lead to useful results. As an example, in the maximum likelihood estimation method, speech is assumed to be a stationary, Gaussian random process. This assumption appears very reasonable for unvoiced sounds and invalid when voiced speech is analyzed. An inverse filter formulation does not require assumptions of Gaussian statistics, but both formulations lead to the identical set of autocorrelation equations. As a second example, the covariance method and Prony's method result in identical covariance equations. In the Prony method, it becomes clear that if the signal is composed of M complex exponentials, $2M$ samples of data suffice to precisely determine the parameters of the exponentials. However, it is not immediately obvious that Prony's method (based upon most numerical analysis formulations) is, in effect, linearly predicting each sample value.

In this chapter, several formulations that lend insight into linear prediction of speech and that lead to results used in later chapters are presented, not necessarily in the historical order of their development. Where possible, both the autocorrelation and covariance formulations are presented together to better illustrate their similarities and differences. Several spectral models which can also be considered as formulations are discussed in Chapter 6.

First, two maximum likelihood approaches are presented, each of which assumes stationary Gaussian statistics for the speech. The maximum likelihood approaches are not treated in detail because of their complexity, but the underlying assumptions are discussed. These maximum likelihood approaches will be shown to result in the covariance method in the case where a conditional maximum likelihood is utilized, or the autocorrelation method as developed by Itakura and Saito [1970], in the case where an approximate maximum likelihood is utilized. Second, an approach based upon minimizing the statistical mean square of a linear predictor error is presented. This approach assumes a stationary and random, but not necessarily Gaussian speech signal, and can lead to either the covariance or autocorrelation method, depending upon how statistical correlations are estimated from the speech data. Deterministic linear prediction and inverse filter formulations were presented in Chapter 1.

Third, the Prony method of exponential curve fitting [Hildebrand, 1956; McDonough, 1963] is discussed and it is shown that the results for an all-pole representation are equivalent to the covariance method developed by Atal and Hanauer [1971b]. Considerable insight can be gained from this formulation in the analysis of voiced speech since the all-pole model of speech is of a complex-exponential nature. The Prony method, however, can be extended through the use of pitch-synchronous analysis and additional computations to include the effects of zeros in the speech model as discussed in Chapter 11. Fourth, a correlation matching approach [Makhoul and Wolf, 1972; Markel and Gray, 1973a] is shown to yield results identical to those of the autocorrelation method with the additional property that a gain term is also obtained.

The final portion of the chapter is devoted to the detailed structure of the inverse filter [Wakita, 1972] and the PARCOR formulation [Itakura and Saito, 1972b]. Considerable insight into the linear prediction process can be gained from this formulation. In addition, the results provide a mathematical framework from which a unified theory for solving either the autocorrelation or covariance equations can be developed.

2.2 Maximum Likelihood

It is assumed that the speech spectrum (over short intervals) is determined by certain parameters whose values for any given interval are unknown. The question is then asked, "What choice of the parameter values makes the probability of occurrence of the actual observations (the speech samples), most likely, i.e., maximizes the probability density of the parameters?" In order to answer such a question, certain assumptions must be made about the speech process.

It is most often assumed that the samples being analyzed are elements of a Gaussian, stationary, random process, generated by passing uncorrelated noise through an all-pole model of the form

$$\frac{1}{A(z)} = \frac{1}{\sum_{i=0}^{M} a_i z^{-i}} \tag{2.1}$$

with $a_0 = 1$. The noise is assumed to have a variance, σ_e^2, which can be estimated. While the assumptions may not appear to be reasonable in light of actual speech statistics, the results obtained are nonetheless equivalent to other formulations. Before proceeding to the mathematics, the specific assumptions for maximum likelihood estimation are discussed.

First, it is assumed that the process is Gaussian. This is a standard assumption in many areas such as information theory, communications theory, and statistics. In part, it is based upon the fact that the Gaussian assumption is often sufficient for tractable mathematics. In part, it is based upon a very liberal view of the central limit theorem, which might be loosely stated, "almost any random process put into almost any linear system will come out almost Gaussian". A final reason for the Gaussian assumption lies in a maximum entropy principle which states that for known values of the first and second moments of a random process, the specific joint probability density which has the largest entropy is the Gaussian probability density.

Second, it is assumed that the process is a stationary process (its statistics do not vary with time). By observation of sampled speech waveforms in Chapter 1, this assumption may appear reasonable over short intervals for unvoiced sounds, such as the /ʃ/ in predic*ti*on, but voiced sounds do not appear to be either random or stationary. It should be noted that the model described in (2.1) having uncorrelated noise passed through it is identical to the model of Chapter 1 for the case of unvoiced sounds, only here it is applied in the analysis of both unvoiced and voiced sounds.

Finally it is assumed that the uncorrelated noise sequence input to $1/A(z)$, $\{e(n)\}$, is Gaussian with zero mean and variance σ_e^2, i.e.,

$$\mathscr{E}[e(n)]=0 \text{ and } \mathscr{E}[e(k)e(n)]=\delta_{kn}\sigma_e^2 \tag{2.2}$$

where $\mathscr{E}$ represents the expectation (stochastic or probabilistic mean) and δ_{kn} is the Kronecker delta. The sequence $\{x(n)\}$ (representing the synthesized speech or a pre-emphasized version of the speech) is obtained by passing the noise sequence $\{e(n)\}$ through the model in (2.1) and results in the equation

$$\sum_{i=0}^{M} a_i x(n-i)=e(n) \qquad \text{with } a_0=1. \tag{2.3}$$

From (2.2) and (2.3) it can be shown that $\{x(n)\}$ is Gaussian and has zero mean and a correlation sequence $\{\varrho(n)\}$, given by

$$\mathscr{E}[x(n)x(l)]=\varrho(n-l). \tag{2.4}$$

The correlation sequence is not simply related to the parameters, and so here will be expressed in a functional form as

$$\varrho(n)=f(a_1, a_2, \ldots, a_M, \sigma_e^2, n). \tag{2.5}$$

Knowing that the sequence $\{x(n)\}$ is Gaussian with zero mean, one can define the joint probability density of the variables $x(0)$, $x(1)$, $x(2)$,..., $x(N-1)$. It is simply the standard Gaussian multivariate probability density, and is a function of the parameters in (2.5) and the sequence of random variables $x(0), x(1), x(N-1)$. The maximum likelihood principle states that if certain values of the random variables are observed, the parameter values which make the observation most likely are those which maximize the probability density. In principle, the probability density is differentiated with respect to each of the parameters, a_1, a_2, ..., a_M, σ_e^2, the derivatives are set equal to zero, and the resulting equations are solved for the parameters. While the approach outlined is conceptually straightforward, it does not lead to fruitful results. In fact, for N greater than 2 the problem becomes extremely non-linear, and to this date it appears that no one has solved the true maximum likelihood problem.

The fact that there have been no successful solutions of a true maximum likelihood analysis of speech does not render the concept inoperable, however. Two approaches can be utilized, both of which modify the maximum likelihood criterion and which lead to the same results as found by the autocorrelation method of linear prediction analysis. The first, due to Itakura and Saito [1970, 1971 a, c] and Itakura [1972a], involves an approximate maximum likelihood leading to the autocorrelation method results. The second, as we shall show, involves a conditional maximum likelihood approach which leads to the covariance method results.

Itakura and Saito utilized approximations based upon the assumption $N \gg M$ (the number of data points greatly exceeds the filter order) to show that the joint probability density for the sequence $\{x(0), x(1), \ldots, x(N-1)\}$ can be approximated by

$$\boxed{p[x(0), x(1), \ldots, x(N-1)] = (2\pi\sigma_e^2)^{-N/2} \exp(-\alpha/2\sigma_e^2)} \tag{2.6}$$

where α is equal to

$$\boxed{\alpha = \sum_{n=-\infty}^{\infty} \left[\sum_{i=0}^{M} a_i x(n-i)\right]^2} \tag{2.7a}$$

with the definition

$$x(n) = 0 \quad \text{for} \quad n < 0 \quad \text{and for} \quad n > N-1. \tag{2.7b}$$

The α of (2.7) is precisely the error energy in the autocorrelation approach to least squares linear prediction. As a result, it is noted that maximizing p in (2.6) is equivalent to first minimizing α in (2.7) as in the autocorrelation method, and then maximizing (2.6) with respect to σ_e^2 by solving $dp/d\sigma_e^2 = 0$. The result is thus equivalent to the autocorrelation method, with the additional property that a gain term

$$\boxed{\sigma_e^2 = \alpha/N} \tag{2.8}$$

is obtained for matching the energy of the model (2.1) to that of the input signal, where the minimum α is used. In the form of a conditional maximum likelihood, the first set of M observations, $x(0), x(1), \ldots, x(M-1)$, is treated as a set of deterministic initial conditions or values and the remaining set of samples $x(M)$, $x(M+1), \ldots, x(N-1)$ is treated as a set of observations of random variables. By not treating $x(0), x(1), \ldots, x(M-1)$ as random, the probability density is defined only for the remaining $N-M$ variables (though it is in fact now a density, conditioned on the initial values). From (2.3) it is seen that the variables $x(M)$, $x(M+1), \ldots, x(N-1)$ are simply a linear transformation of the uncorrelated variables $e(M), e(M+1), \ldots, e(N-1)$. An elementary change of variables and the use of Gaussian statistics lead to the conditional probability density p_c for the $N-M$ observations

$$p_c = (2\pi\sigma_e^2)^{-(N-M)/2} \exp(-\alpha/2\sigma_e^2) \tag{2.9}$$

where α is now equal to

$$\alpha = \sum_{n=M}^{N-1} \left[\sum_{i=0}^{M} a_i x_{n-i} \right]^2 . \tag{2.10}$$

The α of (2.10) is precisely the same as the error energy in the covariance method. Maximization of p_c in (2.9) is effected by first minimizing α, and then maximizing with respect to σ_e^2, with the result for σ_e^2 given by

$$\sigma_e^2 = \alpha/(N-M) \tag{2.11}$$

where α is the minimum value of (2.10), the covariance error minimum.

The maximum likelihood method can thus be used to obtain either the autocorrelation method in the case where an approximate probability density is used (provided $N \gg M$), or the covariance method in the case where a conditional probability density is used. In both cases, a gain term σ_e is obtained in terms of the minimum predictor error α. The logic leading to the maximum likelihood method is based upon stationary random Gaussian speech signal assumptions, and seems most reasonable in the case of unvoiced sounds.

2.3 Minimum Variance

It is now assumed that the speech samples are random and stationary, but no assumption is made concerning the type of statistics they have (such as the Gaussian assumption used in maximum likelihood). From Chapter 1, the linear predictor error sequence $\{e(n)\}$ was defined as

$$e(n) = \sum_{i=0}^{M} a_i x(n-i) \tag{2.12}$$

with $a_0 = 1$. If it is assumed that the samples $\{x(n)\}$ have zero mean, then the error sequence will also have zero mean, and its variance will equal its mean square, which from (2.12) is given by

$$\mathscr{E}[e(n)^2] = \sum_{i=0}^{M} \sum_{j=0}^{M} a_i a_j \mathscr{E}[x(n-i)\,x(n-j)], \tag{2.13}$$

where $\mathscr{E}[\cdot]$ is the expectation operator.

From the stationarity assumption of the samples, the expected value of the product $x(n-i)\,x(n-j)$ will be a function of the difference of the subscripts, given in terms of an autocorrelation sequence, by $\{\varrho(n)\}$

$$\mathscr{E}[x(n-i)\,x(n-j)]=\varrho(i-j)=\varrho(j-i), \tag{2.14}$$

where $\varrho(i-j)$ will be defined below. The predictor error variance can thus be written as the double summation

$$\boxed{\mathscr{E}[e(n)^2]=\sum_{i=0}^{M}\sum_{j=0}^{M} a_i\,\varrho(i-j)\,a_j} \;. \tag{2.15}$$

The criterion for minimizing the predictor error variance is that of choosing the coefficients $a_1, a_2, \ldots, a_M$ so as to minimize (2.15). At this point (2.15) appears similar to both the autocorrelation and covariance methods described in Chapter 1, except that the terms $\varrho(i-j)$ in the summation represent statistical correlations. These statistical correlations can only be approximated from $\{x(n)\}$ since a finite number of samples must be used, and it is the method of estimation that differentiates the results.

If the process is assumed to be ergodic, so that statistical averages equal time averages, then $\varrho(i-j)$ can be expressed as

$$\varrho(i-j)=\mathscr{E}[x(n-i)x(n-j)]=\lim_{L\to\infty}\frac{1}{L}\sum_{n=0}^{L-1} x(n-i)x(n-j). \tag{2.16}$$

With only a finite set of observed samples, $x(0), x(1), \ldots, x(N-1)$, the above expression obviously cannot be directly evaluated. Although numerous approximations can be utilized, only two will be considered here. First, the approximation

$$\boxed{\varrho(i-j)\simeq\frac{1}{N-M}\sum_{n=M}^{N-1} x(n-i)x(n-j)=c_{ij}/(N-M)} \tag{2.17}$$

leads to the covariance method, and second, the approximation

$$\boxed{\varrho(i-j)\simeq\frac{1}{N}\sum_{n=0}^{N-|i-j|-1} x(n)x(n+|i-j|)=r(i-j)/N} \tag{2.18}$$

leads to the autocorrelation method. The two approximations utilize exactly the same set of data points, but in a different manner.

The covariance approximation as indicated by (2.17) has an inconsistency since the left-hand side of that equation is a function of the subscript difference, $i-j$, while the right-hand side is not. This problem arose because of the stationarity assumption in obtaining (2.15) where $\mathscr{E}[x(n-i)\,x(n-j)]$ was replaced by $\varrho(i-j)$. If stationarity is not assumed then there is no way to estimate $\mathscr{E}[x(n-i)\,x(n-j)]$ from sample means, for if the process is not stationary it is not ergodic and ensemble averages are no longer equal to time averages. It has become customary to consider $c_{ij}/(N-M)$ an approximation to $\mathscr{E}[x(n-i)\,x(n-j)]$ in some sense even without a stationary assumption.

The autocorrelation approximation is a biased estimate, since the expected value of the autocorrelation method approximation gives

$$\mathscr{E}[r(i-j)/N] = \varrho(i-j)\left[1 - \frac{|i-j|}{N}\right]. \tag{2.19}$$

The covariance approximation is an unbiased estimate since the expected value of the covariance method approximation gives

$$\mathscr{E}[c_{ij}/(N-M)] = \varrho(i-j). \tag{2.20}$$

Experimentally, the bias of the autocorrelation method estimate, (2.19), appears to have little significance in speech analysis when M (the filter order) is much less than N (the number of data samples being analyzed), since the bias is negligible for $|i-j| \ll N$. The bias is significant, however, when N is only slightly larger than M. This result can be useful for explaining why $N \gg M$ is necessary for accurate spectral estimation in the autocorrelation method.

2.4 Prony's Method

Prony's method has a long history dating from the original development by Prony [1795], [McDonough, 1963], the same year that Gauss presented his least squares estimation theory [Sorenson, 1970]. Least squares estimation was applied to Prony's equations as early as 1924 [Runge and Konig, 1924]. A detailed discussion of Prony's method may be found in McDonough [1963]. Prony's method is important in understanding the linear prediction of speech, since the formulation shows explicitly how the voiced speech model is being represented by complex exponentials in the time domain.

In working with descrete signals, the term *exponential* or *complex exponential* is used in the broad sense to include powers. If, for example, the continuous waveform $\exp(-\alpha t)$ is sampled at the discrete times $t = nT$, $n = 0, 1, \ldots,$ then $\exp(-\alpha nT) = (z)^n$ with $z = \exp(-\alpha T)$.

In Chapter 1 it was noted that the speech model during voicing corresponds to a sequence of unit samples (separated by the pitch period) driving an all-pole filter $1/A(z)$. If transients from preceding pitch periods are ignored, then voiced speech samples during a single pitch period will be proportional to the unit sample response of an all-pole filter. If $\{x(n)\}$ is considered to be the sampled speech data (or possibly pre-emphasized speech data) sequence during a single pitch period, then it can be modeled as a linear combination of M complex exponentials, i.e.,

$$x(n) = \sum_{i=1}^{M} u_i (z_i)^n \tag{2.21}$$

where z_i, $i=1,2,\ldots,M$ defines the roots or zeros of $A(z)$,

$$A(z_i)=0 \quad \text{for} \quad i=1,2,\ldots,M. \tag{2.22}$$

Eq. (2.21) is obtained by simply expressing the model of a single voiced interval in a partial fraction expansion. For a single pitch period, the driving sequence is $e(n)=\delta_{n0}$, so that from (1.8) with $S(z)$ replaced by $X(z)$,

$$X(z)=E(z)/A(z)=1/A(z). \tag{2.23}$$

But $1/A(z)$ can be represented in the partial fraction expansion form

$$1/A(z)=\sum_{i=1}^{M} u_i/(1-z_i z^{-1}). \tag{2.24}$$

The inverse z-transform of (2.23) and (2.24) then gives (2.21).

It should be noted that the u_i and z_i factors in (2.21) may be complex in the case of oscillatory behavior, but they must combine in the summation to give real results since $x(n)$ is real. In (2.21) it is assumed that the roots of $A(z)$ are distinct. This assumption is necessary only for clarity of presentation, and is not restrictive. Eq. (2.21) can therefore be written as a linear combination of real exponentials and damped sinusoids. The general form of (2.21) can be applied even when two of the previously made assumptions are removed. In particular, it can also include the effects of transients from a previous pitch period as well as the effects of zeros due to the nasal tract that might be included in the speech model. The z-transform of $x(n)$ in (2.21) is of the form

$$\begin{aligned} X(z) &= P(z)/A(z) \\ &= \sum_{i=0}^{M-1} p_i z^{-i} \Big/ \sum_{i=0}^{M} a_i z^{-i} \end{aligned} \tag{2.25}$$

where $P(z)$ is a polynomial of degree $M-1$, one less than $A(z)$ with $a_0=1$. The numerator, $P(z)$, can be interpreted as describing either the effects of transients from earlier pitch periods or the effects of zeros in the speech model.

If speech were precisely representable by the model of (2.21), then the unknown parameters u_i and z_i ($2M$ in number) could be obtained by solving the set of $2M$ simultaneous equations indicated by (2.21) for $n=0,1,\ldots,2M-1$. Thus, only $2M$ samples of the speech signal would be needed. The problem is non-linear but it does have a solution. The method of solution is not intuitively obvious as can be noted in the second example to follow, where $M=2$.

Example 1 – 1 real exponential $(M=1)$
For $M=1$, (2.21) gives

$$x(n)=u_1 z_1^n.$$

Taking as observations the first $2M$ samples $x(0)$ and $x(1)$, two simultaneous equations are obtained as

$$x(0)=u_1 \text{ and } x(1)=u_1 z_1.$$

As a result, the unknown parameters are given in terms of the two observations $x(0)$ and $x(1)$ as

$$u_1=x(0) \quad \text{and} \quad z_1=x(1)/x(0).$$

If, for example, $\{x(n)\}$ is made up of one real exponential,

$$x(n)=C\exp(-2\pi BnT), \quad \text{then} \quad u_1=C \quad \text{and} \quad z_1=\exp(-2\pi BT).$$

Example 2 – 2 exponentials (real or complex) $(M=2)$
For $M=2$, (2.21) gives

$$x(n)=u_1 z_1^n+u_2 z_2^n.$$

Taking as observations the first $2M$ samples $x(0)$, $x(1)$, $x(2)$, and $x(3)$, four simultaneous equations are obtained as

$$x(0)=u_1 \quad +u_2 \tag{2.26a}$$

$$x(1)=u_1 z_1+u_2 z_2 \tag{2.26b}$$

$$x(2)=u_1 z_1^2+u_2 z_2^2 \tag{2.26c}$$

$$x(3)=u_1 z_1^3+u_2 z_2^3. \tag{2.26d}$$

At this point there are four equations and four unknowns. One approach to the solution is to attempt to systematically eliminate unknowns. This approach will be illustrated here, but becomes very unwieldy as M increases. As a result, this illustration should be considered as motivation for seeking a better solution algorithm.

It is possible to eliminate u_1 from (2.26) by taking the differences $z_1 x(0)-x(1)$, $z_1 x(1)-x(2)$, and $z_1 x(2)-x(3)$,

$$z_1 x(0)-x(1)=u_2(z_1-z_2) \tag{2.27a}$$

$$z_1 x(1)-x(2)=u_2(z_1-z_2)z_2 \tag{2.27b}$$

$$z_1 x(2)-x(3)=u_2(z_1-z_2)z_2^2. \tag{2.27c}$$

This gives three equations in three unknowns, z_1, z_2, and u_2. Next u_2 can be eliminated in a similar manner to give

$$z_2[z_1 x(0)-x(1)]-[z_1 x(1)-x(2)]=0 \tag{2.28a}$$

$$z_2[z_1 x(1)-x(2)]-[z_1 x(2)-x(3)]=0. \tag{2.28b}$$

There are now two equations in two unknowns, z_1 and z_2. The variable z_2 is easily eliminated in these equations, giving

$$\begin{aligned}&[z_1 x(1)-x(2)][z_1 x(1)-x(2)]\\&-[z_1 x(0)-x(1)][z_1 x(2)-x(3)]=0.\end{aligned} \tag{2.29}$$

At this point a single quadratic equation in the unknown z_1 is obtained. In principle one could solve for z_1 in terms of $x(0), x(1), x(2), x(3)$, use either of (2.28) to obtain z_2, any of (2.27) to obtain u_2, and any of (2.26) to obtain u_1.

The significant result illustrated by these examples is that *if the signal $x(n)$ is composed of precisely M complex exponentials, then $2M$ samples suffice to exactly determine the model parameters.* The second example illustrates the complexity of the problem when approached directly as one of eliminating variables from a set of simultaneous equations. The following approach makes the problem much easier to handle. From (2.25),

$$X(z)A(z)=P(z)$$

or

$$\sum_{i=0}^{M} a_i x(n-i)=\sum_{i=0}^{M-1} p_i \delta_{n,i}$$

so that

$$\boxed{\sum_{i=0}^{M} a_i x(n-i)=0 \quad \text{for} \quad n=M, M+1, \ldots, N-1} \tag{2.30}$$

To account for the possibility that the model may not exactly represent a single pitch period of real speech, an error term $e(n)$ can be introduced so that

$$\boxed{\sum_{i=0}^{M} a_i x(n-i)=e(n) \quad \text{with} \quad a_0=1}\;. \tag{2.31}$$

This equation defines the practical form of Prony's method. The exact solution (2.30) will inevitably fail in the analysis of real speech.

The coefficients a_i can then be obtained by minimizing the squared error α with respect to the coefficients, where

$$\boxed{\alpha = \sum_{n=M}^{N-1} e(n)^2} \quad . \tag{2.32}$$

But this result gives precisely the covariance method of Chapter 1. For the case of an exact model, the minimum value of α will be zero.

Example 3
From Example 1, if $C=1024, B=100$ Hz, and $T=0.1$ ms, the sequence $\{x(n)\}$ becomes samples of the decaying exponential $C\exp(-\pi BnT)$ having the z-transform $X(z)=C/[1-\exp(-\pi BT)z^{-1}]$. The sequences $\{x(n)\}$ and $\{e(n)\}$, obtained by solving for the coefficients $\{a_i\}$ and then substituting into (2.31), are shown in Fig. 2.1A. Since all $e(n)=0$ for $n \geq M=1, \alpha=0$. Similarly, for an increasing exponential as shown in Fig. 2.1B, $\alpha=0$ for $n \geq M=1$.

For Example 2, if $u_1=u_2=C/2$, and $z_1=z_2^*=\exp(-\pi BT+j2\pi FT)$, then

$$x(n)=C[\exp(-\pi BnT)]\cos 2\pi FnT.$$

The z-transform of this signal is of the form

$$X(z)=\frac{C-Crz^{-1}}{1-2r\cos\theta z^{-1}+r^2z^{-2}}=\frac{P(z)}{A(z)}$$

where $r=\exp(-\pi BT)$ and $\theta=2\pi FT$. By choosing the same values for C, B, and T with $F=1.0$ kHz, the resulting sampled complex exponential and error sequence are shown as Fig. 2.1C. This example has a zero at $z=r$ in addition to the two poles z_1 and z_2. Nonetheless, $e(n)=0$ for $n \geq M=2$ and $\alpha=0$. These results can be easily verified. Similar results will be obtained for higher order models. Thus, the very significant theoretical point is emphasized that *if the sampled speech waveform can be precisely represented by the model $P(z)/A(z)$ or $1/A(z)$ within a single pitch period, then the covariance method can extract the parameters of $1/A(z)$ from the sampled data sequence exactly*. Furthermore, only $N=2M$ samples are necessary for the analysis. In practice, $N=4M$ or more may be necessary to obtain reasonable results due to the fact that the speech waveform will generally not precisely fit the model. Prony's method is actually more general than the covariance method definition in that determination of the numerator $P(z)$ is also part of the procedure. By solving the set of simultaneous equations (1.21), the polynomial $A(z)$,

$$A(z)=\sum_{i=0}^{M} a_i z^{-i} \tag{2.33}$$

is defined, with $a_0=1$. But the roots of the polynomial $A(z)$ are from (2.22), the terms utilized to give the linear combination of exponentials in (2.21). Again

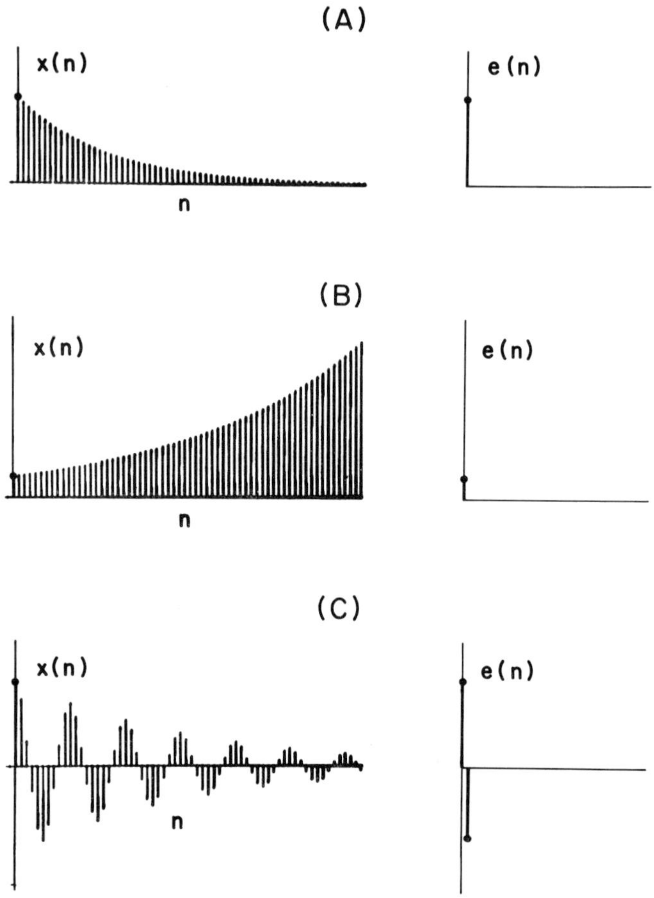

Fig. 2.1 Examples of first and second order models with error sequences.

assuming that the model may not be exact, a residual error $\eta(n)$ can be introduced into (2.21), as

$$x(n) = \sum_{i=1}^{M} u_i (z_i)^n + \eta(n), \tag{2.34}$$

so that an equality expression is obtained. Since the roots z_i are now known, the parameters u_i can be chosen to minimize the sum of the squares of the residual error $\eta(n)$. This minimization is a standard problem in least squares curve fitting.

The essential feature of Prony's method is that the non-linear problem of solving for the $2M$ parameters of the model (2.21) is separated into two straightforward problems. First, a polynomial $A(z)$ defined by the M roots $z_i, i = 1, 2, \ldots, M$ is introduced, and then its coefficients are obtained by solving a set of M linear simultaneous equations. This part of the problem is precisely equivalent to the covariance method. Then, the roots or zeros of $A(z)$ are obtained (a standard numerical analysis problem) and substituted into (2.21), leaving a straightforward least squares curve fitting problem to be solved for defining the M numerator coefficients.

In summary, Prony's method attempts to fit complex exponentials to the speech data, a physically reasonable approach for speech data taken from a single pitch period. The results for the denominator of the speech model, $A(z)$, are identical to the covariance method. Prony's method is more general, however, in that it also allows for the determination of zeros in the speech model. The question of whether the zeros corresponding to the numerator $P(z)$ have physical meaning is discussed in Chapter 11.

2.5 Correlation Matching

In the correlation matching formulation [Makhoul and Wolf, 1972; Markel and Gray, 1973a], a match between the autocorrelation of the input sequence $\{x(n)\}$ and the unit sample response of an all-pole synthesis filter $\sigma/A(z)$ is desired at as many points as possible. Assume a causal, stable synthesis model $H(z)$ of the form

$$\boxed{H(z)=\sigma/A(z)} \quad . \tag{2.35}$$

The constant σ represents a gain factor, and as before, $A(z)$ is given by

$$A(z)=\sum_{i=0}^{M} a_i z^{-i} \quad \text{with} \quad a_0=1. \tag{2.36}$$

Multiplication of both sides of (2.35) and an inverse z-transformation gives

$$\sum_{i=0}^{M} a_i h_{n-i}=\sigma\delta_{n0} \ , \tag{2.37}$$

where the sequence $\{h_n\}$ has the z-transform $H(z)$.

With the assumption that the filter is causal, $h_n=0$ for $n<0$, and stable, the autocorrelation sequence for the unit sample can be expressed as

$$\varrho(i-j)=\sum_{n=-\infty}^{\infty} h_{n-i}h_{n-j}=\sum_{n=0}^{\infty} h_n h_{n+|i-j|}. \tag{2.38}$$

If (2.37) is multiplied by h_{n-j} and summed over all n, then a direct application of (2.38) gives the result

$$\sum_{i=0}^{M} a_i\,\varrho(i-j)=\sigma h_{-j}. \tag{2.39}$$

As the filter is assumed to be causal, the right-hand side of (2.39) is zero for $j>0$, and from (2.37), $h_0=\sigma$, making the right-hand side of (2.39) equal to σ^2 for $j=0$. As a result

$$\sum_{i=0}^{M} a_i \varrho(i-j)=0 \quad \text{for} \quad j>0 \tag{2.40}$$

and

$$\sum_{i=0}^{M} a_i \varrho(i)=\sigma^2. \tag{2.41}$$

To determine the $M+1$ parameters of the synthesis model $H(z)$, $\{\sigma, a_1, a_2, \ldots, a_M\}$, the first $M+1$ autocorrelation samples of the filter unit sample response are chosen to exactly match the first $M+1$ autocorrelation samples of the input data sequence $\{x(n)\}$, i.e.,

$$\varrho(j)=r(j) \quad \text{for} \quad j=0,1,\ldots,M, \tag{2.42}$$

where $\varrho(i-j)$ is given by (2.38) and

$$r(i-j)=\sum_{n=-\infty}^{\infty} x(n-i)x(n-j)=\sum_{n=0}^{N-1-|i-j|} x(n)x(n+|i-j|). \tag{2.43}$$

Combining (2.42) with (2.40) for $m=1,2,\ldots,M$, we have exactly the equations which must be solved in the autocorrelation method. Combining (2.42) with (2.41) results in the expression for matching the energy of the input signal spectrum to the energy of the synthesis model (2.35) unit sample response as

$$\boxed{\sigma^2=\sum_{i=0}^{M} a_i r(i)} \quad . \tag{2.44}$$

Figure 2.2 shows the effect of including the gain term σ in the analysis of both a voiced and an unvoiced speech segment. The model log spectrum closely matches the envelope structure of the signal log spectrum. One of the important properties of this spectral match is that the model $LM[\sigma/A]$ will always have a slightly higher average value than $LM[X]$. This fact will be shown later in Chapter 6.

2.6 PARCOR (Partial Correlation)

A novel approach to the linear prediction of speech was presented in terms of partial correlation coefficients by Itakura and Saito [1969, 1972b, 1972c]. This approach is used to define an internal view of the inverse filter $A(z)$ [Wakita, 1972]

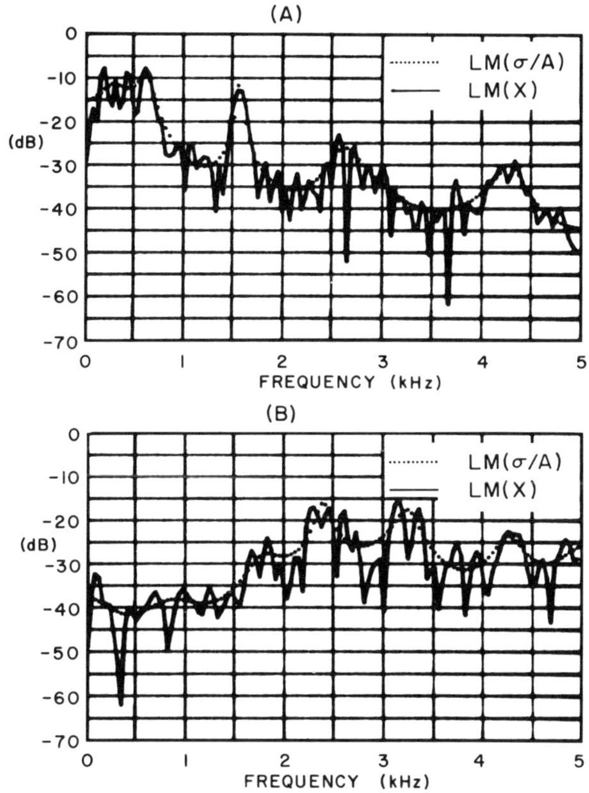

Fig. 2.2 Examples of correlation matching formulation with gain term σ. A) voiced sound. B) unvoiced sound.

and, in addition, to motivate the introduction of inner products for a unified solution procedure in Chapter 3. The properties of this formulation for vocoder applications are discussed in Chapter 10.

Here not just one, but rather M forward and M backward linear predictors are utilized. The m*th* forward prediction error at time n is denoted by $x_m^+(n)$ and the m*th* backward prediction error by $x_m^-(n)$. The forward prediction error results from the difference between the signal and a predicted signal based upon a linear combination of samples at the previous m units of time,

$$
\begin{aligned}
x_m^+(n) &= x(n) - \left[-\sum_{i=1}^{M} a_{mi} x(n-i) \right] \\
&= \sum_{i=0}^{M} a_{mi} x(n-i) \quad \text{with} \quad a_{m0} = 1.
\end{aligned}
\tag{2.45}
$$

The backward prediction error results from the difference between the signal and a backward prediction based upon the m samples to follow, both evaluated at the time $n-m-1$ so that there need be only causal relations involved,

$$\begin{aligned} x_m^-(n) &= x(n-m-1) - \left[-\sum_{i=1}^{m} b_{mi} x(n-i) \right] \\ &= \sum_{i=1}^{m+1} b_{mi} x(n-i) \quad \text{with} \quad b_{m,m+1} = 1. \end{aligned} \tag{2.46}$$

An illustration of the samples used for backward and forward prediction is shown in Fig. 2.3.

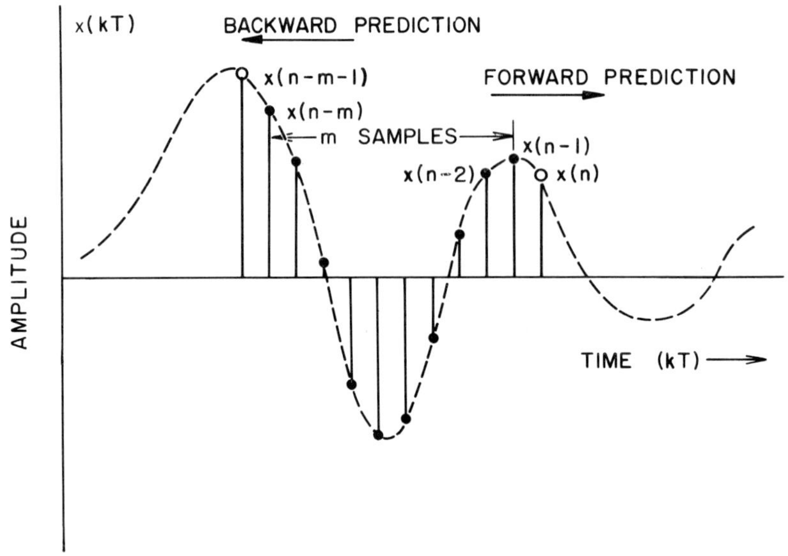

Fig. 2.3 Illustration showing samples used in both forward and backward prediction of order m.

The criterion used in the PARCOR formulation is simultaneous minimization of the total squared error of the forward and backward prediction errors,

$$\alpha_m = \sum_{n=n_0}^{n_1} [x_m^+(n)]^2 \quad \text{and} \quad \beta_m = \sum_{n=n_0}^{n_1} [x_m^-(n)]^2 \tag{2.47}$$

for $m = 1, 2, \ldots, M$.

The minimization can be carried out by taking partial derivatives with respect to the specific parameters and equating them to zero,

$$\frac{\partial \alpha_m}{\partial a_{mi}} = 0 \quad \text{and} \quad \frac{\partial \beta_m}{\partial b_{mi}} = 0$$

for $i=1,2,\ldots,m$. However, this approach does not treat the question of existence or uniqueness of minimizing coefficients, nor does it directly show that the results obtained yield minima rather than points of inflection or maxima. The desired results are now obtained using a well-known orthogonality principle.

2.6.1 Inner Products and an Orthogonality Principle

The forward and backward prediction errors, $x_m^+(n)$ and $x_m^-(n)$, can be considered as the outputs of two filters, $A_m(z)$ and $B_m(z)$, having the common input sequence $\{x(n)\}$, where

$$A_m(z)=\sum_{i=0}^{m} a_{mi}z^{-i} \quad \text{with} \quad a_{m0}=1 \tag{2.48a}$$

and

$$B_m(z)=\sum_{i=1}^{m+1} b_{mi}z^{-i} \quad \text{with} \quad b_{m,m+1}=1. \tag{2.48b}$$

The total squared errors α_m and β_m of (2.47) represent the filter output energies for the time intervals from $n=n_0$ through $n=n_1$.

For notational convenience, we consider a more general situation with filters $F(z)$ and $G(z)$, having real coefficients (real filters). As shown in Fig. 2.4, the filters

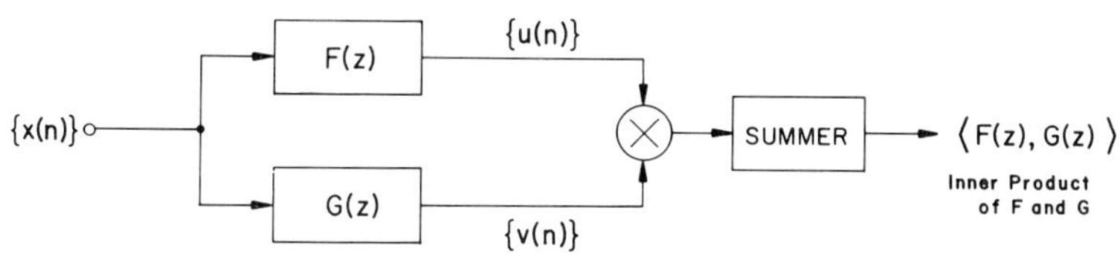

Fig. 2.4 Signal processing interpretation of inner product.

have a common input sequence $\{x(n)\}$. The outputs of the filters are labeled $\{u(n)\}$ and $\{v(n)\}$. These outputs are multiplied and summed from $n=n_0$ through $n=n_1$. The result will be called the inner product of $F(z)$ and $G(z)$ and denoted by $\langle F(z), G(z)\rangle$, with

$$\langle F(z), G(z)\rangle=\sum_{n=n_0}^{n_1} u(n)v(n).$$

Note that the inner product depends upon the input sequence $\{x(n)\}$ and upon the limits of the summation, n_0 and n_1. For now, the inner product can simply be thought of as a notational convenience. For example, if the filters $F(z)$ and $G(z)$ are of the form

$$F(z)=\sum_{i=0}^{\infty} f_i z^{-i} \quad \text{and} \quad G(z)=\sum_{i=0}^{\infty} g_i z^{-i}$$

then a direct calculation shows that

$$\langle F(z), G(z)\rangle = \sum_{i=0}^{\infty} \sum_{j=0}^{\infty} f_i \left[\sum_{n=n_0}^{n_1} x(n-i)x(n-j) \right] g_j, \tag{2.49}$$

and thus

$$\langle z^{-i}, z^{-j}\rangle = \sum_{n=n_0}^{n_1} x(n-i)x(n-j). \tag{2.50}$$

In the covariance method, $n_0 = M$ and $n_1 = N-1$, giving

$$\langle z^{-i}, z^{-j}\rangle = c_{ij} = c_{ji}. \tag{2.51}$$

In the autocorrelation method, where the data samples are non-zero only over the interval $n=0, 1, \ldots, N-1$ with limits $n_0 = -\infty$ and $n_1 = \infty$,

$$\langle z^{-i}, z^{-j}\rangle = r(i-j) = r(j-i) \tag{2.52}$$

is obtained from substitution into (2.49).

In all cases it is important to note that the inner product of two filters represents a number which is not a function of frequency, the transform variable z, or time. It does depend upon the input data sequence, as well as the filter forms, and it does depend upon the summation limits n_0 and n_1. It will be shown later that inner products can also be interpreted in terms of integrations in the z-plane or the frequency domain for the autocorrelation method only. For now the inner product definition must be restricted to Fig. 2.3 or equivalently to the summation of (2.49).

In terms of the inner product definition, the errors to be minimized can be expressed in the form

$$\alpha_m = \langle A_m(z), A_m(z)\rangle \quad \text{and} \quad \beta_m = \langle B_m(z), B_m(z)\rangle. \tag{2.53}$$

The inner product of a filter with itself will be called a norm square, and denoted by the symbol $\|\cdot\|^2$, so that

$$\|F(z)\|^2 = \langle F(z), F(z)\rangle. \tag{2.54}$$

The criterion of minimizing the predictor errors can thus be described as that of minimizing the norm squares of the polynominals $A_m(z)$ and $B_m(z)$, for $m=1, 2, \ldots, M$.

Using the shorthand notation of the inner products, the orthogonality principle will now be developed as it applies to the minimization problem here. If $A_m(z)$ or $B_m(z)$ truly minimize α_m or β_m, then adding $c\,z^{-j}$ (where $j=1, 2, \ldots, m$, and c is any constant) to the polynomial must result in a larger norm square. For example,

$$\|A_m(z) + c\,z^{-j}\|^2 \geq \|A_m(z)\|^2 \quad \text{for} \quad j=1, 2, \ldots, m. \tag{2.55}$$

The constant c can take on any value. If (2.55) is written out in terms of inner products, with the various terms multiplied out, then it is found that

$$2c\langle A_m(z), z^{-j}\rangle + c^2\langle z^{-j}, z^{-j}\rangle \geq 0,$$

for any value of c. If $\langle z^{-j}, z^{-j}\rangle$ is non-zero, choosing c as

$$c = -\langle A_m(z), z^{-j}\rangle / \langle z^{-j}, z^{-j}\rangle$$

leads to

$$-[\langle A_m(z), z^{-j}\rangle]^2 \geq 0.$$

In the unlikely event that $\langle z^{-j}, z^{-j}\rangle$ is zero, c can be chosen as $c = -\langle A_m(z), z^{-j}\rangle$ to obtain the identical result. As this result must hold for $j = 1, 2, \ldots, m$,

$$\langle A_m(z), z^{-j}\rangle = 0 \quad \text{for} \quad j = 1, 2, \ldots, m, \tag{2.56a}$$

must be true. Exactly the same procedure leads to

$$\langle B_m(z), z^{-j}\rangle = 0 \quad \text{for} \quad j = 1, 2, \ldots, m. \tag{2.56b}$$

These relations of (2.56) represent the orthogonality principle as it applies to the minimization of the forward and backward total squared prediction errors,

$$\alpha_m = \|A_m(z)\|^2 \quad \text{and} \quad \beta_m = \|B_m(z)\|^2. \tag{2.57}$$

As stated, they give necessary conditions for a minimum. To show that they are also sufficient is a simple matter of using the orthogonality relations to note that if

$$Q(z) = \sum_{j=1}^{m} q_j z^{-j}$$

then the orthogonality relations of (2.56) lead to the results

$$\|A_m(z) + Q(z)\|^2 = \|A_m(z)\|^2 + \|Q(z)\|^2 \geq \|A_m(z)\|^2 = \alpha_m$$

$$\|B_m(z) + Q(z)\|^2 = \|B_m(z)\|^2 + \|Q(z)\|^2 \geq \|B_m(z)\|^2 = \beta_m.$$

Thus any polynomial differing from $A_m(z)$ or $B_m(z)$ in the coefficients of z^{-j} for $j = 1, 2, \ldots, m$, must have a norm square that is larger than or equal to the norm square of $A_m(z)$ or $B_m(z)$.

The orthogonality relations of (2.56) can also be expressed in terms of the error signals and the delayed input sequence, for from the definitions of the inner products and error signals (2.45) and (2.46),

$$\langle A_m(z), z^{-j}\rangle = \sum_{n=n_0}^{n_1} x_m^+(n)\, x(n-j) = 0 \tag{2.58a}$$

and

$$\langle B_m(z), z^{-j}\rangle = \sum_{n=n_0}^{n_1} x_m^-(n)\, x(n-j) = 0 \tag{2.58b}$$

for $j=1,2,\ldots,m$. These relations have been utilized to discuss the decorrelating process carried out in linear predictive analysis [Wakita, 1972].

2.6.2 The PARCOR Lattice Structure

The next step in the analysis procedure is to use the orthogonality relations (2.56) to solve for the polynomials $A_m(z)$ and $B_m(z)$ in a recursive manner, starting with

$$A_0(z)=1 \quad \text{and} \quad B_0(z)=z^{-1} \tag{2.59}$$

and proceeding with $m=1,2,\ldots,M$.

If $A_{m-1}(z)$ and $B_{m-1}(z)$ have already been obtained, then finding $A_m(z)$ is straightforward. A linear combination of the form $A_{m-1}(z)+k_m B_{m-1}(z)$ will be a polynomial of the proper order, m, with the proper leading coefficient, one, and will be orthogonal to the powers $z^{-1}, z^{-2}, \ldots, z^{-(m-1)}$. If k_m can be chosen so as to make the linear combination orthogonal to z^{-m}, then $A_m(z)$ will have been obtained. Thus,

$$A_m(z)=A_{m-1}(z)+k_m B_{m-1}(z). \tag{2.60}$$

The term k_m is obtained by requiring that $A_m(z)$ be orthogonal to z^{-m}, so that

$$\langle A_m(z), z^{-m}\rangle = 0 = \langle A_{m-1}(z), z^{-m}\rangle + k_m \langle B_{m-1}(z), z^{-m}\rangle. \tag{2.61}$$

By using the fact that both $A_{m-1}(z)$ and $B_{m-1}(z)$ must be orthogonal to the powers $z^{-1}, z^{-2}, \ldots, z^{-(m-1)}$, each of the terms in (2.61) can be expressed in a number of different forms. For example,

$$\langle A_{m-1}(z), z^{-m}\rangle = \langle A_{m-1}(z), B_{m-1}(z)\rangle = \langle 1, B_{m-1}(z)\rangle \tag{2.62a}$$

and

$$\langle B_{m-1}(z), z^{-m}\rangle = \langle B_{m-1}(z), B_{m-1}(z)\rangle = \|B_{m-1}(z)\|^2 = \beta_{m-1}. \tag{2.62b}$$

The constant k_m can be obtained from (2.61) in a number of different forms by utilizing (2.62). One of those can be expressed as

$$\begin{aligned} k_m &= -\frac{1}{\beta_{m-1}} \langle A_{m-1}(z), B_{m-1}(z)\rangle \\ &= -\frac{1}{\beta_{m-1}} \sum_{n=n_0}^{n_1} x_{m-1}^+(n)\, x_{m-1}^-(n). \end{aligned} \tag{2.63}$$

Although the development of Itakura and Saito [1972b, 1972c] was strictly for the autocorrelation method, this development so far encompasses both methods. It will be shown in Chapter 3 that if the denominator of (2.63) is zero (as can occur only in the covariance method) then k_m is arbitrary, and any value can be utilized. The determination of $B_m(z)$ is simplest for the autocorrelation method. For the remainder of this chapter we shall cover only this method, saving the discussion of the covariance method for the following chapter. In the autocorrelation method, (2.52) can be used to restate the orthogonality requirement of (2.56a) in the form

$$\begin{aligned}\langle A_m(z), z^{-j}\rangle &= \sum_{i=0}^{m} a_{mi}\langle z^{-i}, z^{-j}\rangle \\ &= \sum_{i=0}^{m} a_{mi}\, r(i-j) = 0 \quad \text{for} \quad j=1,2,\ldots,m.\end{aligned} \tag{2.64}$$

It can be noted at this point that as $a_{m0}=1$, this equation is identical to the autocorrelation method equations obtained in Chapter 1. This result can be used to determine the coefficients of $B_m(z)$ if we reverse the order of the various subscripts. By defining $l=m+1-j$ and $i=m+1-k$, (2.64) can be written as

$$\sum_{k=1}^{m+1} a_{m,m+1-k}\, r(l-k) = 0 \quad \text{for} \quad l=1,2,\ldots,m.$$

Choosing

$$b_{mk} = a_{m,m+1-k} \quad \text{for} \quad k=1,2,\ldots,m+1, \tag{2.65a}$$

gives the z-transform

$$B_m(z) = z^{-(m+1)} A_m(1/z), \tag{2.65b}$$

which then satisfies the orthogonality requirement

$$\langle B_m(z), z^{-l}\rangle = 0 \quad \text{for} \quad l=1,2,\ldots,m.$$

Thus the $B_m(z)$ polynomials have the same coefficients as the $A_m(z)$ polynomials but with the order reversed. Combining (2.60) and (2.65b) gives a recursion relationship for $B_m(z)$ as

$$B_m(z) = z^{-1}[k_m A_{m-1}(z) + B_{m-1}(z)]. \tag{2.66}$$

The PARCOR lattice is most easily visualized in terms of the use of z-transforms to describe the filters, with $X(z)$ representing the z-transform of the input data sequence $\{x(n)\}$ and defining

$$X_m^+(z) = A_m(z)X(z) \quad \text{and} \quad X_m^-(z) = B_m(z)X(z). \tag{2.67}$$

Then, from (2.60) and (2.66),

$$X_m^+(z) = X_{m-1}^+(z) + k_m X_{m-1}^-(z) \tag{2.68a}$$

$$X_m^-(z) = z^{-1}[k_m X_{m-1}^+(z) + X_{m-1}^-(z)], \tag{2.68b}$$

where $X_m^+(z)$ and $X_m^-(z)$ represent the z-transforms of $x_m^+(n)$ and $x_m^-(n)$, respectively. As a result, (2.68) can be written in the form

$$x_m^+(n) = x_{m-1}^+(n) + k_m x_{m-1}^-(n) \tag{2.69a}$$

$$x_m^-(n) = k_m x_{m-1}^+(n-1) + x_{m-1}^-(n-1). \tag{2.69b}$$

From (2.45) and (2.46) the boundary conditions

$$x_0^+(n) = x(n) \quad \text{and} \quad x_0^-(n) = x(n-1) \tag{2.70}$$

are also obtained. These equations are shown in the form of an analysis filter in Fig. 2.5. The boxes labeled CORR evaluate coefficients k_m using (2.63) or one of its equivalent versions.

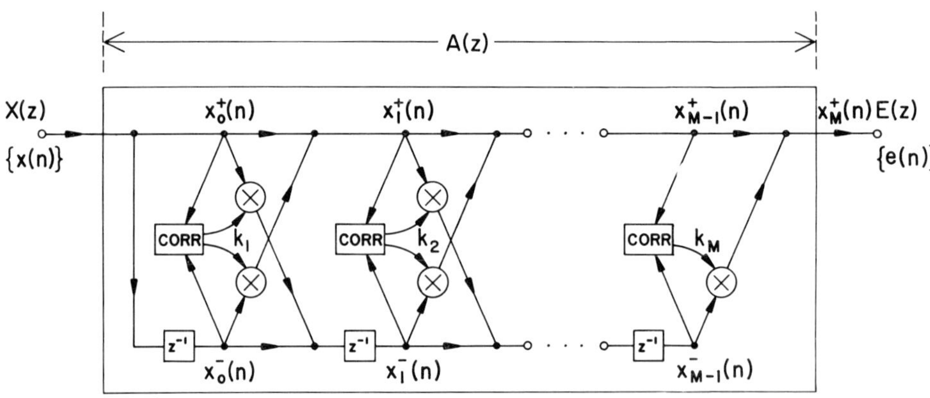

Fig. 2.5 Detailed structure of inverse filter $A(z)$ from PARCOR formulation.

The parameters k_m were defined by Itakura and Saito as partial correlation coefficients (using a different sign convention) since they are equal, but for a sign, to standard statistical correlations in the autocorrelation method. This result follows from (2.63) since the norm squares of $A_m(z)$ and $B_m(z)$ are identical in the autocorrelation method (as can be seen by a direct substitution into the definition (2.49) using (2.52) and which will be shown in more detail in Chapter 3). In the autocorrelation method

$$\alpha_m = \beta_m = \sum_{n=-\infty}^{\infty} [x_m^+(n)]^2 = \sum_{n=-\infty}^{\infty} [x_m^-(n)]^2.$$

As a result,

$$k_m = -\frac{\sum_{n=-\infty}^{\infty} x_{m-1}^{+}(n)\,x_{m-1}^{-}(n)}{\left(\sum_{n=-\infty}^{\infty} [x_{m-1}^{+}(n)]^2 \sum_{n=-\infty}^{\infty} [x_{m-1}^{-}(n)]^2\right)^{1/2}},$$

which represents a correlation coefficient between the forward and backward prediction errors.

In an alternate form, k_m can be computed with the denominator being an arithmetic average of α_{m-1} and β_{m-1}. This approach is discussed in Chapter 10.

3. Solutions and Properties

3.1 Introduction

In the previous chapter, a number of linear prediction formulations were developed. Although different formulations may have appeared to be specifying different problems, it was shown that they all led to the same problem, that of minimizing an expression of the form

$$\alpha = \sum_{i=0}^{M} \sum_{j=0}^{M} a_i c_{ij} a_j, \tag{3.1}$$

where $a_0 = 1$, or equivalently, solving the simultaneous equations

$$\sum_{i=1}^{M} a_i c_{ij} = -c_{0j} \quad \text{for} \quad j = 1, 2, \ldots, M \tag{3.2}$$

for the coefficients a_i. The coefficients c_{ij} were obtained from a correlation of the input data sequence

$$c_{ij} = c_{ji} = \sum_{n=n_0}^{n_1} x(n-i)\, x(n-j). \tag{3.3}$$

In the PARCOR approach of Section 2.6, α, c_{ij}, and a_i do not explicitly appear, but the a_i terms are implicit in the PARCOR lattice since the overall structure defines $A(z)$. In addition, α of (3.1) is minimized by the PARCOR approach, for it is identical to α_M of (2.47).

The limits on the summation of (3.3) differentiate the autocorrelation and covariance methods. In the covariance method the summation is taken from $n_0 = M$ through $n_1 = N-1$. In the autocorrelation method the input samples are first truncated so that $x(n) = 0$ for $n < 0$ and for $n > N-1$, and the summation of (3.3) is taken over all values of n, from $n_0 = -\infty$ through $n_1 = \infty$, though there will be at most a total of N non-zero terms in the summation. In the autocorrelation method the coefficients c_{ij} are a function only of the subscript differences,

$$c_{ij} = r(i-j) \quad \text{(autocorrelation method)}, \tag{3.4}$$

where $r(l)$ is the short-term autocorrelation coefficient defined in Chapter 1.

The coefficients c_{ij} can be directly computed from (3.3) and a general linear simultaneous equation-solving routine can be used to solve (3.2). There are several important reasons, however, for studying these equations in more detail and developing more efficient solution approaches. First, it is desirable to have efficient computation, and in some instances real-time computation of the filter coefficients $\{a_i\}$. It will be shown that a significant decrease in the number of numerical operations can be obtained by using known properties of the coefficients from (3.3).

A second reason for studying the equations in detail is that a number of important properties can be obtained. For example, it will be shown that the equations for both the autocorrelation and covariance methods can be recursively solved for filter polynomials $A_m(z)$ for $m=1, 2, \ldots, M$, where $A(z)=A_M(z)$. This procedure allows the possibility of terminating the solution at an early point if numerical considerations or known speech properties warrant such termination. In addition, for the autocorrelation method a built-in stability test exists to tell whether or not the next recursion step will result in an unstable synthesis filter because of numerical errors.

A third reason for studying the equations is to show that an intermediate set of parameters, k_m, is obtainable from both approaches. In the autocorrelation approach these parameters can be equated to the reflection coefficients of an acoustic tube model of the vocal tract (Chapter 4), provided that the speech is properly pre-emphasized before analysis. As a result, these parameters will be referred to as *reflection coefficients*. They provide necessary and sufficient conditions for stability of the synthesis filter (Chapter 5). The transmission of these parameters will be shown to have a significant advantage over direct transmission of the filter parameters in speech transmission systems (Chapter 10). In the covariance method these parameters are referred to as *generalized reflection coefficients* [Timothy, 1973] and have been utilized for spectral smoothing with a Kalman filter [Matsui, et al., 1972].

The purpose of this chapter is to present a detailed study of algorithms for solving (3.2). The development is based upon an application of the inner product formulation introduced in the preceding chapter. *This approach leads to a unified development for the solutions of both the covariance and autocorrelation methods and lends considerable insight into the solution procedures.*

As it is customary in many areas to use matrix representations for simultaneous equations, equivalent matrix formulations will be given at the beginning and end of the development. For example, (3.1) and (3.2) can be expressed in the form

$$\alpha=\mathbf{a}^t\,\mathbf{C}\mathbf{a}+2\mathbf{a}^t\mathbf{c}+\mathbf{c}^t\mathbf{c} \tag{3.5}$$

and

$$\mathbf{C}\mathbf{a}=-\mathbf{c} \tag{3.6}$$

where the superscript t denotes the transpose, $\mathbf{C}$ is the M by M matrix

$$\mathbf{C}=\begin{pmatrix} c_{11} & c_{12} & \cdot & \cdot & \cdot & c_{1M} \\ c_{21} & c_{22} & \cdot & \cdot & \cdot & c_{2M} \\ \cdot & \cdot & \cdot & \cdot & \cdot & \cdot \\ c_{M1} & c_{M2} & \cdot & \cdot & \cdot & c_{MM} \end{pmatrix} \tag{3.7a}$$

and **a** and **c** are M-dimensional column matrices or vectors whose transposes are given by

$$\mathbf{a}^t=[a_1 \quad a_2 \quad . \; . \; . \quad a_M] \tag{3.7b}$$

and

$$\mathbf{c}^t=[c_{01} \quad c_{02} \quad . \; . \; . \quad c_{0M}]. \tag{3.7c}$$

As **C** is a symmetric matrix, $c_{ij}=c_{ji}$, the solution of (3.6) can be carried out more efficiently than would be the case for a non-symmetric **C**. This approach is referred to as Cholesky decomposition [Clasen, 1966], and can be derived using the classical Gram-Schmidt orthogonalization procedure. In the autocorrelation method the elements of **C** depend only upon their distance from the main diagonal (the difference of their subscripts) so that the matrix is Toeplitz [Grenander and Szego, 1958]. In such a case the solution of (3.6) can be carried out in an even more efficient manner referred to as Levinson's method [Levinson, 1947] or Robinson's method [Robinson, 1967, pp. 274–279]. There are a number of other solution algorithm variations [Morf, 1974] which will not be treated here.

After review of necessary vector space concepts in Section 3.2, the solution algorithms will be developed in Section 3.3. Fortran subroutines and programs which have one-to-one correspondences with various text equations for obtaining solutions are presented. Matrix forms for the autocorrelation and covariance methods are presented in Section 3.4. This presentation draws upon the work of Matsui, et al. [1972], Markel and Gray [1973a], and Timothy [1973].

3.2 Vector Spaces and Inner Products

Through the use of analogies with Euclidian vector spaces, a vector space of polynomials can be utilized for the development of the solution algorithms. It is possible to be even more abstract by introducing a Hilbert space [Narasimha, et al., 1974], but in dealing with the specific speech problems to be considered here this is unnecessary. Rather than utilizing the terminology of abstract vector spaces, we shall restrict our discussion to polynomials in negative powers of z, and point out analogies with ordinary Euclidian spaces.

The first analogy was introduced in Section 2.6.1 where the inner product was defined for polynomials representing the transfer functions of filters having no poles. The inner product was obtained by passing a common input sequence, $\{x(n)\}$, through separate filters, multiplying the filter outputs, and then summing from $n=n_0$ through $n=n_1$. The resulting inner product of two filters or polynomials, $F(z)$ and $G(z)$, given by $\langle F(z), G(z)\rangle$, thus depends upon the input data sequence and also the limits used in the summation, n_0 and n_1. This fact was illustrated in Fig. 2.4.

In (2.49) and (2.50) it was noted that the inner product can be a convenient shorthand notation, for if $F(z)$ and $G(z)$ are defined by

$$F(z)=\sum_{i=0}^{\infty} f_i z^{-i} \quad \text{and} \quad G(z)=\sum_{i=0}^{\infty} g_i z^{-i} \tag{3.8}$$

then

$$\langle F(z), G(z)\rangle = \sum_{i=0}^{\infty}\sum_{j=0}^{\infty} f_i c_{ij} g_j \tag{3.9}$$

with

$$c_{ij}=\langle z^{-i}, z^{-j}\rangle = \sum_{n=n_0}^{n_1} x(n-i)\,x(n-j). \tag{3.10}$$

While the summations of (3.9) are infinite, in most cases they will reduce to finite summations because the polynomials in question will be of finite order. In addition, all summations which must be actually carried out in the solution procedures will reduce to summations over only one index, for one of the polynomials will be a simple power of z, thus allowing the double summation to reduce to a single summation as in the example

$$\langle F(z), z^{-j}\rangle = \sum_{i=0}^{\infty} f_i c_{ij}. \tag{3.11}$$

Using the inner product notation, the total squared error of (3.1) to be minimized can be represented in the very simple form

$$\alpha = \langle A(z), A(z)\rangle \tag{3.12}$$

where $A(z)$ is the polynomial

$$A(z)=\sum_{i=0}^{M} a_i z^{-i} \tag{3.13}$$

with $a_0=1$.

The simultaneous equations (3.2) can be represented in inner product form by inspection of (3.11) and (3.13) as

$$\langle A(z), z^{-j}\rangle = 0 \quad \text{for} \quad j=1,2,\ldots,M. \tag{3.14}$$

Therefore, *the solution for the linear predictor coefficients can be found by forcing the inverse filter, $A(z)$, to be orthogonal to the powers of z^{-1} from z^{-1} through z^{-M}.* This expression forms the basis for the recursive solutions to the problem.

3.2.1 Filter or Polynomial Norms

In Euclidian vector spaces, the dot product of a vector with respect to itself is defined as the square of the length or norm square of that vector. The norm square of a polynomial is defined here as the inner product of that polynomial with respect to itself. Using the notation $||F(z)||$ to denote the norm of $F(z)$, then by definition

$$||F(z)||^2 = \langle F(z), F(z) \rangle. \tag{3.15a}$$

Such norm squares will certainly be non-negative, for they arise from the summation of the squares of filter outputs. By referring to Fig. 2.4, where the filter $F(z)$ has the input sequence $\{x(n)\}$ and output sequence $\{u(n)\}$, it can then be noted that

$$||F(z)||^2 = \sum_{n=n_0}^{n_1} [u(n)]^2 \tag{3.15b}$$

is non-negative (since all filters and signals being considered are real). It can equal zero if and only if $u(n)$, the filter output, is zero for each n in the summation from $n=n_0$ through $n=n_1$. As a result, two elementary properties are obtained:

$$||F(z)||^2 \geq 0 \quad \text{for all} \quad F(z), \tag{3.16}$$

and

$$\text{if} \quad ||F(z)|| = 0 \quad \text{then} \quad \langle F(z), G(z) \rangle = 0. \tag{3.17}$$

The latter property arises from the fact that if the norm of $F(z)$ is zero, it represents a filter whose output is zero for all values of n used in the summation defining the inner product. Following this logic further, it is seen that

$$||F(z)+G(z)|| = ||G(z)|| \quad \text{if} \quad ||F(z)|| = 0, \tag{3.18}$$

so that the addition of a polynomial with zero norm to another polynomial does not change its norm.

This brings to mind an important question: When can a polynomial have a zero norm? In a Euclidian vector space, only the zero vector has zero length. But here, a situation exists where a polynomial might have a zero norm without being zero. If $F(z)$ is such a polynomial, and an input sequence $\{x(n)\}$ is applied, then its output sequence $\{u(n)\}$ must vanish for $n=n_0$ through $n=n_1$. In the autocorrelation method, the input sequence is truncated and the limits on the summation are extended from $n_0 = -\infty$ through $n_1 = \infty$. Thus, as long as there is a single non-zero data point in the input sequence, $F(z)$ can have a zero norm *if and only if* $F(z)$ is itself zero, since a zero norm in this case implies a zero output for all samples. Thus,

$$||F(z)||^2 > 0 \quad \text{for} \quad F(z) \neq 0 \quad \text{(autocorrelation method)}. \tag{3.19}$$

In the covariance method, where the limits on the summation defining the inner product are finite, a zero norm implies only that the filter has a zero output for the samples from $n=n_0$ through $n=n_1$. This event will happen if the input data sequence happens to satisfy a homogeneous difference equation. For example, if the input data sequence satisfies

$$u(n)=\sum_{i=0}^{L} f_i x(n-i)=0 \quad \text{for} \quad n=n_0, n_0+1, \ldots, n_1$$

then the polynomial

$$F(z)=\sum_{i=0}^{L} f_i z^{-i}$$

will have a zero norm from (3.15b). Thus

$$\|F(z)\|^2 \geq 0 \quad \text{for} \quad F(z) \neq 0 \quad \text{(covariance method)}. \tag{3.20}$$

3.2.2 Properties of Inner Products

From the definition of the inner product and the symmetry of $\mathbf{C}$ (i.e., $c_{ij}=c_{ji}$), it is easy to see that the inner product is symmetric,

$$\langle F(z), G(z)\rangle = \langle G(z), F(z)\rangle, \tag{3.21a}$$

and linear,

$$\langle F(z), gG(z)+hH(z)\rangle = g\langle F(z), G(z)\rangle + h\langle F(z), H(z)\rangle. \tag{3.21b}$$

These two properties, and the non-negative value of the norm square, allow us to use the term *inner product* consistently with its usual mathematical meaning. In more general situations, complex coefficients can be used, but they are not necessary here. A development of the inner product which allows complex coefficients for the filters may be found elsewhere [Markel and Gray, 1973a].

In Euclidian vector spaces, the magnitude of the dot product of two vectors is always less than or equal to the product of their magnitudes. The equivalent relation here is one form of the Cauchy-Schwartz inequality which states that

$$|\langle F(z), G(z)\rangle| \leq \|F(z)\| \, \|G(z)\|. \tag{3.22}$$

Equality can hold if and only if the polynomials are linearly dependent in the sense that some linear combination of the two has a zero norm. The left-hand side of (3.22) represents the magnitude of a number, hence the magnitude symbol $|\cdot|$. The right-hand side represents the products of norms.

To prove the Cauchy-Schwartz inequality, we can first note that should $F(z)$ have a zero norm it is automatically satisfied as a result of (3.17). If the norm of $F(z)$ is non-zero, the result follows simply from examination of the norm square of $aF(z)-G(z)$ where

$$a=\langle F(z), G(z)\rangle/\|F(z)\|^2.$$

3.2.3 Orthogonality Relations

In the preceding chapter, an approach for solving (3.2) was outlined in the discussion of the PARCOR formulation. Here the technique will be developed rigorously for both the autocorrelation and covariance methods. For the theoretical development inner products will be used, but the end results will be presented in computational form.

Using vector space terminology, the basic problem in the linear prediction formulation is to find the polynomial $A(z)$ of the form

$$A(z)=\sum_{i=0}^{M} a_i z^{-i} \quad \text{with} \quad a_0=1 \tag{3.23}$$

which minimizes the norm square

$$\alpha=\|A(z)\|^2. \tag{3.24}$$

It was shown in Chapter 2 that a necessary and sufficient condition for this minimization is that $A(z)$ be orthogonal to the powers $z^{-1}, z^{-2}, \ldots, z^{-M}$, i.e.,

$$\langle A(z), z^{-i}\rangle=0 \quad \text{for} \quad i=1,2,\ldots,M. \tag{3.25}$$

The approach for finding $A(z)$ as described in the preceding chapter was to recursively generate polynomials $A_m(z)$ and $B_m(z)$ of the form

$$A_m(z)=\sum_{i=0}^{m} a_{mi} z^{-i} \quad \text{with} \quad a_{m0}=1 \tag{3.26a}$$

and

$$B_m(z)=\sum_{i=1}^{m+1} b_{mi} z^{-i} \quad \text{with} \quad b_{m,m+1}=1 \tag{3.26b}$$

which satisfy the orthogonality relations

$$\langle A_m(z), z^{-i}\rangle=\langle B_m(z), z^{-i}\rangle=0 \quad \text{for} \quad i=1,2,\ldots,m. \tag{3.27}$$

The polynomials $\{A_m(z)\}$ do not form an orthogonal set, for each polynomial contains a z^0 term, and the polynomials are not orthogonal to z^0, from (3.27).

The polynomial $B_m(z)$ from (3.26b) is a linear combination of powers of z from z^{-1} to z^{-m}. Therefore each polynomial is orthogonal to all lower order polynomials $B_{m-1}(z), B_{m-2}(z), B_0(z)$ from (3.27). The polynomials $\{B_m(z)\}$ thus form an orthogonal set, i.e.,

$$\langle B_m(z), B_i(z)\rangle = 0 \quad \text{for} \quad m \neq i. \tag{3.28}$$

One approach to finding the polynomial set $\{B_m(z)\}$ would be a classical Gram-Schmidt orthogonalization of the powers $z^{-1}, z^{-2}, \ldots, z^{-M}$. There is a more efficient approach in the case of the autocorrelation method.

The polynomial set $\{A_m(z)\}$ is most easily found recursively by using the orthogonality requirement of (3.27). As $A_{m-1}(z)$ and $B_{m-1}(z)$ must both be orthogonal to $z^{-1}, z^{-2}, \ldots, z^{-(m-1)}$, $A_m(z)$ can be defined as

$$\boxed{A_m(z) = A_{m-1}(z) + k_m B_{m-1}(z)} \quad . \tag{3.29}$$

The unspecified parameter k_m is chosen so that $A_m(z)$ is orthogonal to z^{-m} as well as to the lower order powers, thus satisfying the orthogonality requirement (3.27), as

$$0 = \langle A_m(z), z^{-m}\rangle = \langle A_{m-1}(z), z^{-m}\rangle + k_m \langle B_{m-1}(z), z^{-m}\rangle. \tag{3.30}$$

The orthogonality relations can be used to simplify many inner product calculations, which otherwise might involve double summations as indicated by (3.9). In particular, by using the fact that both $A_m(z)$ and $B_m(z)$ are orthogonal to the powers $z^{-1}, z^{-2}, \ldots, z^{-m}$, as shown in (3.27), then

$$\langle A_m(z), B_m(z)\rangle = \langle 1, B_m(z)\rangle = \langle A_m(z), z^{-(m+1)}\rangle, \tag{3.31}$$

allowing the inner products to be evaluated using a single summation as in (3.11). Similarly, the squared errors at step m can be evaluated as

$$\alpha_m = \|A_m(z)\|^2 = \langle A_m(z), A_m(z)\rangle = \langle 1, A_m(z)\rangle \tag{3.32a}$$

and

$$\beta_m = \|B_m(z)\|^2 = \langle B_m(z), B_m(z)\rangle = \langle z^{-(m+1)}, B_m(z)\rangle. \tag{3.32b}$$

Therefore, (3.30) can be expressed in the form

$$k_m = -\langle A_{m-1}(z), z^{-m}\rangle / \beta_{m-1}. \tag{3.33}$$

The orthogonality relations can also be used to recursively calculate α_m. Starting with $A_0(z) = 1$ and using (3.29), the inverse filter can be written in terms of the orthogonal polynomials $\{B_i(z)\}$ as

$$\boxed{A_m(z) = 1 + \sum_{i=1}^{m} k_i B_{i-1}(z)} \tag{3.34}$$

for $m > 0$. As the polynomial set $\{B_m(z)\}$ is orthogonal, the norm square of (3.34) can be evaluated as

$$||A_m(z) - 1||^2 = \sum_{i=1}^{m} k_i^2 \beta_{i-1}.$$

But by substituting into the form (3.15a) and applying the linearity property (3.21),

$$\begin{aligned} ||A_m(z) - 1||^2 &= ||A_m(z)||^2 - 2\langle A_m(z), 1\rangle + ||1||^2 \\ &= ||1||^2 - \alpha_m. \end{aligned}$$

Therefore,

$$\alpha_m = ||1||^2 - \sum_{i=1}^{m} k_i^2 \beta_{i-1}. \tag{3.35}$$

Replacing m by $m+1$ in (3.35) and subtracting α_m from α_{m+1} results in

$$\alpha_{m+1} = \alpha_m - k_{m+1}^2 \beta_m. \tag{3.36}$$

3.3 Solution Algorithms

A unified recursive solution for both the autocorrelation and covariance methods is presented in this section. The procedure results in an m*th* order inverse filter $A_m(z)$, a coefficient k_m, and a total squared error α_m at each recursion step m, for $m = 1, 2, \ldots, M$.

The final results are presented in Figs. 3.1 and 3.2 by way of the Fortran subroutines AUTO and COVAR, for the autocorrelation and covariance method implementations, respectively. The equation numbers in the text, corresponding to specific Fortran statements, are indicated on the programs. Table 3.1 shows the relationship between text and subroutine variables.

Table 3.1. Listing of Fortran and text variables

TEXT	*COVAR*	*AUTO*
N	N	N
$x(n)$ for $n = 0, 1, \ldots, N-1$	$X(n+1)$	$X(n+1)$
M	M	M
a_{mi} for $0 \le i \le m \le M$	$A(i+1)$	$A(i+1)$
α_m for $m = 0, 1, \ldots, M$	*ALPHA*	*ALPHA*
k_m for $m = 1, 2, \ldots, M$	$GRC(m)$	$RC(m)$
c_{ij} for $0 \le j \le i \le M$	$C(i+1, j+1)$	$R(i-j+1)$
$r(i)$ for $i = 0, 1, \ldots, M$	—	$R(i+1)$
β_m for $m = 0, 1, \ldots, M-1$	$BETA\ (m+1)$	*ALPHA*
b_{mi} for $1 \le i \le m+1 \le M$	$B(m+1, i)$	$B(i)$
γ_{mi} for $0 \le i \le m-1 \le M-2$	*GAM*	—

	FORTRAN PROGRAM	TEXT EQUATION
	SUBROUTINE AUTO(N,X,M,A,ALPHA,RC) DIMENSION X(1),A(1),RC(1) DIMENSION R(21),B(20) MP=M+1	–
     100	DO 100 K=1,MP R(K)=0. NK=N-K+1 DO 100 NP=1,NK NPK=NP+K-1 R(K)=R(K)+X(NP)*X(NPK)	(3.39)
	ALPHA=R(1)	(3.41a)
	RC(1)=-R(2)/R(1)	(3.42)
	A(1)=1. A(2)=RC(1)	(3.43)
	ALPHA=ALPHA-RC(1)*RC(1)*ALPHA	(3.44)
	MF=M DO 400 MINC=2,MF M=MINC-1	–
  200	DO 200 J=1,MINC JB=MINC-J+1 B(J)=A(JB)	(3.53)
	M=M+1	–
   300	S=0. DO 300 IP=1,M MIP=M-IP+2 S=S+R(MIP)*A(IP) RC(M)=-S/ALPHA	(3.55)
 350	DO 350 IP=2,M A(IP)=A(IP)+RC(M)*B(IP-1) A(M+1)=RC(M)	(3.56)
	ALPHA=ALPHA-RC(M)*RC(M)*ALPHA	(3.57)
 500 400	IF(ALPHA.LE.0.)WRITE(5,500) FORMAT(' INSUFFICENT ACCURACY'/) CONTINUE RETURN END	–

Fig. 3.1. Fortran subroutine AUTO for implementing the autocorrelation method of linear prediction.

3.3.1 Correlation Matrix

The first step in the solution is to obtain the coefficients c_{ij} as given by (3.3). In the covariance method $n_0 = M$ and $n_1 = N-1$, so that

$$c_{ij} = \sum_{n=M}^{N-1} x(n-i)x(n-j) \tag{3.37}$$

becomes the computation expression. This summation is evaluated only for $j = 0, 1, \ldots, i$, and $i = 0, 1, \ldots, M$, since $\mathbf{C}$ is symmetric ($c_{ij} = c_{ji}$). Thus $M(M+1)/2$ coefficients are calculated. In Chapter 9, an efficient computational method for evaluating these coefficients is presented. In the autocorrelation method, the number of calculations is decreased since $n_0 = -\infty$ and $n_1 = \infty$ with $x(n) = 0$ for $n < 0$ and $n > N-1$ giving

$$c_{ij} = c_{0,\,i-j} = c_{i-j,\,0} = r(i-j). \tag{3.38}$$

FORTRAN PROGRAM		TEXT EQUATION
	SUBROUTINE COVAR(N,X,M,A,ALPHA,GRC) DIMENSION X(1),A(1),GRC(1) DIMENSION C(21,21),B(20,20),BETA(20) MP=M+1	–
       100	DO 100 IP=1,MP DO 100 JP=1,IP C(IP,JP)=0. DO 100 NP=MP,N NPI=NP+1-IP NPJ=NP+1-JP C(IP,JP)=C(IP,JP)+X(NPI)*X(NPJ)	(3.37)
	B(1,1)=1.	(3.40)
	ALPHA=C(1,1) BETA(1)=C(2,2)	(3.41)
	GRC(1)=-C(2,1)/C(2,2)	(3.42)
	A(1)=1. A(2)=GRC(1)	(3.43)
	ALPHA=ALPHA-GRC(1)*GRC(1)*BETA(1)	(3.44)
	MF=M DO 400 MINC=2,MF M=MINC-1	–
	B(MINC,MINC)=1.	(3.51a)
  130 150 190 200	DO 200 IP=1,M IF(BETA(IP)) 600,200,130 GAM=0. DO 150 J=1,IP GAM=GAM+C(M+2,J+1)*B(IP,J) GAM=GAM/BETA(IP) DO 190 JP=1,IP B(MINC,JP)=B(MINC,JP)-GAM*B(IP,JP) CONTINUE	(3.50) & (3.51b)
  250	BETA(MINC)=0. DO 250 J=1,MINC BETA(MINC)=BETA(MINC)+C(M+2,J+1)*B(MINC,J)	(3.52)
	M=M+1 IF(BETA(M)) 600,360,260	–
260 300	S=0. DO 300 IP=1,M S=S+C(M+1,IP)*A(IP) GRC(M)=-S/BETA(M)	(3.55)
 350	DO 350 IP=2,M A(IP)=A(IP)+GRC(M)*B(M,IP-1) A(M+1)=GRC(M)	(3.56)
360	CONTINUE	–
	ALPHA=ALPHA-GRC(M)*GRC(M)*BETA(M)	(3.57)
 400 600	IF(ALPHA) 600,600,400 CONTINUE RETURN END	–

Fig. 3.2 Fortran subroutine COVAR for implementing the covariance method of linear prediction.

The necessary $M+1$ autocorrelation coefficients are evaluated from

$$c_{0,k}=r(k)=\sum_{n=-\infty}^{\infty} x(n)x(n-k)=\sum_{n=0}^{N-1-k} x(n)x(n+k) \qquad (3.39)$$

where $k=0,1,\ldots,M$.

The equations for evaluating (3.37) and (3.39) are implemented in the *DO* loops of the FORTRAN subroutines which terminate with statement number 100 as shown in Figs. 3.1 and 3.2.

3.3.2 Initialization

At the start of the solution procedure, (3.26) gives the initial conditions for $m=0$ as

$$A_0(z)=1 \quad \text{and} \quad B_0(z)=z^{-1}$$

or

$$a_{00}=1 \quad \text{and} \quad b_{01}=1. \tag{3.40}$$

Direct evalution of the inner products in (3.31) and (3.32) using (3.9) gives

$$\alpha_0=\langle A_0(z), A_0(z)\rangle=c_{00} \tag{3.41a}$$

and

$$\beta_0=\langle B_0(z), B_0(z)\rangle=c_{11}. \tag{3.41b}$$

In the autocorrelation method, both c_{00} and c_{11} equal $r(0)$ from (3.4). In addition,

$$\langle A_0(z), B_0(z)\rangle=\langle 1, z^{-1}\rangle=c_{01}=c_{10},$$

so that (3.33) with $m=1$ gives the result

$$k_1=-c_{10}/\beta_0=-c_{10}/c_{11}. \tag{3.42}$$

Using (3.29), $A_1(z)$ can be evaluated as

$$A_1(z)=A_0(z)+k_1 B_0(z)=1+k_1 z^{-1}$$

so that

$$a_{10}=1 \quad \text{and} \quad a_{11}=k_1. \tag{3.43}$$

Using (3.36) with $m=0$,

$$\alpha_1=\alpha_0-k_1^2\beta_0, \tag{3.44}$$

which completes the initialization procedure. Recursion relationships for obtaining $A_m(z)$ for $m=2, 3, \ldots, M$, will now be developed. At completion, the inverse filter and total squared error are given by

$$A(z)=A_M(z) \quad \text{and} \quad \alpha=\alpha_M.$$

3.3.3 Gram-Schmidt Orthogonalization

Assume that step $m-1$ has been completed. This implies that the terms $B_i(z)$ and β_i for $i=0, 1, \ldots, m-1$, are known, and that $A_m(z)$ and α_m are known. From (3.29) and (3.30) it is seen that to complete step m (replacing m by $m+1$ in (3.29)), it is necessary to know only $B_m(z)$. The classical Gram-Schmidt orthogonalization procedure can be used to recursively obtain $B_m(z)$ for both the covariance and autocorrelation methods, as will now be shown.

What is desired is a polynomial of the form (3.26b)

$$\begin{aligned} B_m(z) &= \sum_{i=1}^{m+1} b_{mi} z^{-i} \\ &= z^{-(m+1)} + \sum_{i=1}^{m} b_{mi} z^{-i}, \end{aligned} \tag{3.45}$$

which satisfies the orthogonality relation (3.27)

$$\langle B_m(z), z^{-j} \rangle = 0 \quad \text{for} \quad j=1, 2, \ldots, m \tag{3.46a}$$

and thus

$$\langle B_i(z), B_j(z) \rangle = 0 \quad \text{for} \quad i \neq j, \tag{3.46b}$$

where $b_{m,m+1}=1$. As the lower-ordered polynomials, $B_i(z)$ for $i=0, 1, \ldots, m-1$, are assumed to be known, a linear combination of them with the term $z^{-(m+1)}$ can define $B_m(z)$ in the form

$$B_m(z) = z^{-(m+1)} - \sum_{i=0}^{m-1} \gamma_{mi} B_i(z). \tag{3.47}$$

The coefficients γ_{mi} are determined by applying the orthogonality requirements of (3.46). Taking inner products of (3.47) with respect to $B_n(z)$ and applying the orthogonality relations of (3.46b) for $n=0, 1, \ldots, m-1$, results in

$$\langle B_n(z), B_m(z) \rangle = 0 = \langle z^{-(m+1)}, B_n(z) \rangle - \gamma_{mn} \beta_n \tag{3.48}$$

with

$$\beta_n = \langle B_n(z), B_n(z) \rangle.$$

If β_n is zero, (3.48) is automatically satisfied since both sides of the equation are zero as seen from (3.17). Therefore,

$$\gamma_{mn} = \begin{cases} \langle z^{-(m+1)}, B_n(z) \rangle / \beta_n & \text{for} \quad \beta_n \neq 0 \\ \text{arbitrary} & \text{for} \quad \beta_n = 0 \end{cases}. \tag{3.49}$$

If $\beta_n \neq 0$ for $n=0, 1, \ldots, m-1$, then (3.49) can be written in computational form by direct evaluation of the inner product as

$$\gamma_{mn} = \frac{1}{\beta_n} \sum_{j=1}^{n+1} c_{m+1,j} b_{nj} \tag{3.50}$$

for $n=0, 1, \ldots, m-1$. $B_m(z)$ can then be found from (3.47), which, when expressed in computational form, gives

$$b_{m,m+1} = 1 \tag{3.51a}$$

$$b_{mj} = -\sum_{i=0}^{m-1} \gamma_{mi} b_{ij} = -\sum_{i=j-1}^{m-1} \gamma_{mi} b_{ij} \;. \tag{3.51b}$$

Using (3.32b) to obtain β_m, results in

$$\beta_m = \langle z^{-(m+1)}, B_m(z) \rangle = \sum_{j=1}^{m+1} c_{m+1,j} b_{mj} \;. \tag{3.52}$$

The evaluation of $B_m(z)$ and β_m at step m is now complete. The new $A_{m+1}(z)$ and α_{m+1} are evaluated next. First, the simplified process (valid only for the autocorrelation method) is presented.

3.3.4 Levinson Recursion

For the autocorrelation method, the Gram-Schmidt procedure produces results that can be equivalently obtained much more efficiently by other means. It was noted in Chapter 2 that when the coefficients c_{ij} depend only upon the difference of the subscripts, then the polynomials $B_m(z)$ are related to the polynomials $A_m(z)$ by

$$B_m(z) = z^{-(m+1)} A_m(1/z) \tag{3.53}$$

and therefore,

$$b_{mi} = a_{m,m+1-i} \quad \text{for} \quad i=1, 2, \ldots, m+1.$$

Application of (3.53) reduces the necessary number of operations in the solution of the autocorrelation equations to the order of M^2, denoted as $O(M^2)$, whereas the full Gram-Schmidt procedure requires $O(M^3)$ operations. The efficient recursive solution procedure is due to Levinson [1947]. From (3.53), it can be shown that the norm squares of $B_m(z)$ and $A_m(z)$ are identical, so that

$$\beta_m = \alpha_m.$$

Thus, the simple replacement as indicated by (3.53) produces the same results as the Gram-Schmidt orthogonalization used in the autocorrelation approach, only it does it much more efficiently.

3.3.5 Updating $A_m(z)$

To complete step m, (3.29) has its index incremented by one to give

$$A_{m+1}(z)=A_m(z)+k_{m+1}\,B_m(z). \tag{3.54}$$

Eq. (3.30) with m incremented by one then gives

$$0=\langle A_m(z), z^{-(m+1)}\rangle+k_{m+1}\,\beta_m.$$

If β_m is non-zero, the coefficient k_{m+1} is obtained as

$$k_{m+1}=-\frac{1}{\beta_m}\sum_{i=0}^{m} c_{m+1,\,i}\,a_{m,\,i}. \tag{3.55}$$

In the autocorrelation method, $\beta_m=\alpha_m$ and $c_{m+1,i}=r(m+1-i)$. Once k_{m+1} has been computed, the filter coefficients are evaluated from (3.54) as

$$a_{m+1,0}=1 \tag{3.56a}$$

$$a_{m+1,i}=a_{mi}+k_{m+1}\,b_{mi} \quad \text{for} \quad i=1,\ldots,m \tag{3.56b}$$

$$a_{m+1,m+1}=k_{m+1}. \tag{3.56c}$$

The term α_{m+1} is directly evaluated from (3.36) as

$$\alpha_{m+1}=\alpha_m-k_{m+1}^2\,\beta_m. \tag{3.57}$$

Step m of the process is now complete.

Since the initial conditions for $m=0$ have been obtained in Section 3.3.2, and since it has been shown how to complete step m given that step $m-1$ is completed, the recursive procedure for both the covariance and autocorrelation methods is now specified. If any of the norm squares α_m or β_m are zero or negative, the Fortran subroutines terminate the process, since the results are either in error (in the case of negative norm squares) or may be exceeding the accuracy of the computer (in the case of a zero norm square). With floating-point arithmetic, a zero norm square is highly unlikely. It is more likely that round-off or truncation errors would result in a very small norm square rather than one which is zero. A zero norm can theoretically occur only in the covariance method. In the autocorrelation method, the parameters $\{k_m\}$ are referred to as the *reflection coefficients* since they define the reflection coefficients of an acoustic tube model of the vocal tract as shown in Chapter 4. In the covariance method the parameters $\{k_m\}$ are referred to as the *generalized reflection coefficients* since the same recursion form in (3.33) is used for evaluating k_m in both methods. In Chapters 4 and 5, the reflection coefficients will be shown to theoretically satisfy certain important properties that the generalized reflection coefficients do not satisfy. The reason for using the same symbol with a different name is that, as a special case of the covariance method, if the input data sequence $\{x(n)\}$ is truncated so that $x(n)=0$ for $n<M$ and $n>N-M-1$, then the results will be identical to those of the autocorrelation method.

3.3.6 A Test Example

An example usage of the Fortran programs AUTO and COVAR is shown in Fig. 3.3. A test signal $x(n)$ is defined by

$$x(n)=2(.99)^{n-4}-(.99)^{2(n-4)} \qquad n=M, M+1, \ldots, N-M-1$$

where $N=24$ and $M=4$. To illustrate the special condition where the generalized reflection coefficients of the covariance method equal the reflection coefficients of

```
C
      DIMENSION X(100),A(21),RC(21)
      DATA X/100*0./
      Z=1.
      DO 10 J=5,20
      X(J)=2.*Z-Z*Z
   10 Z=.99*Z
C
      M=4
      N=24
      CALL COVAR(N,X,M,A,ALPHA,RC)
      MP=M+1
      WRITE (5,20) (I,A(I),RC(I),I=1,MP)
      WRITE (5,30) ALPHA
C
      M=4
      CALL AUTO(N,X,M,A,ALPHA,RC)
      WRITE (5,20) (I,A(I),RC(I),I=1,MP)
      WRITE (5,30) ALPHA
   20 FORMAT (I3,2E16.8)
   30 FORMAT(E16.8,/)
C
      END
```

J	A(J)	RC(J)
1	0.10000000E 01	-0.93784922E 00
2	-0.96669060E 00	0.33252385E-01
3	0.96835429E-04	0.34218445E-01
4	0.90468675E-04	0.35259910E-01
5	0.35259910E-01	0.00000000E 00
ALPHA =	0.18936189E 01	

J	A(J)	RC(J)
1	0.10000000E 01	-0.93784922E 00
2	-0.96669060E 00	0.33252385E-01
3	0.96824893E-04	0.34218431E-01
4	0.90513378E-04	0.35259861E-01
5	0.35259861E-01	0.00000000E 00
ALPHA =	0.18936189E 01	

Fig. 3.3 Test program with results for AUTO and COVAR.

the autocorrelation method, $x(n)$ was set to zero for $n=0, 1, 2, 3$ and $n=20, 21, 22, 23$. It is seen that for this case, identical results are obtained within the computer accuracy used. If the example is rerun without $x(n)$ artificially set to zero (by setting the *DO LOOP* index from 1 to 24), it will be seen that dramatically different results are obtained. The covariance method will produce a zero norm result $\|A(z)\| = ALPHA = 0$ within the computer accuracy since $\{x(n)\}$ is composed of a linear combination of two exponentials. Whenever a portion of voiced speech being analyzed with the covariance is composed of only complex exponential behavior (as assumed in the linear speech production model within a single pitch period), similar behavior will be observed. This property as it applies to formant frequency estimation is discussed in Chapter 7.

3.4 Matrix Forms

In the preceding sections, solutions to (3.2) have been obtained without explicitly using matrices. Here some of the matrix identities that are inherent in the solution process will be presented. In matrix notation, (3.34) can be expressed as

$$\mathbf{a} = \mathbf{B}\mathbf{k} \tag{3.58}$$

where $\mathbf{a}^t$ is given by (3.7b), $\mathbf{k}^t$ by

$$\mathbf{k}^t = [k_1 k_2 \quad \dots \quad k_M] \tag{3.59}$$

and $\mathbf{B}$ by

$$\mathbf{B} = \begin{pmatrix} 1 & b_{11} & b_{21} & \cdot & \cdot & \cdot & b_{M-1,1} \\ 0 & 1 & b_{22} & \cdot & \cdot & \cdot & b_{M-1,2} \\ 0 & 0 & 1 & \cdot & \cdot & \cdot & b_{M-1,3} \\ \cdot & \cdot & \cdot & \cdot & \cdot & \cdot & \cdot \\ 0 & 0 & 0 & \cdot & \cdot & \cdot & 1 \end{pmatrix} . \tag{3.60}$$

The m*th* column of the triangular matrix $\mathbf{B}$ contains the coefficients of the polynomial $B_m(z)$.

As the polynomials $B_m(z)$ and $B_n(z)$, $n \neq m$, are orthogonal to each other,

$$\langle B_m(z), B_n(z) \rangle = \sum_{i=1}^{m+1} \sum_{j=1}^{n+1} b_{mi} c_{ij} b_{nj} = \delta_{mn} \beta_n .$$

In matrix form this equation becomes

$$\mathbf{B}^t \mathbf{C} \mathbf{B} = \boldsymbol{\beta} \tag{3.61}$$

where $\boldsymbol{\beta}$ is the diagonal matrix

$$\boldsymbol{\beta} = \begin{pmatrix} \beta_0 & 0 & . & . & . & . & 0 \\ 0 & \beta_1 & . & . & . & . & 0 \\ . & . & . & . & . & . & . \\ 0 & 0 & . & . & . & . & \beta_{M-1} \end{pmatrix}. \tag{3.62}$$

The determinants of $\mathbf{C}$ and $\boldsymbol{\beta}$ are therefore related by

$$|\mathbf{C}| = |\boldsymbol{\beta}| = \prod_{m-0}^{M-1} \beta_m . \tag{3.63}$$

This determinant has been applied by Strube [1974] in the estimation of the moment of glottal closure, and can be easily obtained from the subroutine COVAR (see Table 3.1).

The solution of (3.6) in matrix form can be described by first using (3.58) to write

$$\mathbf{C}\,\mathbf{B}\mathbf{k} = -\mathbf{c}.$$

If this expression is premultiplied by $\mathbf{B}^t$ and (3.61) is applied, then

$$\boldsymbol{\beta}\mathbf{k} = -\mathbf{B}^t\mathbf{c}.$$

As $\boldsymbol{\beta}$ is diagonal, it is easily invertible (if it is not singular) giving

$$\mathbf{a} = \mathbf{B}\mathbf{k} = -\mathbf{B}\boldsymbol{\beta}^{-1}\mathbf{B}^t\mathbf{c}$$

for the inverse filter coefficients, or

$$\mathbf{k} = -\boldsymbol{\beta}^{-1}\mathbf{B}^t\mathbf{c}$$

for the generalized reflection coefficients.

The term decomposition refers to the decomposition of $\mathbf{C}^{-1}$, if it exists, which from (3.61) is given by

$$\mathbf{C}^{-1} = \mathbf{B}\boldsymbol{\beta}^{-1}\mathbf{B}^t. \tag{3.64}$$

Thus the inverse of $\mathbf{C}$ can be explicitly described as the product of an upper triangular matrix, a diagonal matrix, and a lower triangular matrix. In this case, the triangular matrices are normalized so that their diagonal elements are unity, as seen from (3.60). This decomposition is usually termed triangular decomposition, but in the case of a symmetric matrix $\mathbf{C}$, the upper and lower triangular matrices are transposes of each other, and the process is called Cholesky decomposition.

The final total squared error of (3.5) is easily expressed in terms of the matrices as

$$\alpha = c_{00} - \mathbf{a}^t\,\mathbf{C}\mathbf{a} = c_{00} - \mathbf{k}^t\,\mathbf{B}^t\,\mathbf{C}\,\mathbf{B}\mathbf{k} = c_{00} - \mathbf{k}^t\boldsymbol{\beta}\mathbf{k}. \tag{3.65}$$

The matrix representation has been introduced to present the unified solution from a different point of view, even though matrices are not explicitly necessary to carry out the solutions.

4. Acoustic Tube Modeling

4.1 Introduction

Acoustic tube models of speech production have been studied for a number of years [Chiba and Kajiyama, 1941; Dunn, 1950; Stevens, et al., 1953; Fant, 1960; Kelly and Lochbaum, 1962]. In these studies, it was shown that from a given tube shape, the resonance frequencies could be obtained. The important inverse problem of determining a unique tube shape from resonance characteristics has received considerable attention in the studies by Schroeder and Mermelstein [1965], Mermelstein [1967], Schroeder [1967], Heinz [1967], and Sondhi and Gopinath [1971].

Interest in the linear prediction model within this context stems from the fact that it allows *the parameters of these acoustic tube models to be estimated directly from the acoustical speech waveform.*

The first attempt at directly computing an acoustic tube model of the vocal tract from the speech waveform was due to Atal [1970b]. He demonstrated that the formant frequencies and bandwidths are sufficient to uniquely determine the areas of an acoustic tube having a specified number of sections. He also demonstrated that a transfer function with M poles is always realizable as the transfer function of an acoustic tube consisting of M cylindrical sections of equal length [Atal and Hanauer, 1971b]. Thus a unique discrete tube shape can be reconstructed from a given-order transfer function polynomial. Wakita [1972] showed that the same acoustic tube model is equivalently represented from the inverse filter $A(z)$ obtained by linear prediction of the acoustical speech waveforms. He also demonstrated the important experimental result that *if the speech is properly pre-emphasized, and if boundary conditions of the acoustic tube are properly chosen, then very reasonable vocal tract shapes can be directly estimated using the autocorrelation method of linear prediction.*

The purposes of this chapter are to 1) briefly review the derivation of the acoustic tube model, 2) show how the parameters of the model are directly obtained using linear prediction of the speech waveform, and 3) show experimental results that tend to validate the method proposed by Wakita. Several other topics related to acoustic tube modeling of the vocal tract are also discussed.

This chapter is based almost exclusively upon the studies of Wakita [1972–1975], Wakita and Gray [1974–1975], Atal [1971a], and Atal and Hanauer [1971b]. Using hindsight, the equations developed by these researchers are reformulated to avoid matrix manipulation and to lend insight into the solution process.

The material is presented in sufficient detail to be essentially self-contained from the starting equations governing pressure (the acoustic analog of voltage) and volume velocity (the acoustic analog of current).

The acoustic tube to be developed for modeling the vocal tract is undoubtedly very simplistic. However, it will be seen that 1) the resulting mathematics are tractable, 2) the parameters of the model can be obtained directly from linear prediction of the speech waveform, and 3) the acoustic tube model can result in very reasonable estimates of the human vocal tract shape (in terms of area functions).

4.2 Acoustic Tube Derivation

The basic assumptions used in deriving the acoustic tube model of the vocal tract are as follows [Atal and Hanauer, 1971b; Wakita, 1972, 1973b]:

1) The vocal tract is assumed to consist of a set of M interconnected sections of equal length. Each individual section is of uniform area.

2) The transverse dimension of each section is small enough compared with a wavelength so that the sound propagation through an individual section can be treated as a plane wave.

3) The sections are rigid so that internal losses due to wall vibration, viscosity, and heat conduction are negligible.

4) The normal assumptions leading to elementary wave propagation equations are valid [see, for example, Morse and Ingard, 1968, p. 243].

5) The model is linear and uncoupled from the glottis.

6) The effects of the nasal tract can be ignored.

The assumptions concerning boundary conditions for the acoustic tube are discussed later in the chapter. The particular choice of boundary conditions is critical for the acoustic tube to be a valid model of the vocal tract shape. In addition, proper boundary conditions are necessary to have the acoustic tube model be equivalent to the linear prediction model $1/A(z)$.

A mid-sagittal X-ray section of a human vocal tract is shown in Fig. 4.1A. A schematic version of this photograph was shown earlier as Fig. 1.1. The outline of the vocal tract is shown with a dashed line centered along the vocal tract. A schematic model of the vocal tract is shown in Fig. 4.1B as a series of 8 uniform cylindrical sections [Dunn, 1950]. Each section has a constant area $\mathscr{A}_m$, whose value is estimated as the average area of the m*th* non-uniform section of the vocal tract. Alternate representations of this model are shown in Figs. 4.1C and D. These figures show a two-dimensional representation of the area $\mathscr{A}_m$ versus distance from the glottis to the lips, with Fig. 4.1D simply having a reversed axis for area (the area $\mathscr{A}_m$ for a physical system must certainly be positive valued). The form of Fig. 4.1D will be used in most instances as suggested by Wakita, since in the physical system the hard palate or roof of the mouth is relatively immovable with respect to the articulators such as the tongue and lips. For example, some vowels require a small lip opening while others require a rather large opening. The resultant graph is referred to as the *discrete area function representation* or, for simplicity, as the *area function.*

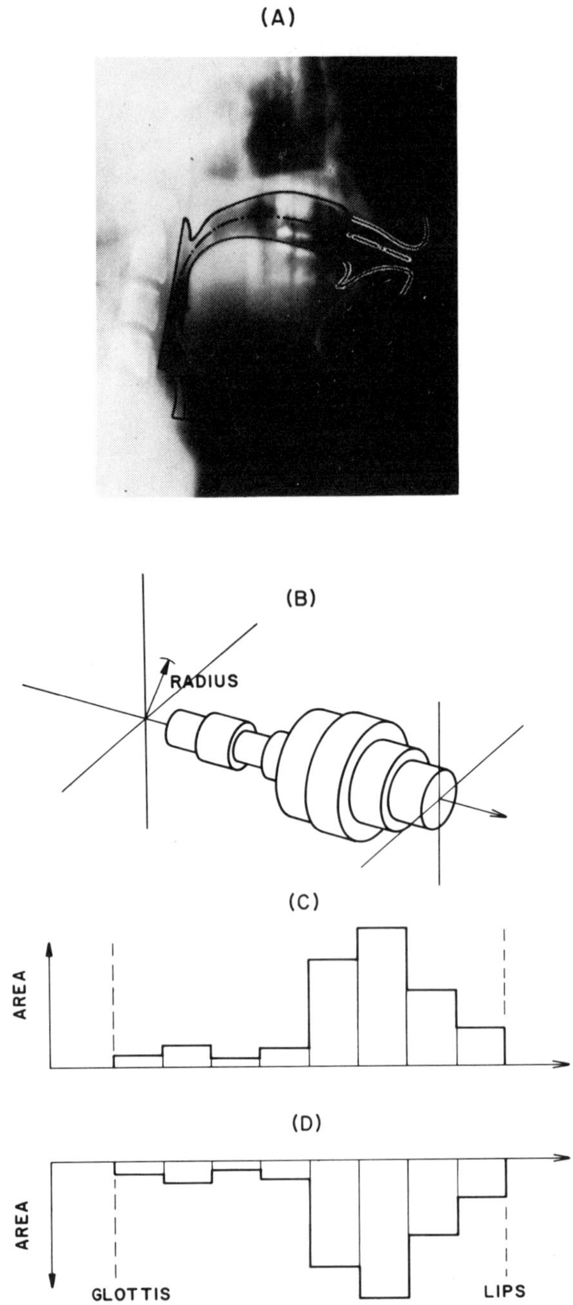

Fig. 4.1 Vocal tract representations. A) a mid-sagittal X-ray section. B) series of uniform cylindrical sections. C) discrete area function. D) area function with reversed axis.

4.2.1 Single Section Derivation

The equations governing pressure and volume velocity in an acoustic tube which satisfy the above stated assumptions are well known in acoustics [Morse and Ingard, 1968, p. 243] as the momentum equation

$$\frac{\partial p_m(x,t)}{\partial x} = \frac{-\varrho}{\mathscr{A}_m(x)} \frac{\partial u_m(x,t)}{\partial t}, \tag{4.1}$$

and the continuity of mass equation

$$\frac{\partial u_m(x,t)}{\partial x} = \frac{-\mathscr{A}_m(x)}{\varrho c^2} \frac{\partial p_m(x,t)}{\partial t}. \tag{4.2}$$

The variables $p_m(x,t)$ and $u_m(x,t)$ define the pressure and volume velocity, respectively, within the m*th* acoustic tube section as a function of time t and distance x measured from the center of the tube. The term ϱ defines the air density and c is the velocity of propagation or speed of sound for air. These equations combine to give the classical Webster horn equation as follows. From (4.1) and (4.2)

$$\begin{aligned}\frac{\partial^2 p_m(x,t)}{\partial x \partial t} &= \frac{-\varrho}{\mathscr{A}_m(x)} \frac{\partial^2 u_m(x,t)}{\partial t^2} \\ &= -\varrho c^2 \frac{\partial}{\partial x}\left[\frac{1}{\mathscr{A}_m(x)} \frac{\partial u_m(x,t)}{\partial x}\right].\end{aligned} \tag{4.3}$$

The Webster horn equation for the volume velocity is then

$$\boxed{\frac{\partial}{\partial x}\left[\frac{1}{\mathscr{A}_m(x)} \frac{\partial u_m(x,t)}{\partial x}\right] = \frac{1}{c^2 \mathscr{A}_m(x)} \frac{\partial^2 u_m(x,t)}{\partial t^2}}. \tag{4.4}$$

If $\mathscr{A}_m(x)$ is defined as a constant over the total length of the section, i.e.,

$$\mathscr{A}_m(x) = \mathscr{A}_m, \tag{4.5}$$

Webster's horn equation reduces to the one-dimensional wave equation

$$\frac{\partial^2(\cdot)}{\partial x^2} = \frac{1}{c^2} \frac{\partial^2(\cdot)}{\partial t^2}, \tag{4.6}$$

which holds for both the pressure and volume velocity arguments $(\cdot)$. Since $\mathscr{A}_m(x)$ is assumed to be constant over the length of the section, the solution to the one-dimensional wave equation for volume velocity and pressure is expressed as a linear combination of forward and reverse traveling waves (indicated by the superscripts $+$ and $-$ respectively) in the form

$$u_m(x,t) = u_m^+(t - x/c) - u_m^-(t + x/c), \tag{4.7a}$$
and
$$p_m(x,t) = p_m^+(t - x/c) + p_m^-(t + x/c), \tag{4.7b}$$

where x/c has units of time. The forward-traveling wave moves in the direction from the glottis to the lips, while the reverse-traveling wave moves in the direction from the lips to the glottis. These relationships are shown for the volume velocity in Fig. 4.2. The volume velocity $u_m(x, t)$ at any location x and time t within the m*th* section is the difference between the forward-traveling wave $u_m^+(t - x/c)$ and

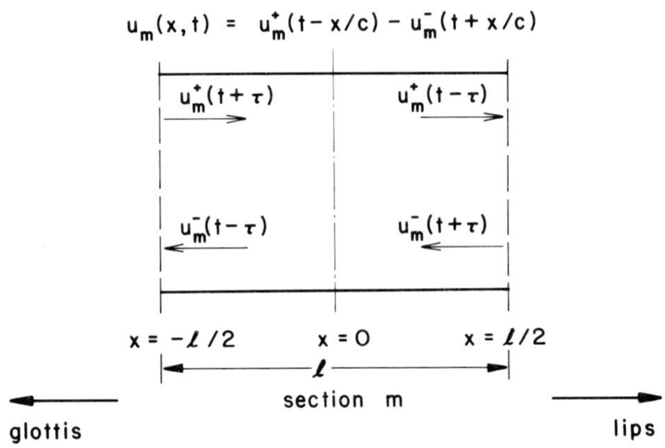

Fig. 4.2 Forward and reverse traveling wave definitions for volume velocity in section *m*.

the reverse-traveling wave $u_m^-(t + x/c)$ from (4.7a). The center of the section is defined as $x = 0$. The section length is defined as l so that the left and right ends are located at $x = -l/2$ and $x = +l/2$, respectively. If τ is defined as the time for a wave to propagate halfway along a single section, then

$$\tau = l/2c. \tag{4.8}$$

For example, the forward-traveling volume velocity wave at the left end of the tube is $u_m^+[t - (-l/2c)] = u_m^+(t + \tau)$.

The equations governing pressure and volume velocity can be used to describe $p_m(x, t)$ in terms of volume velocity components. Eq. (4.7) is substituted into (4.1) and (4.2). The partial derivatives with respect to t are then eliminated by applying the relationship

$$\frac{\partial f(t \pm x/c)}{\partial t} = \pm c \frac{\partial f(t \pm x/c)}{\partial x}, \tag{4.9}$$

which is true for any differentiable function $f(\cdot)$. Integrating with respect to t, and solving for $p_m^+(t-x/c)$ and $p_m^-(t+x/c)$ gives

$$p_m^+(t-x/c)=\frac{\varrho c}{\mathscr{A}_m}u_m^+(t-x/c)+C_1 \tag{4.10a}$$

and

$$p_m^-(t+c/x)=\frac{\varrho c}{\mathscr{A}_m}u_m^-(t+x/c)+C_2, \tag{4.10b}$$

where C_1, C_2 are constants of integration. Therefore, to within a constant, the pressure can be related to the velocity components using (4.7b) and (4.10) as

$$\boxed{p_m(x,t)=\frac{\varrho c}{\mathscr{A}_m}[u_m^+(t-x/c)+u_m^-(t+x/c)]} \quad . \tag{4.11}$$

The constant term can be omitted since it is pressure variation, and not absolute pressure in the absence of traveling waves, which is important.

4.2.2 Continuity Conditions

The volume velocity at the right end of section m ($x=+l/2$) is obtained from (4.7a) as

$$u_m(l/2,t)=u_m^+(t-\tau)-u_m^-(t+\tau). \tag{4.12}$$

In a similar manner the volume velocity at the left end of section $m-1$ is found to be

$$u_{m-1}(-l/2,t)=u_{m-1}^+(t+\tau)-u_{m-1}^-(t-\tau). \tag{4.13}$$

These continuity conditions are illustrated in Fig. 4.3. In a physical system, the pressure and volume velocity between two sections must be continuous, i.e.,

$$u_m(l/2,t)=u_{m-1}(-l/2,t) \tag{4.14a}$$

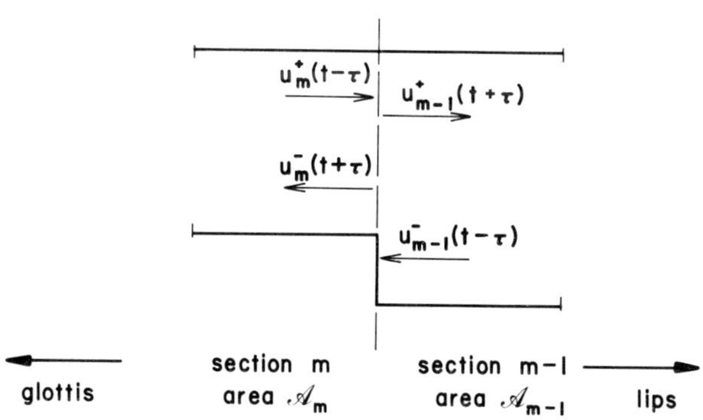

Fig. 4.3 Continuity conditions between section $m-1$ and section m for volume velocity.

and

$$p_m(l/2,t)=p_{m-1}(-l/2,t). \tag{4.14b}$$

A direct substitution of (4.7) and (4.11) into (4.14a) and (4.14b) gives

$$u_m^+(t-\tau)-u_m^-(t+\tau)=u_{m-1}^+(t+\tau)-u_{m-1}^-(t-\tau) \tag{4.15a}$$

and

$$u_m^+(t-\tau)+u_m^-(t+\tau)=[u_{m-1}^+(t+\tau)+u_{m-1}^-(t-\tau)]\,[\mathscr{A}_m/\mathscr{A}_{m-1}]. \tag{4.15b}$$

In the absence of a reverse-traveling wave in section $m-1$, i. e., $u_{m-1}^-(t-\tau)=0$, $-u_m^-(t+\tau)$ can be considered the reflection of $u_m^+(t-\tau)$ at the junction in Fig. 4.3. By combining Eqs. (4.15) the result

$$-u_m^-(t+\tau)=\frac{\mathscr{A}_{m-1}-\mathscr{A}_m}{\mathscr{A}_{m-1}+\mathscr{A}_m}u_m^+(t-\tau)$$

is obtained. Thus a *reflection coefficient* μ_m can be defined as

$$\mu_m=\frac{\mathscr{A}_{m-1}-\mathscr{A}_m}{\mathscr{A}_{m-1}+\mathscr{A}_m}. \tag{4.16}$$

Therefore,

$$\frac{\mathscr{A}_m}{\mathscr{A}_{m-1}}=\frac{1-\mu_m}{1+\mu_m}. \tag{4.17}$$

Note that if the two sections $\mathscr{A}_m$ and A_{m-1} have identical areas there is no reflection ($\mu_m=0$). Next the volume velocity terms at the right edge of section m are computed as a function of the terms of the left edge of section $m-1$. By taking one half the sum and difference of (4.15) and applying (4.17), the results

$$u_m^+(t-\tau)=\frac{u_{m-1}^+(t+\tau)-\mu_m u_{m-1}^-(t-\tau)}{1+\mu_m} \tag{4.18a}$$

and

$$u_m^-(t+\tau)=\frac{-\mu_m u_{m-1}^+(t+\tau)+u_{m-1}^-(t-\tau)}{1+\mu_m} \tag{4.18b}$$

are obtained. Next (4.18a) is solved for the forward-traveling volume velocity at the left edge of section $m-1$, $u_{m-1}^+(t+\tau)$. The result is substituted into (4.18b) to obtain the reverse-traveling wave at the right edge of section $m, u_m^-(t+\tau)$. The results

$$u_{m-1}^+(t+\tau)=\mu_m u_{m-1}^-(t-\tau)+(1+\mu_m)u_m^+(t-\tau) \tag{4.19a}$$

and

$$u_m^-(t+\tau)=(1-\mu_m)u_{m-1}^-(t-\tau)-\mu_m u_m^+(t-\tau) \tag{4.19b}$$

satisfy the form of the Kelly-Lochbaum structure [Kelly and Lochbaum, 1962].

The location of each of the volume velocity terms in (4.15) is shown in Fig. 4.4A. This figure shows the *physical model* at the junction between sections m and $m-1$. Eqs. (4.19) relate the volume velocity components at the right of section m and the left of section $m-1$. From Fig. 4.4A it is noted that the only difference between the volume velocity terms at the left and right edges of section

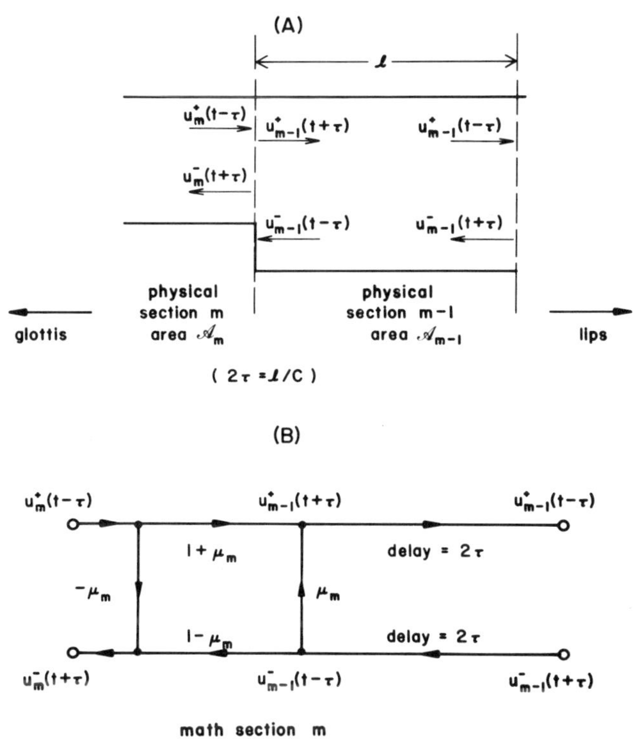

Fig. 4.4 Correspondence between a physical and mathematical section. A) physical model between section m and $m-1$. B) mathematical model for section m.

$m-1$ is the time delay $2\tau=2l/2c$. Therefore, it is necessary to incorporate only the time delay 2τ to uniquely describe the relationship between the volume velocity at the right edge of sections m and $m-1$. Eqs. (4.19) including the delays form the *mathematical model* for section m, and are shown in flow graph form in Fig. 4.4B. All that remains to uniquely define the recursion from the glottis to the lips is to determine the boundary conditions.

4.2.3 Boundary Conditions

The physical and mathematical models of the M-section acoustic tube are shown in Fig. 4.5. At the glottis, $u_{M-1}^{+}(t+\tau)$ and $u_{M-1}^{-}(t-\tau)$ must be related, while at the lips $u_0^{+}(t-\tau)$ and $u_0^{-}(t+\tau)$ must be related. In the mathematical model, Fig. 4.5B, it can be seen that the mathematical section M is incomplete in the sense that the portion which defines the reflection μ_{M-1} at the glottal junction is not yet specified. Completion of this section completes the glottal boundary conditions.

The location of the lips in the acoustic tube model by definition occurs at $x=l/2$ in section zero and is also illustrated in Fig. 4.5. The termination at the lips can be defined in several ways. Following Wakita [1972, 1973b], the acoustic tube is assumed to be open at the lips so that the pressure $p_0(l/2, t)$ equals zero. This condition is analogous to a short-circuited transmission line where the voltage must be zero. In actual fact there must be a non-zero radiation impedance or there could be no pressure wave leaving the lips. The transfer function obtained assuming zero radiation impedance is almost identical to that obtained for small impedance as shown in Section 4.6.1, (4.77). From (4.11), the boundary condition is easily obtained as

$$\boxed{u_0^{+}(t-\tau) = -u_0^{-}(t+\tau)} \quad . \tag{4.20}$$

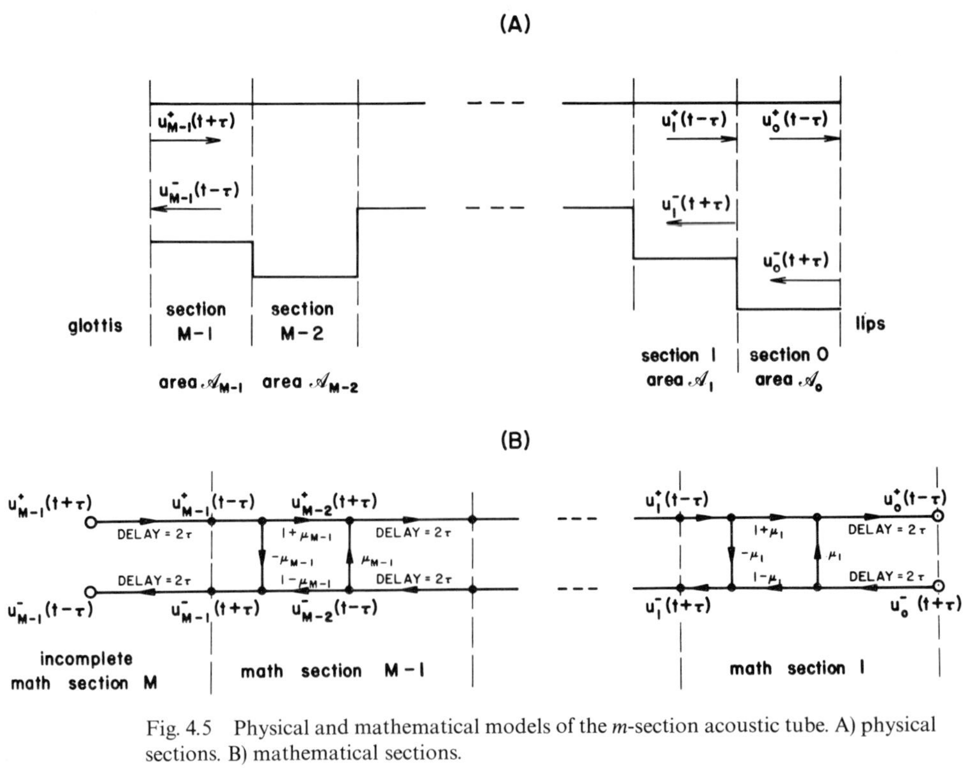

Fig. 4.5 Physical and mathematical models of the m-section acoustic tube. A) physical sections. B) mathematical sections.

The volume velocity output of the acoustic tube $u_L(t)$ (measured at the corresponding lip location $x=l/2$ in section $m=0$) is from (4.7a)

$$u_L(t)=u_0^+(t-\tau)-u_0^-(t+\tau). \tag{4.21}$$

Substituting (4.20) into (4.21) results in

$$\boxed{u_L(t)=2u_0^+(t-\tau)}\ . \tag{4.22}$$

Determination of the boundary condition at the glottis is somewhat more involved. Two equivalent models for terminating the glottis are shown in Fig. 4.6. In Fig. 4.6A, the acoustic tube is driven by a volume velocity source $u_G(t)$ whose source

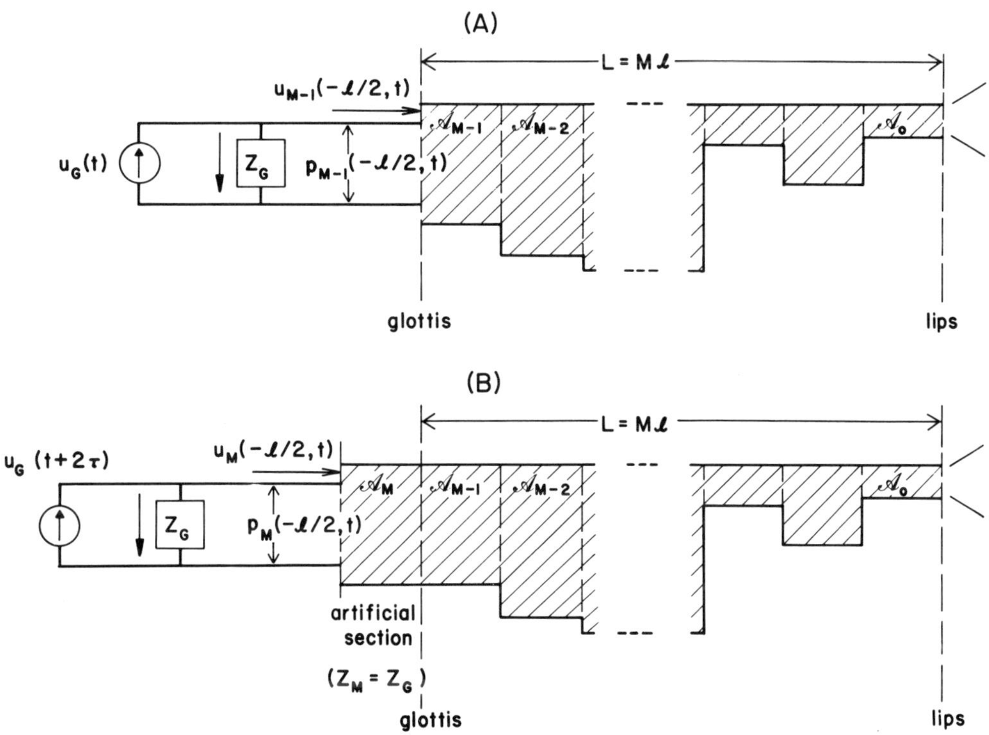

Fig. 4.6 Physical models for terminating the glottis. A) volume velocity source with impedance Z_G. B) artificial section matched to volume velocity source Z_G.

impedance is Z_G. In Fig. 4.6B, the $M-1th$ physical section (whose left edge defines the glottis) is connected to an artificial section whose impedance is matched to the volume velocity source impedance Z_G. The addition of the artificial section M has the effect of placing all reflections at the glottis between sections $M-1$ and M, rather than at the connection between section $M-1$ and the volume velocity source. The two models are precisely equivalent. The delay due to the M*th* section is compensated for by defining the source as $u_G(t+2\tau)$.

The location of the glottis in the acoustic tube model occurs by definition at $x=-l/2$ in section $M-1$, as illustrated in Fig. 4.6. From (4.11) and (4.7a), the pressure and volume velocity are given by

$$p_{M-1}(-l/2,t)=\frac{\varrho c}{\mathscr{A}_{M-1}}[u^+_{M-1}(t+\tau)+u^-_{M-1}(t-\tau)] \tag{4.23a}$$

and

$$u_{M-1}(-l/2,t)=u^+_{M-1}(t+\tau)-u^-_{M-1}(t-\tau). \tag{4.23b}$$

The driving function to the glottis is defined as a volume velocity source $u_G(t)$ which drives the glottis through an acoustic impedance Z_G. This term is analogous to electrical impedance when volume velocity and pressure are analogous to current and voltage, respectively. From Fig. 4.6A (using electrical analogies if desired), the volume velocity of the source must equal the volume velocity going through the impedance Z_G plus that going into the acoustic tube. Assuming Z_G is real,

$$u_G(t)=\frac{1}{Z_G}p_{M-1}(-l/2,t)+u_{M-1}(-l/2,t). \tag{4.24}$$

Substituting (4.23) into (4.24) results in

$$\begin{aligned}u_G(t)=&\frac{\varrho c}{Z_G\mathscr{A}_{M-1}}[u^+_{M-1}(t+\tau)+u^-_{M-1}(t-\tau)]\\&+u^+_{M-1}(t+\tau)-u^-_{M-1}(t-\tau).\end{aligned} \tag{4.25}$$

Using the standard definition of the characteristic impedance of section m,

$$Z_m=\varrho c/\mathscr{A}_m\,, \tag{4.26}$$

a glottal area

$$\mathscr{A}_M=\varrho c/Z_G \tag{4.27}$$

and a reflection coefficient

$$\mu_M=\frac{\mathscr{A}_{M-1}-\mathscr{A}_M}{\mathscr{A}_{M-1}+\mathscr{A}_M}=\frac{Z_G-Z_{M-1}}{Z_G+Z_{M-1}} \tag{4.28}$$

are defined. As a result of these definitions,

$$\frac{\varrho c}{Z_G\mathscr{A}_{M-1}}=\frac{Z_{M-1}}{Z_G}=\frac{1-\mu_M}{1+\mu_M}. \tag{4.29}$$

Substitution of (4.29) into (4.25) gives

$$u_G(t)=\frac{2u_{M-1}^{+}(t+\tau)-2\mu_M u_{M-1}^{-}(t-\tau)}{1+\mu_M}. \tag{4.30}$$

With the exception of a scale factor, this equation is identical to (4.18a) with m replaced by M, and indeed becomes identical if we define

$$u_M^{+}(t-\tau)=u_G(t)/2. \tag{4.31}$$

The velocity $u_M^{+}(t-\tau)$ represents a forward-traveling wave in an artificial section added to the vocal tract model to accommodate reflections at the glottis. To complete the mathematical model, $u_M^{-}(t+\tau)$ is defined by using (4.18b) with $m=M$, and in a similar manner to the derivation of (4.19), the results

$$u_{M-1}^{+}(t+\tau)=\mu_M u_{M-1}^{-}(t-\tau)+(1+\mu_M)u_M^{+}(t-\tau) \tag{4.32}$$

and

$$u_M^{-}(t+\tau)=(1-\mu_M)u_{M-1}^{-}(t-\tau)-\mu_M u_M^{+}(t-\tau) \tag{4.33}$$

are obtained. This mathematical model of the glottal termination is shown in Fig. 4.7.

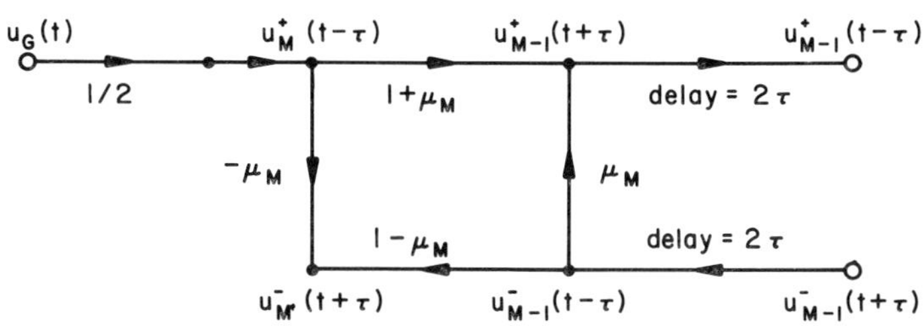

Fig. 4.7 Mathematical model showing glottal termination.

4.3 Relationship Between Acoustic Tube and Linear Prediction

The relationship between the volume velocity terms from section m to section $m-1$ have now been developed in detail along with a specific set of lip and glottal boundary conditions. What is desired next is a relationship that can be used, starting from the volume velocity waveform at the lips, $u_L(t)$, or equivalently from the volume velocity components at the right edge of the physical section $m=0$, to compute the volume velocity components of the acoustic tube down to the glottis.

Wakita's [1972] result that this recursive relationship is equivalent to the recursive design of the inverse filter is now developed. Referring to Fig. 4.4 and (4.18), the volume velocity components at the right edge of the physical section m are given by

$$u_m^+(t-\tau)=\frac{u_{m-1}^+(t+\tau)-\mu_m u_{m-1}^-(t-\tau)}{1+\mu_m} \tag{4.34}$$

and

$$u_m^-(t+\tau)=\frac{u_{m-1}^-(t-\tau)-\mu_m u_{m-1}^+(t+\tau)}{1+\mu_m}. \tag{4.35}$$

An absolute time reference will be included in the equations by relating all times to the lips with a delay t_m, where

$$t_m=2(m+1)\tau \tag{4.36}$$

represents the time for a wave to pass from the left edge of the physical section m to the lips. In addition, it is seen from (4.34) and (4.35) that products of $1/(1+\mu_m)$ arise as m increases. To accommodate both of these factors, the time delay and the product terms, as well as a sign change, are introduced to make results consistent with the inverse filter equations. New variables

$$y_m^+(t)=c_m u_m^+(t+\tau-t_m) \tag{4.37a}$$

and

$$y_m^-(t)=-c_m u_m^-(t-\tau-t_m) \tag{4.37b}$$

are defined, where the coefficients c_m are given by

$$c_m=\prod_{i=1}^{m}(1+\mu_i) \quad \text{for} \quad m>0, \tag{4.38}$$

with $c_0=1$.

Replacing t by $t+2\tau-t_m$ in (4.34) and (4.35) and multiplying these equations by

$$c_m=(1+\mu_m)c_{m-1} \tag{4.39}$$

then gives the results

$$c_m u_m^+(t+\tau-t_m)=c_{m-1}[u_{m-1}^+(t+3\tau-t_m)-\mu_m u_{m-1}^-(t+\tau-t_m)]$$

$$c_m u_m^-(t+3\tau-t_m)=c_{m-1}[u_{m-1}^-(t+\tau-t_m)-\mu_m u_{m-1}^+(t+3\tau-t_m)].$$

These results can be expressed in terms of the new variables by noting that

$$t+3\tau-t_m=t+\tau-t_{m-1} \quad \text{and} \quad t+\tau-t_m=t-\tau-t_{m-1}.$$

With the definition

$$T = 4\tau = 2l/c, \tag{4.40}$$

so that T defines twice the time needed for a wave to propagate through a single section, the equations

$$y_m^+(t) = y_{m-1}^+(t) + \mu_m y_{m-1}^-(t) \tag{4.41a}$$

$$y_m^-(t+T) = \mu_m y_{m-1}^+(t) + y_{m-1}^-(t) \tag{4.41b}$$

for $m = 1, 2, \ldots, M$ are obtained.

The boundary condition at the lips and the volume velocity in terms of the new variables arise directly from substitution of (4.37) into (4.20) and (4.22). As $c_0 = 1$, the boundary condition

$$y_0^+(t) = y_0^-(t+T) \tag{4.42}$$

is obtained, resulting in the volume velocity leaving the lips as

$$u_L(t) = 2y_0^+(t). \tag{4.43}$$

The boundary condition at the glottis is found by replacing t by $t - 2M\tau$ in (4.31) and multiplying by c_M. As

$$t - 2M\tau = t - t_M + 2\tau,$$

(4.37a) can be applied to (4.31) with the result

$$y_M^+(t) = \frac{c_M}{2} u_G(t - 2M\tau). \tag{4.44}$$

The delay factor $2M\tau = MT/2$ represents the time it takes for a wave to propagate from the glottis to the lips.

If the terms $y_m^+(t)$ and $y_m^-(t)$ from (4.37) are sampled at the times $t = nT$, the z-transforms are directly obtained from (4.41) as

$$\boxed{\begin{aligned} Y_m^+(z) &= Y_{m-1}^+(z) + \mu_m Y_{m-1}^-(z) \qquad (4.45\text{a}) \\ z\,Y_m^-(z) &= \mu_m Y_{m-1}^+(z) + Y_{m-1}^-(z) \qquad (4.45\text{b}) \end{aligned}}$$

where $Y_m^+(z)$ and $Y_m^-(z)$ are the z-transforms of $y_m^+(nT)$ and $y_m^-(nT)$, respectively. The boundary conditions are given from (4.42) and (4.44) by

$$Y_0^+(z) = z\,Y_0^-(z) = U_L(z)/2 \tag{4.46a}$$

and

$$Y_M^+(z) = [c_M U_G(z) z^{-M/2}]/2. \tag{4.46b}$$

By rewriting (4.45) in the form

$$\boxed{\begin{aligned} Y_{m-1}^{+}(z) &= Y_m^{+}(z) - \mu_m Y_{m-1}^{-}(z) \\ Y_m^{-}(z) &= z^{-1}[\mu_m Y_{m-1}^{+}(z) + Y_{m-1}^{-}(z)] \end{aligned}} \qquad \begin{aligned} &(4.47\text{a}) \\ &(4.47\text{b}) \end{aligned}$$

for $m = M, M-1, \ldots, 1$, a structure known as the lattice or two-multiplier filter [Itakura and Saito, 1971b; Gray and Markel, 1973] is obtained. Eqs. (4.47) define the speech production or synthesis equations of the acoustic tube model.

From Chapter 2, (2.60) and (2.66), it was shown that the inverse filter $A(z)$ from the autocorrelation method can be recursively obtained from

$$\boxed{\begin{aligned} A_m(z) &= A_{m-1}(z) + k_m B_{m-1}(z) \\ z B_m(z) &= k_m A_{m-1}(z) + B_{m-1}(z) \end{aligned}} \qquad \begin{aligned} &(4.48\text{a}) \\ &(4.48\text{b}) \end{aligned}$$

with initial conditions

$$A_0(z) = 1 \quad \text{and} \quad z B_0(z) = 1 \tag{4.49}$$

so that $A_0(z) = z B_0(z)$. If $A_0(z)$ and $B_0(z)$ are multiplied by $Y_0^{+}(z)$ then

$$Y_0^{+}(z) A_0(z) = z\, Y_0^{+}(z) B_0(z).$$

Comparing this result with (4.46a), one sees that the boundary conditions of the linear prediction filter and the acoustic tube equations are related by

$$Y_0^{+}(z) = Y_0^{+}(z) A_0(z) \tag{4.50a}$$

and

$$Y_0^{-}(z) = Y_0^{+}(z) B_0(z). \tag{4.50b}$$

If μ_m is set equal to k_m, then multiplication of both sides of (4.47) by $Y_0^{+}(z)$ gives, by direct comparison with the acoustic tube relations (4.45),

$$Y_m^{+}(z) = Y_0^{+}(z) A_m(z) \tag{4.51a}$$

$$Y_m^{-}(z) = Y_0^{+}(z) B_m(z), \tag{4.51b}$$

for $m = 1, \ldots, M$.

$$\text{At} \quad m = M \quad (\text{with } A(z) = A_M(z)),$$

$$Y_M^{+}(z) = Y_0^{+}(z) A(z).$$

But from (2.67) (with $X_0^+(z)=X(z)$),

$$X_M^+(z)=X_0^+(z)A_M(z)$$

or

$$E(z)=X(z)A(z),$$

so that

$$\boxed{\frac{Y_M^+(z)}{Y_0^+(z)}=\frac{X_M^+(z)}{X_0^+(z)}=\frac{E(z)}{X(z)}=A(z)} \quad . \tag{4.52}$$

Therefore, if the number of coefficients M in the inverse filter equals the number of sections M in the acoustic tube model, then *the reflection coefficients μ_m, which uniquely define the area ratios of the acoustic tube model of the vocal tract, can be obtained directly by linear prediction analysis of the speech waveform.*

Since the z-transforms of $y_m^+(t)$ and $y_m^-(t)$ were based upon sampling at intervals $t=nT$ where $T=4\tau$ from (4.40), the acoustic speech waveform must be sampled according to the same relationship. Therefore, the relation between sampling frequency $f_s=1/T$, number of sections M, length of the acoustic tube $L=Ml$, and speed of sound is given from (4.40) as

$$T=2l/c=2Ml/Mc$$

or

$$\boxed{f_s=Mc/2L} \quad . \tag{4.53}$$

In modeling the vocal tract with the acoustic tube, L is defined as the vocal tract length.

In terms of the speech production model in Chapter 1, the acoustical speech waveform $s(t)$ at the lips is considered as a first approximation to be the derivative of the volume velocity at the lips. Therefore, within a constant scaling factor, the z-transform of the speech measured at a distance d from the lips is given by

$$S(z)=L(z)U_L(z)z^{-f_s d/c}=2L(z)Y_0^+(z)z^{-f_s d/c}, \tag{4.54}$$

where the lip radiation filter $L(z)$ is approximated by

$$L(z)=1-\mu z^{-1} \tag{4.55}$$

with $0<\mu\leq 1$ near unity.

From Chapter 1, $S(z)$ is approximated in terms of the vocal tract transfer function $V(z)$ by

$$\begin{aligned} S(z)&=L(z)V(z)G(z)E(z)\\ &\simeq\frac{V(z)E(z)}{L(z)} \end{aligned} \tag{4.56}$$

since the glottal shaping term $G(z)$ is approximated by $G(z)=1/L^2(z)$. If the inverse filter input is described as a pre-emphasized version of the speech, i.e.,

$$X(z)=P(z)S(z),$$

then the inverse filter transfer function from the above equations is

$$\frac{E(z)}{V(z)} \simeq \frac{L(z)}{V(z)P(z)} = A(z) \tag{4.57}$$

Therefore, by choosing the pre-emphasis filter equal to the lip radiation factor, i.e.,

$$P(z)=L(z)\simeq 1-z^{-1},$$

the result

$$\boxed{V(z)\simeq 1/A(z)} \tag{4.58}$$

is obtained. Therefore, *by performing linear prediction analysis on the pre-emphasized speech signal, an inverse filter is obtained whose reciprocal* $1/A(z)$ *is an estimate of the vocal tract transfer function.* To have this relationship, the reflection coefficients of the acoustic tube model are defined by $\mu_m = k_m$, $m=1, 2, \ldots, M$. The area functions of the estimated vocal tract shape are then computed from (4.17) as

$$\mathscr{A}_{m-1} = \frac{1+\mu_m}{1-\mu_m} \mathscr{A}_m \tag{4.59}$$

for $m=M, M-1, \ldots, 1$, keeping in mind that $\mathscr{A}_M$ is an artificial area defined by (4.27). Having no absolute reference value, $\mathscr{A}_M$ is usually assumed to be unity.

The relationships between the input and output of the inverse filter and the volume velocities at the lips and glottis are now derived. Using (4.54), the input $X(z)=X_0^+(z)$ can be equivalently written as

$$\begin{aligned} X(z)=X_0^+(z)=P(z)S(z)=P(z)L(z)U_L(z)z^{-f_s d/c} \\ =2P(z)L(z)Y_0^+(z)z^{-f_s d/c}. \end{aligned} \tag{4.60}$$

Using (4.52) the inverse filter output can be written as

$$E(z)=X_M^+(z)=A(z)X_0^+(z)=Y_M^+(z)X_0^+(z)/Y_0^+(z).$$

Substituting for $Y_M^+(z)$ from (4.46 b), and the ratio $X_0^+(z)/Y_0^+(z)$ from (4.60), gives the result

$$E(z)=c_M U_G(z)P(z)L(z)z^{-M/2}z^{-f_s d/c}.$$

Therefore since $M=2L/c$ from (4.53), the glottal volume velocity and lip volume velocity estimates are related to the acoustical speech waveform and the inverse filter output by

$$U_L(z) = \frac{S(z) z^{f_s d/c}}{L(z)} \qquad (4.61a)$$

$$U_G(z) = \frac{z^{(L+d) f_s/c} E(z)}{c_M L(z) P(z)} \quad . \qquad (4.61b)$$

The various relationships derived in this section are summarized on the flow graph of Fig. 4.8.

Given that the reflection coefficients $\mu_m = k_m$, $m = 1, 2, \ldots, M$, have been obtained from linear prediction analysis, the flow graph explicitly shows how to compute the error signal $E(z)$ from the input $X(z) = P(z) S(z)$. (The computation of the error signal and its properties for real speech are covered in Chapter 8.) Within a scale factor, the volume velocity at the lips is modeled as the approximate integration of the speech signal. The delayed glottal volume velocity within a scalc factor is modeled by the approximate double integration $1/(1-\mu z^{-1})^2$ of the error signal. The delay term $z^{-M/2}$ shows that the estimate from the error signal is delayed by $M/2$ samples or $f_s L/c$ from (4.53), due to the sound propagation delay from the glottis to the lips.

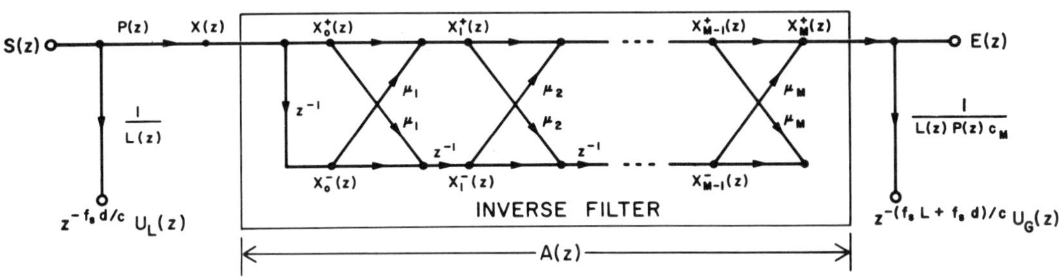

Fig. 4.8 Flow graph showing various relationships starting from the acoustic speech signal $S(z)$.

From a purely computational point of view, if μ is near unity, the double integration for estimation of $U_G(z)$ can cause large bias buildup. A preferable procedure might be to perform the analysis by choosing $P(z) = 1/L(z)$, determining the filter $A(z)$, passing $S(z)$ through $A(z)$ to obtain $E(z)/L(z)$, and then performing one additional integration to estimate the delayed volume velocity waveform.

4.4 An Algorithm, Examples, and Evaluation

The procedure developed by Wakita for directly estimating the vocal tract area functions from speech has been based upon application of the autocorrelation method. The recursion based upon the $A_m(z)$ and $B_m(z)$ polynomials requires that $B_m(z)=z^{-(m+1)}A_m(1/z)$ (see Chapter 3) which occurs only in the autocorrelation method. Nonetheless, it is possible to estimate vocal tract area functions from both methods. An algorithm that allows estimation of acoustic tube area functions directly from speech is now presented, several examples are given, and then an evaluation of the accuracy of direct vocal tract area function estimation using the Ishizaka-Flanagan [1972] speech production model is considered.

4.4.1 An Algorithm

An algorithm for estimating vocal tract area functions directly from the acoustic speech waveform is shown in Fig. 4.9. This algorithm builds upon the results of the previous chapters where details of performing either the autocorrelation method or covariance method have been presented.

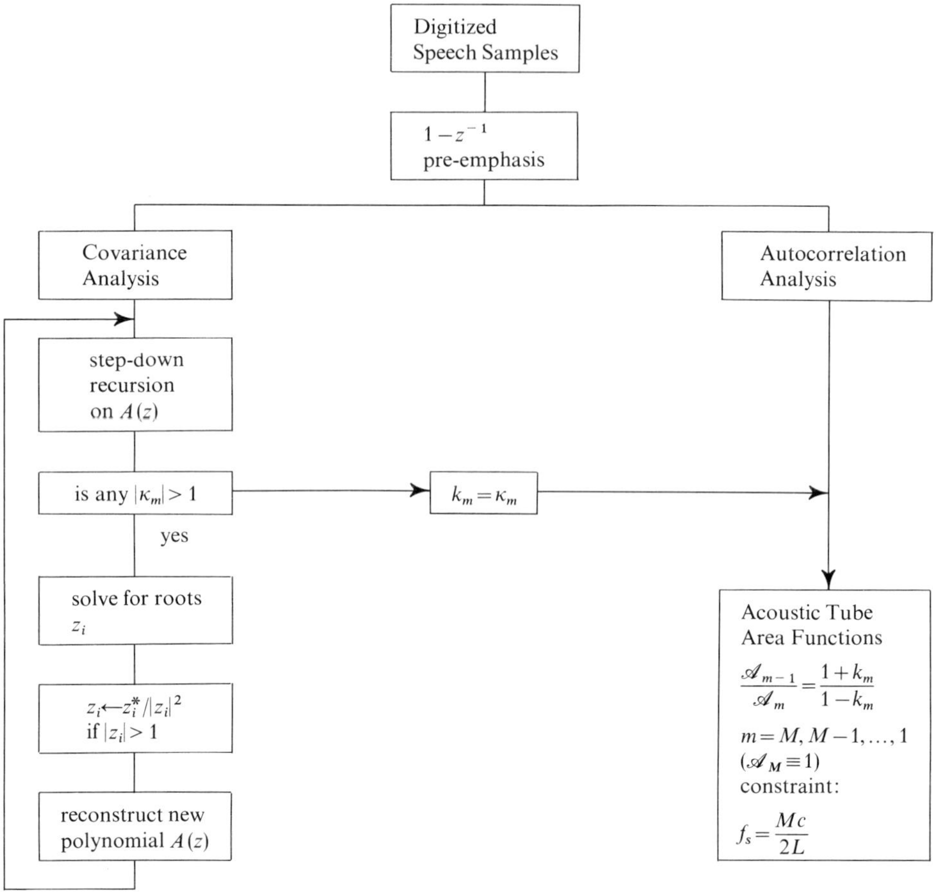

Fig. 4.9 An algorithm for obtaining vocal tract area function estimates from digitized speech samples.

Generally, without better knowledge, a vocal tract length of about 17 cm [Fant, 1960] is assumed for male speakers. If the sampling frequency f_s is chosen to include at least three formants, then $f_s \geq 6$ kHz. From (4.52), M is then computed where the speed of sound is $c = 34$ cm/ms.

The speech waveform is then sampled and digitized at the rate f_s and stored for further processing. Valid estimation of vocal tract area functions requires the assumption of voiced non-nasal speech and preprocessing to approximately eliminate the glottal and lip radiation characteristics.

By choosing an interval of voiced non-nasalized speech less than 20–25 ms in duration, and then differencing the data to obtain the processed signal $X(z)$, the assumptions will generally be satisfied. Next, either the autocorrelation method or the covariance method is applied to obtain the resultant inverse filter $A(z)$. The autocorrelation method requires only use of the solution algorithm in Chapter 3 to obtain k_m for $m = 1, 2, \ldots, M$. The resulting filter $1/A(z)$ defines the estimate of the vocal tract transfer function. This filter is theoretically stable, as will be shown in the next chapter, since $|k_m| < 1$ and thus the area functions are positive valued and finite.

The procedure for estimating area functions using the covariance method is taken from Atal and Hanauer [1971b]. From the pre-emphasized speech waveform (without windowing the data), the coefficients $\{a_i\}$ of an inverse filter $A(z)$ are obtained. This filter polynomial is tested to see if any roots lie outside the unit circle in the z-plane (equivalently, to test whether or not $1/A(z)$ is unstable). For efficiently performing this test without having to perform polynomial root-solving, a step-down procedure (see Chapter 5) is employed. The step-down procedure is precisely the reverse of the step-up procedure used for implementation of the autocorrelation method. The output of this procedure is a set of parameters $\{\kappa_m\}$. If $|\kappa_m| \geq 1$ for any $m = 1, 2, \ldots, M$, then not only is $1/A(z)$ unstable, but in addition, negative areas are obtained. The roots of $A(z)$ are then computed and any roots outside the unit circle are replaced by their reciprocals to ensure that the resulting root will lie within the unit circle. These roots are then used to reconstruct a new polynomial $A(z)$. The step-down recursion is performed again and this time the definition $k_m = \kappa_m$ is made since all κ_m terms will now be bounded by unity (since the new $1/A(z)$ is stable). The area functions $\mathscr{A}_m$ for $m = M, M-1, \ldots, 0$ are then computed. This procedure was used by Atal and Hanauer [1971b] to ensure positive valued acoustic tube area functions for parameter quantization in a linear prediction speech analysis synthesis system. No claim was made as to the applicability of the method for estimating *vocal tract* area functions. If the original polynomial has roots outside the unit circle, then the resulting area functions probably have no physical relationship to actual vocal tract area functions. However, in cases where speech data can be accurately modeled in a single pitch period by complex exponentials, the covariance method may give more accurate results than the autocorrelation method. (This point has been discussed in Chapter 2 with Prony's method and will again be discussed in Chapter 8 in connection with accuracy in formant estimation). Vocal tract estimation using the autocorrelation method is foolproof in the sense that each analysis will result in an acoustic tube having positive valued area functions without any modification.

4.4.2 Examples

An example for the autocorrelation method analysis [Wakita, 1972] is shown in Fig. 4.10. A set of seven reflection coefficients and normalized areas was obtained from the analysis of a spoken vowel /i/ as in *e*ve after pre-emphasis as shown in

(A)

J	REFL(J) $= \mu_{J-1}$	AREA(J) $= \mathcal{A}_{J-1}$	POSITION
1	0.66992	3.24655	← LIPS
2	-0.13410	0.64172	·
3	-0.29675	0.84048	·
4	-0.58978	1.54982	·
5	0.04873	6.00632	
6	0.40587	5.44810	
7	0.39438	2.30240	← GLOTTIS

(B)

```
C
C          CALCULATE AREA FUNCTIONS
C          AREA(1):LIPS...AREA(M):GLOTTIS
C
           SM=1.
           DO 1 J=1,M
           JJ=M+1-J
           SM=SM*(1.+REFL(JJ))/(1.-REFL(JJ))
1          AREA(JJ)=SM
```

(C)

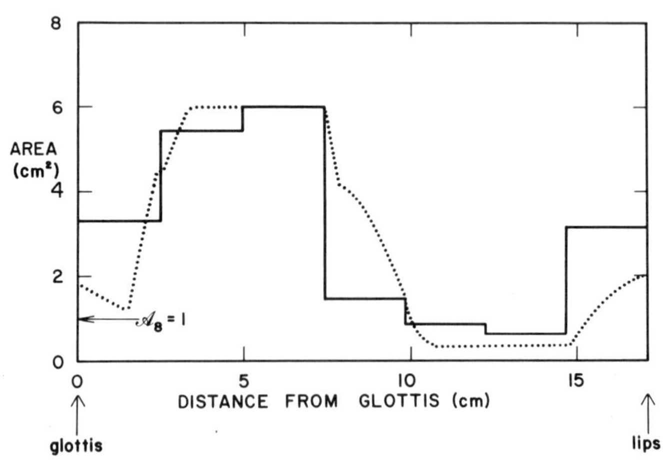

Fig. 4.10 An area function calculation example. A) coefficient listing. B) computational program. C) resulting area function (solid) and measured area function (dotted) for a similar sound.

Fig. 4.10A. A Fortran program for computing the normalized areas (using $\mathscr{A}_7 = 1$) is shown in Fig. 4.10B. The resulting estimate of the vocal tract area function estimate is shown in Fig. 4.10C as a solid line along with an experimentally measured vocal tract area function for a Russian vowel /i/ from Fant [1960]. The opening at the lips is in a relatively open condition while the main constriction occurs in the front portion of the vocal tract, about 3 cm from the lips in this example. A large area occurs at the back of the mouth (approximately 5 cm from the glottis). This shape is very similar to the X-ray tracings made by Fant. The difference made by performing an analysis without pre-emphasis is shown in Fig. 4.11 [Wakita, 1972]. The resulting area functions have no apparent relation to actual vocal tract behavior, whereas with pre-emphasis the results appear quite reasonable. Since the pre-emphasis factor $(1 - z^{-1})$ is only an approximation, the conclusion is drawn that if methods can be developed for more accurately estimating the glottal and lip radiation characteristics, then even more accurate vocal tract area function estimates should be possible.

An analysis example from continuous speech is given in Fig. 4.12 for the 320 ms portion "they give" from the utterance "Did they give all their goods away?" spoken by an adult male. The speech was sampled at 10 kHz and analysis was made for a frame interval of 10 ms with a Hamming window, shifted in 10 ms increments. The area functions in Fig. 4.12B are shown every 20 ms. The dynamic movement of the area functions during the vowel portions /e/ of the*y* and /ɪ/ of g*i*ve in Fig. 4.12C is reasonably well represented. The relative lip opening areas for the voiced portions seem quite reasonable. Although the meaning of area functions for consonants is not clear, it is interesting to note that the interval preceding the /g/ of *g*ive shows

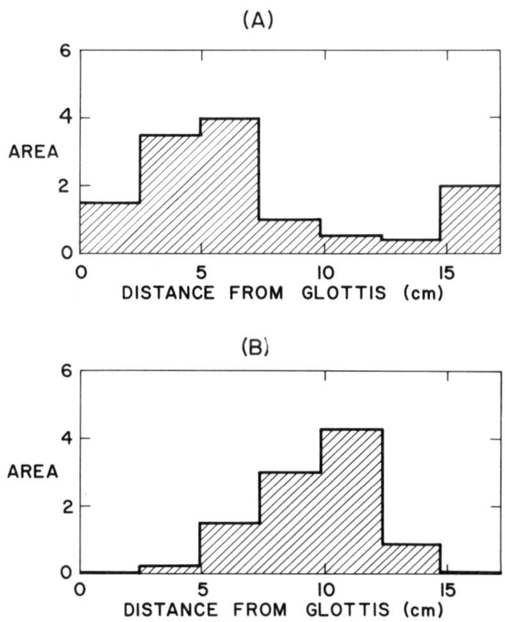

Fig. 4.11 Comparison of area functions. A) pre-emphasis. B) no pre-emphasis. [From Wakita, 1972]

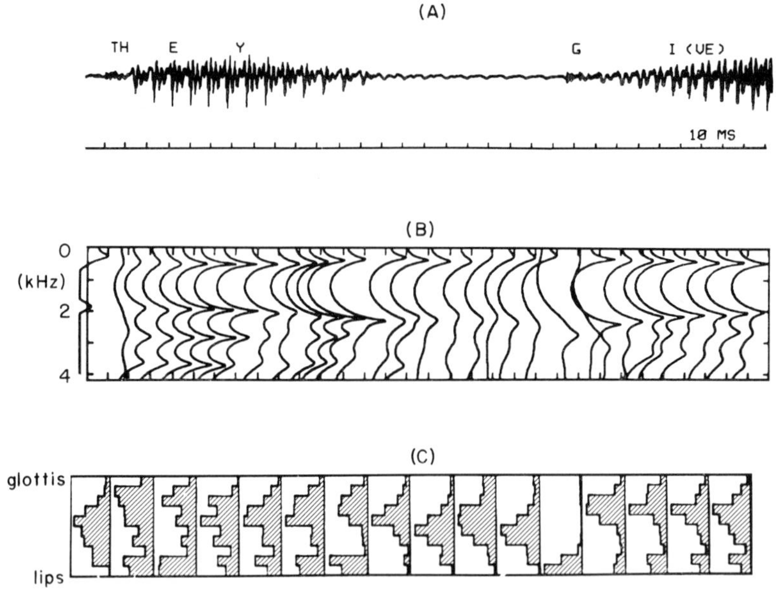

Fig. 4.12 Analysis example. A) acoustic waveform. B) spectral envelopes. C) area functions.

very small lip opening. In addition, the appearance of tract closure with lip opening for the initiation of the /g/ would be expected from physiological considerations, and indeed is indicated to be of this form.

4.4.3 Evaluation of the Procedure

Although reasonable area functions have been obtained using Wakita's method in several studies [Wakita, 1972; Nakajima, 1974; Crichton and Fallside, 1974b], the question of accuracy in the method still remains. A qualitative evaluation can be made by comparing results against X-ray tracings from similar utterances. However, the accuracy of X-ray tracings is somewhat questionable since many assumptions and estimations regarding the lateral dimension of the vocal tract are also necessary.

A major problem in estimating vocal tract area functions is how to eliminate the effects of other factors affecting the resonance frequencies and bandwidths of the vocal tract. It is known that effects of the glottal waves and radiation can be reasonably well eliminated by fixed pre-emphasis of the speech since their characteristics vary relatively slowly in the frequency domain.

An adaptive pre-emphasis of the speech wave has also been shown to give reasonable area functions for various speakers including men, women, and children [Nakajima, 1974]. Wakita [1974c] has performed an experiment to quantitatively determine the accuracy of his method, based upon the analysis of five vowels synthesized from the Ishizaka-Flanagan [1972] speech production model. This procedure has the advantage that the area functions used for the synthesis are known. In addition, the model appears to be the most accurate representation of true vocal tract characteristics presently available. The laryngeal model considers the most essential physiological factors, such as the subglottal pressure, the vocal cord tension, and the equilibrium area of the vocal cord opening. Within the vocal tract, losses due to wall vibration, viscosity, and heat conduction are included in the model. An analysis of the synthetic vowels (by taking into account the effect of losses on resonance frequencies) resulted in the area function estimates shown in Fig. 4.13. A root mean square error (NRMSE) normalized by the maximum area of each reference vowel is shown by the solid line in Fig. 4.14 compared with the NRMSE obtained for Fant's area functions (dotted lines). The results appear quite reasonable with the largest percentage error of about 13 percent for /ɑ/ (see Wakita [1974c] for details).

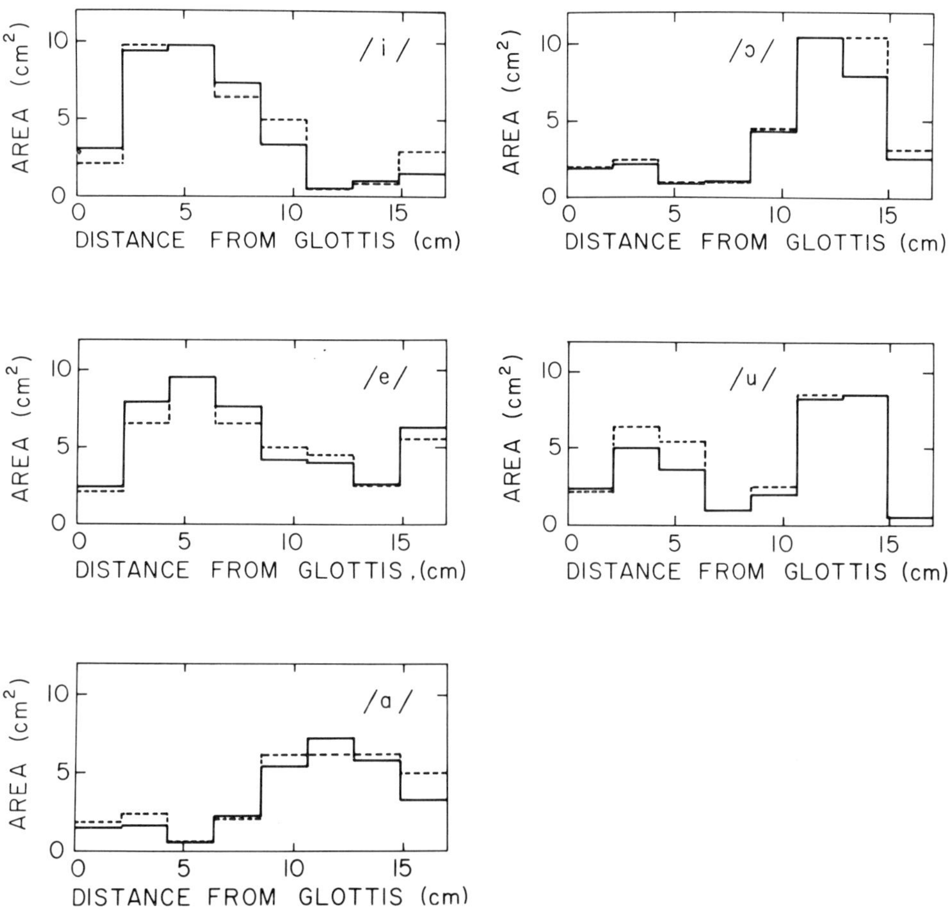

Fig. 4.13 Area function estimates obtained by accounting for formant losses in Ishizaka-Flanagan model. [From Wakita, 1974c]

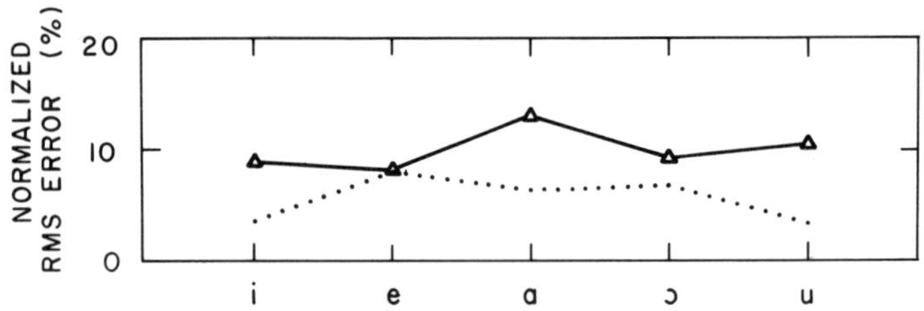

Fig. 4.14 Comparison of normalized root mean square errors. [From Wakita, 1974c]

4.5 Estimation of Lip Impedance

There has been considerable interest in the problem of determining a unique vocal tract shape directly from acoustical measurements, to avoid the drawbacks of X-ray measurements such as the limited safe dosage and the laborious task of processing data. In this context, Schroeder and Mermelstein [1965, 1967] first presented the problem as a first-order perturbation analysis of the Webster horn equation. The work was then extended by Mermelstein [1967] and by Heinz [1967]. Mermelstein [1967] presented the problem as an inverse eigenvalue solution for nearly uniform tracts, based upon the fact that two sets of eigenvalues for different boundary conditions give a unique area function. He chose lip impedance zero and pole frequencies (input impedance at the lips looking into the vocal tract) as two sets of eigenvalues and found a method for determining the area function by applying the perturbation theory iteratively so that the first few lip impedance singularities matched with the measured values of these singularities. In Schroeder's [1967] method an impedance tube was employed to measure lip impedance zeros and poles, since no information on lip impedance poles was believed to be available from the acoustic speech waveform. Wakita [1972, 1973a] has shown, however, that from the acoustic tube model of the vocal tract and its relationship to the linear prediction filter, pole and zero locations of the lip impedance can be directly estimated.

4.5.1 Lip Impedance Derivation

The impedance at the lips seen looking into the vocal tract is found by setting the glottal volume velocity source $u_G(t)$ to zero. A source can then be applied at the lips to determine the lip impedance by taking the ratio of the pressure transform to the negative volume velocity transform at the lips.

The pressure at the lips is given from (4.11) with $m=0$ and $x=l/2$ as

$$p_L(t)=p_0(l/2,t)=\varrho c[u_0^+(t-\tau)+u_0^-(t+\tau)]/\mathscr{A}_0. \tag{4.62a}$$

The volume velocity entering the lips is found in the same manner from (4.7a) as

$$-u_L(t) = -u_0(l/2, t) = -[u_0^+(t-\tau) - u_0^-(t+\tau)]. \tag{4.62b}$$

In terms of the variables $y_m^+(t)$ and $y_m^-(t)$ from (4.37) and (4.38), the lip pressure and lip volume velocity are

$$p_L(t) = \varrho c[y_0^+(t) - y_0^-(t+T)]/\mathscr{A}_0 \tag{4.63a}$$

and

$$-u_L(t) = -y_0^+(t) - y_0^-(t+T). \tag{4.63b}$$

From the z-transform of these equations, the input impedance seen looking into the lips is given by

$$\begin{aligned} Z_{in}(z) &= \frac{P_L(z)}{-U_L(z)} \\ &= \frac{\varrho c}{\mathscr{A}_0} \frac{Y_0^+(z) - z\,Y_0^-(z)}{-Y_0^+(z) - z\,Y_0^-(z)} \end{aligned}$$

or,

$$\boxed{Z_{in}(z) = -Z_0 \frac{Y_0^+(z) - z\,Y_0^-(z)}{Y_0^+(z) + z\,Y_0^-(z)}} \tag{4.64}$$

where $Z_0 = \varrho c/\mathscr{A}_0$ defines the characteristic impedance of the zeroth acoustic tube section from (4.26). The negative sign is used for $u_L(t)$ since $u_L(t)$ is now interpreted as an input to the vocal tract instead of an output.

The lip impedance (4.64) has been derived in terms of the reflection coefficients μ_m by Wakita [1972] and Wakita and Gray [1975]. A slight variation on earlier approaches is given here which shows $Z_{in}(z)$ in terms of $A(z)$. The terms $Y_m^+(z)$ and $Y_m^-(z)$ from (4.45) will be used in the derivation. The boundary conditions, however, must be modified, since lip impedance is defined under the conditions $u_G(t)=0$. Since $u_G(t)=0$, from (4.46b),

$$Y_M^+(z) = 0. \tag{4.65}$$

There is no longer any boundary condition at the lips since a source $-u_L(t)$ has been introduced to drive the vocal tract.

A new set of polynomials is defined from the $A_m(z)$ and $B_m(z)$ polynomials of (4.48) which differs only in the sign choice for the reflection coefficients. In particular, with $\mu_m = -k_m$,

$$A_m^r(z) = A_{m-1}^r(z) - \mu_m B_{m-1}^r(z) \tag{4.66a}$$

$$z\,B_m^r(z) = -\mu_m A_{m-1}^r(z) + B_{m-1}^r(z) \tag{4.66b}$$

with initial conditions

$$A_0^r(z) = 1, \; z\,B_0^r(z) = 1. \tag{4.67}$$

In a similar fashion where $A(z)$ denotes the final polynomial $A_M(z)$, $A^r(z)$ denotes the final polynomial $A_M^r(z)$. The final result in terms of these variables is

$$\boxed{Z_{in}(z)=Z_0\frac{A(z)}{A^r(z)}} \quad . \tag{4.68}$$

The locations of the lip impedance zeros are given by the roots of $A(z)$, or equivalently, by the formant frequencies since $A(z)=1/V(z)$ in the vocal tract model. The locations of the lip impedance poles are obtained from the roots of the polynomial $A^r(z)$ which has the same recursion as $A(z)$ except that all reflection coefficients have their signs reversed.

In order to prove this result, it will first be shown that the following equations satisfy the necessary recursions (4.45) for the acoustic tube:

$$Y_m^+(z)=A_m(z)\,A_M^r(z)-A_m^r(z)\,A_M(z) \tag{4.69a}$$

$$Y_m^-(z)=B_m(z)A_M^r(z)+B_m^r(z)A_M(z). \tag{4.69b}$$

A direct substitution of (4.69) into (4.45) using (4.48) for $A_m(z)$ and $B_m(z)$ and (4.66) for $A_m^r(z)$ and $B_m^r(z)$ shows that (4.69) satisfies (4.45). The boundary condition $Y_M^+(z)=0$ is obviously satisfied by substituting $m=M$ into (4.69a). For $m=0$, (4.69) gives

$$Y_0^+(z)=A_M^r(z)-A_M(z) \tag{4.70a}$$

and

$$Y_0^-(z)=z^{-1}[A_M^r(z)+A_M(z)]. \tag{4.70b}$$

The lip impedance result immediately follows by substitution of (4.70) into (4.64). To test the lip impedance results, Wakita [1972] used the X-ray data for five Russian vowels as obtained by Fant [1960]. In order to obtain the reflection coefficients, Fant's area functions were first converted to eight-section acoustic tube models. The area of each section was represented by the average area of the original area function within that section. The seven reflection coefficients for the five Russian vowels thus obtained are shown in Table 4.1, along with the measured vocal tract length for the vowel. A sound velocity of 34 cm/ms was utilized, so that the sampling period required in the estimation of lip impedance poles and zeros is given by

$$T(\text{ms})=\frac{2L}{Mc}=\frac{L(\text{cm})}{136\,\text{cm/ms}}\,. \tag{4.71}$$

Since the true glottal impedance is unknown, different choices for μ_8 (or Z_G) were assumed. The lip impedance singularities were then computed in the following manner. Starting with $m=1$, the recursion equations (4.66) and (4.67) were used to

Table 4.1. Reflection coefficients for five Russian vowels. From Wakita and Gray [1975b].

Reflection Coefficient	/a/	/e/	/i/	/o/	/u/
μ_1	−0.212	0.381	0.600	0.529	0.845
μ_2	0.0	−0.200	0.037	0.237	0.144
μ_3	0.312	−0.212	−0.708	0.292	0.596
μ_4	0.355	−0.189	−0.382	0.451	0.200
μ_5	0.333	−0.044	−0.066	0.259	0.032
μ_6	−0.260	0.178	0.055	−0.500	−0.685
μ_7	−0.081	0.456	0.480	0.091	0.250
μ_8	assumed 0.5 to 1.0				
Vocal Tract Length (cm)	17.0	16.5	16.5	18.5	19.5

obtain $A^r(z) = A^r_M(z)$. A *step-up* procedure is used to accomplish these recursions. Chapter 5 includes a Fortran program STEPUP which can be used for obtaining both $A(z)$ and $A^r(z)$ given the reflection coefficients $\mu_m, m = 1, 2, \ldots, M$, (before calling the routine to compute $A^r(z)$, simply reverse the signs of the reflection coefficients).

The lip impedance poles and zeros (in terms of the frequencies of the impedance minima and maxima) are shown in Table 4.2. The zero and pole frequencies of the lip impedance were obtained by solving for the roots of the polynomials $A(z)$ and $A^r(z)$. In this case, the zero and pole frequencies were computed for a com-

Table 4.2. The lip impedance zero frequencies (F_i) and pole frequencies (P_i) in Hz for five Russian vowels computed from the reflection coefficients compared with Mermelstein's results for the same vowels. From Wakita and Gray [1975b].

Parameter	Vowel														
	/i/			/e/			/a/			/o/			/u/		
	Eq. (4.68) (Hz)	Mermelstein (Hz)	Difference (%)	Eq. (4.68) (Hz)	Mermelstein (Hz)	Difference (%)	Eq. (4.68) (Hz)	Mermelstein (Hz)	Difference (%)	Eq. (4.68) (Hz)	Mermelstein (Hz)	Difference (%)	Eq. (4.68) (Hz)	Mermelstein (Hz)	Difference (%)
F_1	234	224	4.5	431	421	−2.4	678	669	1.3	545	523	4.2	292	250	18.0
P_1	819	784	4.5	1079	1047	3.1	922	863	6.8	727	689	5.5	552	536	3.0
F_2	2107	2256	−6.6	1908	1971	−3.2	1193	1139	4.7	961	911	5.5	654	601	8.8
P_2	2145	2287	−6.2	2124	2153	−1.3	2111	2111	0	2122	2184	−2.8	2243	2291	−2.1
F_3	3088	3166	−2.4	2859	2908	−1.7	2570	2487	3.3	2328	2350	−0.9	2278	2308	−1.2
P_3	3388	3549	−4.5	3111	3114	−0.1	3162	3074	2.9	3124	3392	−7.9	108	3575	−13.0
F_4	3682	—	—	3736	—	—	3617	—	—	3329	—	—	3297	—	—

pletely lossless case ($\mu_8 = 1$) in order to compare the results with those obtained by Mermelstein [1967]. A sound velocity of 34 cm/ms was used. The results shown in

Table 4.2 are in good agreement. For the zero frequencies, $\{F_i\}$, the deviation was less than 8.8%, except for F_1 of /u/ in which the deviation was 18.0%. The large deviation for F_1 of /u/ is mainly due to the rough quantization of Fant's area function. The deviation for the pole frequencies, $\{P_i\}$, was less than 6.8%, except for P_3 of /o/ and /u/ in which Mermelstein's result did not fall between F_3 and F_4 of /o/ and /u/ obtained by the linear prediction analysis.

Wakita assumed $\mu_8 = 1$ or $Z_G = \infty$, but by assuming different values, we found the results to be only slightly changed by decreasing the assumed value of μ_8 to values as low as 0.5 (values of $Z_G > 3Z_{M-1}$). The frequency locations of the impedance minima and maxima remain relatively fixed in this range, while the bandwidths decrease towards zero as μ_8 is allowed to approach unity.

4.6 Further Topics

In this section, two additional acoustic tube topics are discussed. The first topic is concerned with methods of artificially introducing losses into the acoustic tube model. The second topic is concerned with a theoretical proof of the acoustical tube model stability, based upon energy considerations.

4.6.1 Losses in the Acoustic Tube Model

The acoustic tube model discussed in this chapter is completely lossless except for the glottal impedance, whereas in a more accurate model, distributed losses in the vocal tract due to the effects of wall vibration, viscous losses, and heat conduction would be considered [Flanagan, 1972]. At present, there appears to be no way of incorporating these losses into a simple analysis model for the vocal tract. One recent method suggested by Wakita [1974c] is to use a formant correction chart based upon experimentally observed shifts from synthesis models such as the Ishizaka-Flanagan model that includes the above-mentioned losses.

There are two other elementary ways in which losses can be introduced in the acoustic tube model of this chapter (other than glottal impedance). The simplest way is to assume that in each section waves decay as they are propagating and that the decay is the same for each section. The effect here is to cause each delay (z^{-1}) to be replaced by a delay and an attenuation factor (βz^{-1} with $0 < \beta < 1$). The net result is to replace each z in the transfer function or in the impedance expression by z/β. This type of damping has the effect of moving all the poles and zeros within the unit circle closer to the center (the bandwidths are increased). In the procedure for analyzing vocal tract reflection coefficients, this effect could be compensated for if there were some way to estimate the damping factor β.

Another way in which damping can be introduced into the model is through the boundary condition applied at the lips, where it was previously assumed that the radiation impedance seen by the lips was zero (since the pressure there was equal to zero). Instead, a non-zero radiation impedance $Z_r = Z_r(z)$ is assumed so that the boundary condition becomes

$$Z_r = Z_0 \frac{Y_0^+(z) - z\,Y_0^-(z)}{Y_0^+(z) + z\,Y_0^-(z)} = \frac{P_L(z)}{U_L(z)} \tag{4.72}$$

from (4.64), where Z_0 is given by (4.26). When $Z_r = 0$, this equation gives the earlier boundary condition $Y_0^+(z) = z\,Y_0^-(z)$.

As in Section 4.3, it is possible to satisfy (4.45) with expressions of the form

$$Y_m^+(z) = P(z)\,A_m(z) + Q(z)\,A_m^r(z) \tag{4.73a}$$

and

$$Y_m^-(z) = P(z)\,B_m(z) - Q(z)\,B_m^r(z), \tag{4.73b}$$

a fact which can be verified by direct substitution. $P(z)$ and $Q(z)$ can be arbitrary as far as the recursion relations are concerned. Their effect shows up only in terms of boundary conditions.

At the lips the expression

$$(Z_r/Z_0) = \frac{2\,Q(z)}{2\,P(z)} = Q(z)/P(z) \tag{4.74}$$

is obtained by substituting (4.73) into (4.72). The z-transform of the lip volume velocity is then obtained from (4.63b) and (4.73) as

$$U_L(z) = 2\,P(z). \tag{4.75}$$

The boundary condition at the glottis is

$$Y_M^+(z) = P(z)\,A(z) + Q(z)\,A^r(z) = 1/2\,c_M\,z^{-M/2}\,U_G(z) \tag{4.76}$$

from (4.46b) and (4.73), with $A(z) = A_M(z)$ and $A^r(z) = A_M^r(z)$. Combining (4.74) through (4.76) gives the net result for the transfer function from the glottal volume velocity to lip volume velocity as

$$\boxed{\frac{U_L(z)}{U_G(z)} = c_M\,z^{-M/2}\,\frac{1}{A(z) + (Z_r/Z_0)A^r(z)}}\,. \tag{4.77}$$

When (Z_r/Z_0) is very small, (4.77) reduces to $V(z) = 1/A(z)$ within a scale factor and delay term. The important realization is that if one knew the reflection coefficients and Z_r/Z_0, then the vocal tract transfer function could be determined. Although it is not clear how (4.77) can be used to derive $A(z)$ from the speech waveform for the case of non-zero lip radiation impedance, (4.77) is a more accurate synthesis model which can incorporate the network model of lip radiation impedance derived by Flanagan [1972, p. 36].

4.6.2 Acoustic Tube Stability

Stability of the acoustic tube is of considerable importance if a speech synthesis model is developed from the acoustic tube equations (such as (4.47)). In the next chapter, the stability of speech synthesis models is discussed in detail. Stability is discussed here in terms of the physics of the acoustic tube.

The kinetic energy per unit volume in the acoustic tube is simply one half of the density times the square of the velocity, $1/2\ \varrho[u_m(x,t)/\mathscr{A}_m]^2$. This result is intuitively satisfying, for if it is multiplied by the volume then it becomes simply one half the mass times velocity squared. The potential energy per unit volume is given by $[p_m(x,t)]^2/2\varrho c^2$. This result is not so intuitive, though it does seem reasonable that it is proportional to the square of the pressure. Using Fig. 4.2 as a guide, the energy in section m can be found by integrating over the volume of that section. As there is only an x-axis variation in pressure and velocity for spatial dependence, this integration becomes a multiplication by the cross-sectional area $\mathscr{A}_m$ and a single integration over x, giving

$$E_m(t)=\frac{\mathscr{A}_m}{2}\int_{-l/2}^{l/2}\left\{\varrho[u_m(x,t)/\mathscr{A}_m]^2+\frac{[p_m(x,t)]^2}{\varrho c^2}\right\}dx \tag{4.78}$$

as the energy in section m.

The rate of change of energy is found by differentiating (4.78) with respect to time, t. The evaluation can be performed by applying (4.1) and (4.2), to show that

$$\begin{aligned}\frac{1}{\mathscr{A}_m^2}\frac{\partial}{\partial t}[\varrho u_m^2(x,t)]&=\frac{2\varrho u_m(x,t)}{\mathscr{A}_m^2}\frac{\partial u_m(x,t)}{\partial t}\\&=-\frac{2}{\mathscr{A}_m}u_m(x,t)\frac{\partial p_m(x,t)}{\partial x}\end{aligned} \tag{4.79a}$$

and

$$\begin{aligned}\frac{\partial}{\partial t}[p_m^2(x,t)/\varrho c^2]&=\frac{2}{\varrho c^2}p_m(x,t)\frac{\partial p_m(x,t)}{\partial t}\\&=-\frac{2}{\mathscr{A}_m}p_m(x,t)\frac{\partial u_m(x,t)}{\partial x}.\end{aligned} \tag{4.79b}$$

Differentiating (4.78) with respect to t and applying (4.79) yields the result

$$\begin{aligned}\frac{dE_m(t)}{dt}&=-\int_{-l/2}^{l/2}\left[u_m\frac{\partial p_m(x,t)}{\partial x}+p_m\frac{\partial u_m}{\partial x}\right]dx\\&=-u_m(x,t)p_m(x,t)\Big|_{x=-l/2}^{x=l/2}.\end{aligned} \tag{4.80}$$

The product of volume velocity and pressure thus represents a power flow (rate of change of energy) similar to the Poynting vector used in electromagnetic field theory. From (4.7a) and (4.11), this can be written as

$$\frac{dE_m(t)}{dt} = -\frac{\varrho c}{\mathscr{A}_m}\{[u_m^+(t-x/c)]^2 - [u_m^-(t+x/c)]^2\}\Big|_{x=-l/2}^{x=l/2}$$
$$= \frac{\varrho c}{\mathscr{A}_m}\{-[u_m^+(t-\tau)]^2 + [u_m^-(t+\tau)]^2 + [u_m^+(t+\tau)]^2 - [u_m^-(t-\tau)]^2\}. \tag{4.81}$$

If m is replaced by $m+1$ in (4.15) and the left- and right-hand sides of each of the equations are multiplied, the result

$$[u_{m+1}^+(t-\tau)]^2 - [u_{m+1}^-(t+\tau)]^2 = \{[u_m^+(t+\tau)]^2 - [u_m^-(t-\tau)]^2\}\,[\mathscr{A}_{m+1}/\mathscr{A}_m] \tag{4.82}$$

is obtained. This result is applied to (4.81) along with the definition of the impedance of section m, $Z_m = \varrho c/\mathscr{A}_m$, to obtain

$$\frac{dE_m(t)}{dt} = Z_m\{[u_m^+(t-\tau)]^2 - [u_m^-(t+\tau)]^2\} - Z_{m-1}\{[u_{m-1}^+(t-\tau)]^2 - [u_{m-1}^-(t+\tau)]^2\}. \tag{4.83}$$

This equation relates the rate of change of energy in a section to the forward and reverse volume velocities. If these energies are summed over all of the physical sections, $m=0,1,\ldots,M-1$, then cancellation occurs in (4.83) and the rate of change of total energy becomes

$$\frac{d}{dt}\sum_{m=0}^{M-1} E_m(t) = -Z_0\{[u_0^+(t-\tau)]^2 - [u_0^-(t+\tau)]^2\} + Z_M\{[u_M^+(t-\tau)]^2 - [u_M^-(t+\tau)]^2\}. \tag{4.84}$$

Application of the boundary conditions (4.20) and (4.31) then results in

$$\frac{d}{dt}\sum_{m=0}^{M-1} E_m(t) = \frac{1}{4}\{Z_M[u_G(t)]^2 - Z_M[u_M^-(t+\tau)]^2\} \tag{4.85}$$

where Z_M is the characteristic impedance of the artificial Mth section M, equal to the glottal impedance Z_G.

If the driving function is removed, $u_G(t)=0$, (4.85) shows that the total acoustical energy can only decay, and as it can never be negative, it must have some limit. From (4.85), $u_M^-(t+\tau)$ must therefore go to zero as t goes to infinity. Thus as t goes to infinity, in the absence of a driving function, both $u_M^+(\cdot)$ and $u_M^-(\cdot)$ must approach zero. From the recursion relations of (4.19) it is seen that if the reflection coefficients are less than unity, $u_m^+(t+\tau)$ and $u_m^-(t-\tau)$ must also approach zero as t goes to infinity, for $m=M-1, M-2, \ldots, 0$. Therefore, removing the driving function $u_G(t)$ results in an energy in the acoustic tube which can only decay, and which will in fact approach zero t goes to infinity.

5. Speech Synthesis Structures

5.1 Introduction

In Chapter 4, an acoustic tube model for speech production was developed. The reflection coefficients defining the model were demonstrated to be obtainable directly from the speech waveform using linear prediction. It was shown that by cascading the Kelly-Lochbaum sections, or by introducing a transformation of variables, mathematical structures were obtainable for performing speech synthesis.

In this chapter, a general theory is developed for speech synthesis structures in terms of the reflection coefficients of the acoustic tube model under the assumption that the condition $f_s = Mc/2L$ of Chapter 4 (4.52) has been satisfied. In that case $\mu_m = k_m, m = 0, \ldots, M-1$. These structures are new to the field of digital signal processing and are a specific contribution from the work of researchers in speech [Itakura and Saito, 1971b; Gray and Markel, 1973, 1975a]. The structures will be shown to be completely general in the sense that any digital filter of the form

$$G(z) = P(z)/A(z) \tag{5.1}$$

can be efficiently implemented, where the numerator polynomial $P(z)$ and the denominator polynomial $A(z)$ are of the form

$$P(z) = P_M(z) = \sum_{m=0}^{M} p_{Mm} z^{-m} \tag{5.2}$$

and

$$A(z) = A_M(z) = \sum_{m=0}^{M} a_{Mm} z^{-m}, \tag{5.3}$$

with $a_{M0} = 1$. If $P(z) = 1$, the filter $G(z) = 1/A(z)$ becomes the all-pole synthesis filter that has found wide usage in speech synthesis. The synthesis procedures will allow incorporation of zeros in the synthesis filter, however, since if accurate methods can be obtained for analyzing zero behavior in speech (particularly for nasalized vowels), higher quality synthetic speech may be obtainable.

There are several reasons for extending the mathematics leading to the lattice form that resulted from the Kelly-Lochbaum acoustic tube sections of Chapter 4.

An academic reason is to see what other kinds of interesting mathematical representations of acoustic tubes are possible. The results of academic studies have, however, led to very practical applications. In this chapter it will be seen that a number of tradeoffs can be made between the number and types of computer operations necessary to implement the synthesis filter. A study of the fixed-point accuracy of the various filters to be developed is beyond the scope of this book but has been covered elsewhere [Gray and Markel, 1975a; Markel and Gray, 1974b, 1975a, 1975b].

In any discussion of speech synthesis using the form $G(z)=P(z)/A(z)$ or $G(z)=1/A(z)$, the question of stability must be considered. Depending on how the parameters were obtained, the speech synthesis filter may be stable or unstable. The practical effect of instability is that the results may be meaningless, e.g., the spectral peaks picked from $1/A(z)$ as formants may be completely erroneous.

As an integral part of the theory, a recursive procedure for efficiently testing the stability of $G(z)$ is presented. The procedure is referred to as a *step-down* recursion. It is based upon the inverse of a *step-up* procedure, which can be used to obtain $A(z)=A_M(z)$ in the autocorrelation method from the reflection coefficients. It is shown that a necessary and sufficient condition for the synthesis filter to be stable is that all of the reflection coefficients of the step-down procedure be bounded by unity.

The topic of stability will be considered first, and then additional properties will be developed for the generation of speech synthesis structures. Finally, in addition to the standard direct form synthesis filter, four different speech synthesis structures will be developed directly from the acoustic tube reflection coefficients.

5.2 Stability

The stability of any filter having the form $1/A(z)$ or $P(z)/A(z)$ from (5.2) and (5.3) is determined entirely by the roots of the denominator polynomial $A(z)$ when the filter coefficients are constant. If any root lies on or outside the unit circle $|z|=1$, the filter is said to be unstable in the sense that it will have either a constant oscillation or exponentially increasing oscillation for a unit sample input [Jury, 1964, p. 80]. In speech synthesis, the filter coefficients will generally be modified every few ms. The study of stability of time-varying coefficient filters is extremely difficult since one can no longer directly apply z-transform theory and the above polynomial root location criteria. Fortunately, the coefficient variations that occur in speech synthesis do not generally cause stability problems outside the scope of what can be treated by assuming constant coefficients. That is, if a particular analysis frame has resulted in an inverse filter $A(z)$ with roots outside the unit circle, a perceptually noticeable effect may be obtained in the synthesis, even if other analysis frames have roots within the circle. Conversely, if all analysis frames result in polynomials having roots within the unit circle, then, from a practical point of view, it is unlikely that updating the synthesis filter in a time-varying manner will cause any instabilities. Interest here will be focused on the stability of filters with constant coefficients. In order to study the stability of the speech synthesis filter $1/A(z)$ (or for that matter, any filter of the form $P(z)/A(z)$), it is convenient to first derive several simple properties.

5.2.1 Step-up Procedure

The necessary equations for generating the all-pole synthesis structure $1/A(z) = 1/A_M(z)$ in terms of the acoustic tube reflection coefficients, $k_m, m=1, 2, \dots, M$, are given from (4.48) as

$$A_m(z) = A_{m-1}(z) + k_m B_{m-1}(z) \tag{5.4a}$$

and

$$zB_m(z) = k_m A_{m-1}(z) + B_{m-1}(z) \tag{5.4b}$$

for $m=1, 2, \dots, M$. The initial conditions are given from (4.49) as

$$A_0(z) = 1 \quad \text{and} \quad zB_0 = 1. \tag{5.4c}$$

The polynomials $A_m(z)$ and $B_m(z)$ were previously defined from (2.48) and (2.65) as

$$A_m(z) = \sum_{i=0}^{m} a_{mi} z^{-i} \qquad (a_{m0} = 1) \tag{5.5}$$

and

$$\begin{aligned} B_m(z) &= z^{-(m+1)} A_m(1/z) \\ &= z^{-(m+1)} \sum_{i=0}^{m} a_{mi} z^{i} \\ &= \sum_{i=1}^{m+1} a_{m,m+1-i} z^{-i}. \end{aligned} \tag{5.6}$$

Therefore, using (5.4a),

$$\sum_{i=0}^{m} a_{mi} z^{-i} = \sum_{i=0}^{m-1} a_{m-1,i} z^{-i} + k_m \sum_{i=1}^{m} a_{m-1,m-i} z^{-i} \tag{5.7}$$

for $m=1, \dots, M$.

Equating like powers of z, a *step-up procedure* results for recursively obtaining the synthesis filter $1/A(z)$, from a knowledge of only the reflection coefficients $k_m, m=1, 2, \dots, M$. By inspection of (5.7), a computational expression can be written as

$$a_{mi}=\begin{cases} a_{m-1,i} & i=0 \\ a_{m-1,i}+k_m a_{m-1,m-i} & i=1,2,\ldots,m-1 \\ k_m & i=m \end{cases}$$
$$\text{for } m=1,2,\ldots,M \quad \text{with} \quad a_{00}=1 \tag{5.8}$$

The synthesis filter is then given by $1/A(z)=1/A_M(z)$ where

$$A_M(z)=\sum_{i=0}^{M} a_{Mi}z^{-i}. \tag{5.9}$$

A Fortran subroutine STEPUP for accomplishing this transformation is given in Fig. 5.1. The program inputs are the reflection coefficients $RC(I)=k_I$, $I=1,2,\ldots,M$, and M, the order of the filter (equivalently, the number of acoustic tube sections). The output is $a_{M,I-1}=A(I)$, $I=1,2,\ldots,M+1$.

```
C
      SUBROUTINE STEPUP(A,RC,M)
      DIMENSION A(1),RC(1),B(21)
      A(1)=1.
      A(2)=RC(1)
      DO 30 MINC=2,M
      DO 10 J=1,MINC
      JB=MINC-J+1
   10 B(J)=A(JB)
      DO 20 IP=2,MINC
   20 A(IP)=A(IP)+RC(MINC)*B(IP-1)
      A(MINC+1)=RC(MINC)
   30 CONTINUE
      RETURN
      END
```

Fig. 5.1 Fortran subroutine STEPUP for performing the step-up recursion procedure.

5.2.2 Step-down Procedure

It is also possible to perform the inverse of the step-up procedure, that is, given the synthesis filter $1/A(z)$, it is possible to obtain a *step-down procedure* for determining the corresponding acoustic tube reflection coefficients $\mu_m=k_m$ for $m=1,2,\ldots,M$. Substituting $B_{m-1}(z)$ from (5.4b) into (5.4a) gives

$$A_m(z)=A_{m-1}(z)+k_m[zB_m(z)-k_m A_{m-1}(z)]. \tag{5.10}$$

Solving for $A_{m-1}(z)$ in terms of $A_m(z)$ gives

$$A_{m-1}(z)=\frac{A_m(z)-zk_m B_m(z)}{1-k_m^2},$$

or

$$A_{m-1}(z) = \frac{A_m(z) - z^{-m} k_m A_m(1/z)}{1 - k_m^2} \tag{5.11}$$

for $|k_m| \neq 1$. This restriction of k_m will be considered in detail shortly. Substitution of (5.5) and (5.6) gives the coefficient relationship

$$\sum_{i=0}^{m-1} a_{m-1,i} z^{-i} = \frac{\sum_{i=0}^{m} a_{mi} z^{-i} - k_m \sum_{i=0}^{m} a_{m,m-i} z^{-i}}{1 - k_m^2}.$$

Thus, the computational expression is

$$\boxed{\begin{array}{l} a_{m-1,i} = \dfrac{a_{mi} - k_m a_{m,m-i}}{1 - k_m^2} \\ \text{with} \quad k_m = a_{mm} \\ \text{for} \quad m = M, M-1, \ldots, 1, \quad \text{and} \quad i = 0, 1, \ldots, m-1, \\ \text{and} \quad |k_m| < 1 \end{array}} \;. \tag{5.12}$$

Note that the highest ordered coefficients on the right side of (5.12) combine to give a zero result as they must, since $A_{m-1}(z)$ is, by definition, a polynomial of one order lower, i.e., $a_{mm} - k_m a_{m0} = 0$ since $a_{mm} = k_m$ from (5.8), and $a_{m0} = 1$. Also, the lowest ordered coefficients on the right side combine to give a unity coefficient $a_{m-1,0}$ as they must, by definition, since

$$a_{m-1,0} \frac{a_{m0} - k_m a_{mm}}{1 - k_m^2} = 1.$$

A Fortran subroutine STEPDN is presented in Fig. 5.2 for performing the step-down procedure based upon (5.12). The inputs are $A(I) = a_{I-1} = a_{M, I-1}$ $I = 1, 2, \ldots,$

```
C
      SUBROUTINE STEPDN(A,RC,M)
      DIMENSION A(1),RC(1),B(21)
      MP1=M+1
      ALPHA=1.
      DO 30 J=1,M
      MR=M+1-J
      MRP1=MR+1
      D=1.-A(MRP1)*A(MRP1)
      ALPHA=ALPHA/D
      DO 10 K=1,MR
      MM=MR+2-K
10    B(K)=A(MM)
      DO 20 K=1,MR
20    A(K)=(A(K)-A(MRP1)*B(K))/D
      RC(MR)=A(MRP1)
30    IF (ABS(RC(MR)) .GE. 1.) WRITE (5,40)
40    FORMAT('  1/A(Z) IS UNSTABLE',/)
      RETURN
      END
```

Fig. 5.2 Fortran subroutine STEPDN for performing the step-down recursion procedure.

$M+1$, and M, and the output is $k_I=RC(I), I=1,2,\ldots,M$. If $|k_I|\geq 1$, the procedure is stopped. If the filter $1/A(z)$ was initially designed using the autocorrelation method of linear prediction, it will be shown later that k_m is theoretically bounded by unity. It will be shown in the next section that this step-down procedure with the test for $|k_m|\geq 1, m=M, M-1,\ldots,1$, provides an efficient recursive procedure for testing the stability of *any* filter of the form $1/A(z)$ or $P(z)/A(z)$. Essentially this same procedure was used by Atal and Hanauer [1971b] to evaluate the stability of the synthesis filter $1/A(z)$ obtained from the covariance method (see Fig. 4.10, Chapter 4). The result was apparently first shown by Grenander and Szego [1958].

An Example

An example of the step-up and step-down procedures is shown in Fig. 5.3. From the given set of reflection coefficients $RC(I)=k_I, I=1,2,\ldots,8$, subroutine STEPUP is used to construct the synthesis filter denominator $A(z)$ having coefficients $a_i, i=1,2,\ldots,8$, and $a_0=1$ related to the Fortran variable $A(I)$ as $a_{I-1}=A(I)$, $I=1,2,\ldots,9$. The subroutine STEPDN is then used to construct a set of reflection coefficients $k_I=RC(I), I=1,2,\ldots,8$. The results illustrate the fact that if

```
C
C         TEST DATA FOR STEPUP
C         AND STEPDOWN PROCEDURE
C
          DIMENSION A(9),RC(8)
          DATA RC/-.94217,  .92386, -.56198,
     1            -.09454,  .20218,  .53595,
     2            -.32922, -.05899/
          M=8
          MP1=M+1
C
C         INPUT ARRAY = RC(1),....,RC(8)
C
          CALL STEPUP(A,RC,M)
          WRITE (5,10) (I,A(I),I=1,MP1)
C
          CALL STEPDN(A,RC,M)
          WRITE (5,10) (I,RC(I),I=1,M)
C
   10     FORMAT(1X,I5,F12.5)
          END
```

I	A(I)	I	RC(I)
1	1.00000	1	-0.94217
2	-2.34644	2	0.92386
3	1.65697	3	-0.56198
4	-0.00599	4	-0.09454
5	0.32305	5	0.20218
6	-1.48213	6	0.53595
7	1.15463	7	-0.32922
8	-0.18966	8	-0.05899
9	-0.05899		

Fig. 5.3 An example using STEPUP and STEPDN.

$|k_l| < 1$, then STEPUP and STEPDN are exact inverses of each other except for computational roundoff error.

5.2.3 Polynomial Properties

The polynomial $A_m(z)$ has the form

$$A_m(z) = \sum_{i=0}^{m} a_{mi} z^{-i} \qquad (a_{m0} = 1). \tag{5.13}$$

If the roots of $A_m(z)$ are defined as z_{ml}, $l = 1, 2, \ldots, m$, so that $A_m(z_{ml}) = 0$, then an equivalent form is

$$A_m(z) = \prod_{l=1}^{m} (1 - z_{ml} z^{-1}) \tag{5.14a}$$

or

$$= \prod_{l=1}^{m} (1 - z_{ml}^{*} z^{-1}). \tag{5.14b}$$

Since $A_m(z)$ is assumed to be real, the roots must appear in complex conjugate pairs, or be real roots.

Product of roots
From (5.4a) and (5.6), $A_m(z)$ can be written as

$$\begin{aligned} A_m(z) &= A_{m-1}(z) + k_m B_{m-1}(z) \\ &= \sum_{i=0}^{m-1} a_{m-1,i} z^{-i} + k_m \sum_{i=0}^{m-1} a_{m-1,i} z^{-(m-i)} \end{aligned} \tag{5.15}$$

From (5.15) and (5.14a), the coefficients of the highest power z^{-m} are, by inspection, k_m and $(-z_{m1})(-z_{m2}) \ldots (-z_{mm})$, respectively, so that

$$\boxed{k_m = \prod_{l=0}^{m} (-z_{ml})} \quad . \tag{5.16}$$

Stability bound on $|k_m|$
The magnitude of (5.16) satisfies

$$|k_m| = \left|\prod_{l=1}^{m} (-z_{ml})\right| \le \prod_{l=1}^{m} |z_{ml}|. \tag{5.17}$$

If $|k_m| \ge 1$, at least one of the roots of $A_m(z)$ is outside the unit circle $|z| = 1$. Thus

$$|k_m| < 1 \tag{5.18}$$

is a necessary condition for $1/A_m(z)$ to be stable.

Alternate expressions for k_m
Substitution of the *lth* root of $A_{m-1}(z)$, $z_{m-1,l}$ in (5.11) gives the result

$$A_{m-1}(z_{m-1,l}) = 0 = \frac{A_m(z_{m-1,l}) - z_{m-1,l}\, k_m B_m(z_{m-1,l})}{1-k_m^2}\,.$$

As long as $|k_m| \neq 1$,

$$k_m = \frac{A_m(z_{m-1,l})}{z_{m-1,l} B_m(z_{m-1,l})}\,. \tag{5.19}$$

A new variable $F_m(z)$ is defined from (5.19) and (5.6) as

$$\boxed{F_m(z) = \frac{A_m(z)}{z B_m(z)} = \frac{z^m A_m(z)}{A_m(1/z)}}\,. \tag{5.20}$$

In terms of this variable,

$$\boxed{k_m = F_m(z_{m-1,l}) \qquad k_m \neq 1}\,. \tag{5.21}$$

Similarly, by substituting the roots of $A_m(z)$ into (5.15), an equivalent expression for k_m is obtained as

$$A_m(z_{ml}) = 0 = A_{m-1}(z_{ml}) + k_m B_{m-1}(z_{ml})$$

or by applying (5.20),

$$\boxed{k_m = -z_{ml} F_{m-1}(z_{ml})}\,. \tag{5.22}$$

5.2.4 A Bound on $|F_m(z)|$

In determining necessary and sufficient conditions for the stability of $A(z)$, a bound on the magnitude of $F_m(z)$ is found to be very useful. From (5.20) and (5.14), $F_m(z)$ can be written in the following forms:

$$F_m(z) = \frac{z^m A_m(z)}{A_m(1/z)} = z^m \prod_{l=1}^{m} \frac{(1 - z_{ml} z^{-1})}{(1 - z_{ml}^* z)} = \prod_{l=1}^{m} \frac{(z - z_{ml})}{(1 - z_{ml}^* z)}$$

with a magnitude squared value

$$|F_m(z)|^2 = \prod_{l=1}^{m} \left| \frac{z - z_{ml}}{1 - z_{ml}^* z} \right|^2 . \tag{5.23}$$

What is desired is a bound on the magnitude of $F_m(z)$ in terms of the magnitude of z, given that $A_m(z)$ has all its roots inside the unit circle (i.e., $|z_{ml}| < 1$, $l = 1, 2, \ldots, m$).

The magnitude square of the terms in the product is most easily obtained by multiplying each term by its complex conjugate, i.e.,

$$\begin{aligned}
\left| \frac{(z - z_{ml})}{(1 - z z_{ml}^*)} \right|^2 &= \frac{(z - z_{ml})(z^* - z_{ml}^*)}{(1 - z z_{ml}^*)(1 - z^* z_{ml})} \\
&= \frac{z z^* - z_{ml} z^* - z_{ml}^* z + z_{ml} z_{ml}^*}{1 - z_{ml} z^* - z_{ml}^* z + z_{ml} z_{ml}^* z z^*} .
\end{aligned}$$

The difference between the numerator and denominator can be written as

$$\begin{aligned}
|z - z_{ml}|^2 - |1 - z_{ml}^* z|^2 &= z z^* - 1 + z_{ml} z_{ml}^* - z_{ml} z_{ml}^* z z^* \\
&= |z|^2 - 1 + |z_{ml}|^2 - |z_{ml}|^2 |z|^2 \\
&= (|z|^2 - 1)(1 - |z_{ml}|^2).
\end{aligned}$$

Assuming $A_m(z)$ has all its roots within the unit circle, i.e., $|z_{ml}| < 1$ for $l = 1, 2, \ldots, m$, this difference must therefore have the same sign as the difference $|z|^2 - 1$. As the difference represents the difference between the numerator and denominator for individual terms in the product of (5.23), the product must be greater than, less than, or equal to one according to whether $|z|$ is greater than, less than, or equal to one, i.e.

$$\boxed{\begin{aligned}
|F_m(z)| &> 1 \quad \text{if} \quad |z| > 1 \\
&< 1 \quad \text{if} \quad |z| < 1 \\
&= 1 \quad \text{if} \quad |z| = 1 \\
\text{with} \quad & |z_{ml}| < 1, \quad l = 1, 2, \ldots, m
\end{aligned}} \quad . \tag{5.24}$$

This result will be used to show that if $A_m(z)$ has its roots within the unit circle, then $A_{m-1}(z)$ must have its roots within the unit circle.

5.2.5 Necessary and Sufficient Stability Conditions

Eq. (5.18) states that a necessary condition for the stability of $1/A_m(z)$ is that the single term k_m satisfies $|k_m| < 1$. This requirement does not yet guarantee stability of $1/A_m(z)$, it guarantees only that the product of the roots of $A_m(z)$ has a magnitude less than one, not that all roots have magnitudes less than one.

From (5.21), if $|k_m| < 1$, then

$$|F_m\,(z_{m-1,l})| < 1.$$

By applying the bound on $|F_m(z)|$ for $z = z_{m-1,l}$ in (5.24), the result

$$|z_{m-1,l}| < 1 \tag{5.25}$$

is obtained, i.e., *if $A_m(z)$ has its roots within the unit circle, then $A_{m-1}(z)$ has its roots within the unit circle.* This result can be stated in several other equivalent ways: 1) If $1/A_m(z)$ is stable, $1/A_{m-1}(z)$ is stable; 2) stability of $1/A_m(z)$ is *sufficient* to assume the stability of $1/A_{m-1}(z)$; and 3) stability of $1/A_{m-1}(z)$ is *necessary* for the stability of $1/A_m(z)$.

Next it is shown that $|k_m| < 1$ and stability of $1/A_{m-1}(z)$ together form not only necessary but also sufficient conditions for the stability of $1/A_m(z)$. The step-up recursion evaluated at the roots z_{ml} of $A_m(z)$ from (5.22) gives

$$k_m = -z_{ml}F_{m-1}(z_{ml}).$$

If $1/A_{m-1}(z)$ is assumed stable, i.e., $|z_{m-1,l}| < 1$, and

$$\begin{aligned} |k_m| &= |z_{ml}F_{m-1}(z_{ml})| \\ &\le |z_{ml}|\,|F_{m-1}(z_{ml})| < 1, \end{aligned} \tag{5.26}$$

then $|z_{ml}|$ must be bounded by unity from (5.24).

The conditions that $|k_m| < 1$ and that $1/A_{m-1}(z)$ is stable are thus sufficient to ensure that $1/A_m(z)$ is stable. Recursively applying this result to $A_m(z)$ for $m = M$ down to $m = 1$, with $A_M(z) = A(z)$, then gives both necessary and sufficient conditions for the stability of $1/A(z)$ strictly in terms of the reflection coefficients as

$$\boxed{|k_m| < 1 \quad \text{for} \quad m = M, M-1, \ldots, 1}\;. \tag{5.27}$$

The step-down procedure is used to recursively compute these terms. If at any point k_m has a magnitude equal to or greater than one, $1/A(z)$ has been shown to be unstable due to its having at least one root on or outside the unit circle, and the procedure stops.

Example. Assume that $A(z) = A_2(z) = 1 + 1.75z^{-1} - .5z^{-2}$. From the step-down procedure

$$k_2 = -.5$$

and

$$A_1(z)=\frac{A_2(z)-k_1 z^{-2} A_2(1/z)}{1-k_1^2}$$

$$=\frac{1+1.75z^{-1}-.5z^{-2}+.5(z^{-2}+1.75z^{-1}-.5)}{.75}$$

$$=1+3.5z^{-1}.$$

Since $k_1=3.5$, $A(z)$ is known to have at least one root on or outside the unit circle and $1/A(z)$ is unstable.

5.2.6 Application of Results

It is important to realize that the necessary and sufficient conditions for stability just derived are completely general in the sense that *any* polynomial can be efficiently tested for the existence of roots on or outside the unit circle. If the polynomial has a non-unity leading coefficient, all terms are divided by that value, which leaves the root locations unchanged.

If the covariance method is applied for linear prediction analysis, the step-down recursion procedure should be applied to the polynomial for each frame of data. If $1/A(z)$ is unstable, the results obtained by further processing, e.g., generation of an acoustic tube for parameter quantization, performing spectral analysis, etc., must be carefully interpreted. Replacing all roots outside the unit circle by their reciprocals will assure stability, as suggested by Atal and Hanauer [1971b], but the original linear prediction formulation is no longer being solved. Some other ad hoc procedure is necessary if a root lies precisely on the unit circle. It should also be emphasized that although both the covariance and the autocorrelation equations were shown in Chapter 3 to be solvable using the general recursion

$$A_m(z)=A_{m-1}(z)+k_m B_{m-1}(z),$$

for $m=1,2,\ldots,M$, with $A(z)=A_M(z)$, only in the case of the autocorrelation method do the reflection coefficients of the step-down recursion procedure equal the reflection coefficients $\{k_m\}$ obtained in the analysis. The reflection coefficients from the step-down procedure will not, in general, equal the generalized reflection coefficients $\{k_m\}$ from the covariance analysis. *In the covariance method, the generalized reflection coefficients are not sufficient to correctly determine* $A(z)$. The reason is that $B_m(z)$ is uniquely determined from $A_m(z)$ as $B_m(z)=z^{-(m+1)}A_m(1/z)$ in the autocorrelation method, whereas in the covariance method $B_m(z)$ requires knowledge of the actual covariance coefficients.

Replacement of the reflection coefficients by the generalized reflection coefficients defines a new filter $A'(z)$ whose norm is greater than that of $A(z)$. The authors are aware of two studies in which the generalized reflection coefficients have been applied to speech processing, one where an efficient voice transmission system based upon the covariance method was being developed [Welch, 1974] and one where spectral smoothing was desired based upon a Kalman filter [Matsui, et al., 1972].

If sufficient numerical accuracy is maintained, then *the stability test is unnecessary in the autocorrelation method for* $|k_m| < 1$, $m = 1, 2, \ldots, M$, *is theoretically guaranteed.* From (3.57) with $\beta_m = \alpha_m$ for the autocorrelation method,

$$\alpha_m = \alpha_{m-1}(1 - k_m^2)$$

for $m = 1, \ldots, M$. Since α_m is the total squared error in the analysis (a positive quantity), $k_m^2 = |k_m|^2 < 1$ must always be satisfied!

5.3 Recursive Parameter Evaluation

In this section, properties of the transfer function

$$G(z) = P(z)/A(z) \tag{5.28}$$

will be developed using inner products. The results will then be used for recursively designing speech synthesis structures from the acoustic tube reflection coefficients. The polynomial $A(z)$ can be any Mth order polynomial of the form (5.5) having all its roots inside the unit circle (so that $1/A(z)$ is stable). By applying the step-down procedure of the previous section, this requirement has been shown to be equivalent to showing that $|k_m| < 1$ for $m = 1, 2, \ldots, M$.

5.3.1 Inner Product Properties

With reference to Fig. 5.4, an inner product is once again defined, only now the common input to the transfer functions $P(z)$ and $Q(z)$ is simply the unit sample response of the filter $1/A(z)$. The product of $P(z)$ and $1/A(z)$ gives $G(z)$ as shown

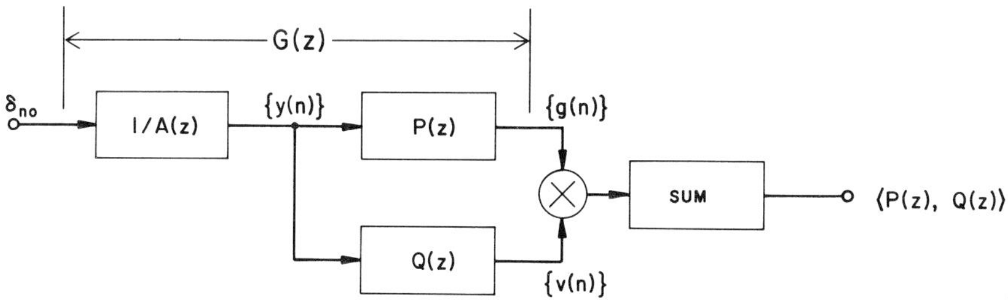

Fig. 5.4 A signal processing definition of the inner product $\langle P(z), Q(z)\rangle$ with $G(z) = P(z)/A(z)$.

with the unit sample response $\{g(n)\}$. The inner product is then computed from the figure as

$$\langle P(z), Q(z)\rangle = \sum_{n=0}^{\infty} g(n)\, v(n). \tag{5.29}$$

In terms of the transfer function coefficients and the input $y(n)$, satisfying $y(n)=0$ for $n<0$, the inner product is

$$\langle P(z), Q(z)\rangle = \sum_{n=0}^{\infty} \sum_{i=0}^{\infty} p_i y(n-i) \sum_{j=0}^{\infty} q_j y(n-j)$$

or

$$\boxed{\langle P(z), Q(z)\rangle = \sum_{i=0}^{\infty} \sum_{j=0}^{\infty} p_i r_y(i-j) q_j} \tag{5.30}$$

where

$$r_y(i-j) = \sum_{n=0}^{\infty} y(n-i)\, y(n-j)$$

$$= r_y(j-i).$$

For $i=0$, the autocorrelation

$$r_y(j) = \sum_{n=0}^{\infty} y(n)\, y(n-j)$$

has a z-transform

$$R_y(z) = Y(z)\, Y(1/z)$$

or

$$\boxed{R_y(z) = \frac{1}{A(z)\, A(1/z)}}\,. \tag{5.31}$$

The cross-correlation defined by

$$u(k) = \sum_{n=0}^{\infty} g(n)\, v(n-k)$$

has the z-transform

$$U(z) = G(z)\, V(1/z).$$

By applying the z-transform inversion integral

$$u(k) = \oint U(z)\, z^{-k} \frac{dz}{2\pi j z} \tag{5.32}$$

for $k=0$, the inner product is equivalently written as

$$\langle P(z), Q(z)\rangle = \oint G(z)\, V(1/z) \frac{dz}{2\pi j z}.$$

Equivalent expressions in terms of the transfer functions $P(z)$ and $Q(z)$ are then

$$\langle P(z), Q(z)\rangle = \oint P(z)\, Y(z)\, Y(1/z)\, Q(1/z) \frac{dz}{2\pi j z} \tag{5.33a}$$

$$= \int_{-\pi}^{\pi} \frac{P(e^{j\theta})\, Q(e^{-j\theta})}{|A(e^{j\theta})|^2} \frac{d\theta}{2\pi} \tag{5.33b}$$

$$= \int_{-\pi}^{\pi} P(e^{j\theta})\, R_y(e^{j\theta})\, Q(e^{-j\theta}) \frac{d\theta}{2\pi} \tag{5.33c}$$

$$= \sum_{i=0}^{\infty} \sum_{j=0}^{\infty} p_i\, r_y(i-j)\, q_j \tag{5.33d}$$

The second expression is obtained by substituting $z=\exp(j\theta)$ and choosing the unit circle $|z|=1$ for the contour. The third expression is obtained by substituting (5.31). The fourth expression is simply a restatement of (5.30). The inner product can therefore be written as a computational expression in either the sampled data domain (5.33d), the frequency domain (5.33b) and (5.33c), or the complex z-transform domain (5.33a).

Basic Properties

From (5.29) or Fig. 5.4, it is seen that the order of the inner product terms is irrelevant, i.e.,

$$\langle P(z), Q(z)\rangle = \langle Q(z), P(z)\rangle. \tag{5.34}$$

From (5.33b) it can also be seen that

$$\langle P(z), Q(z)\rangle = \langle 1, P(1/z)\, Q(z)\rangle \tag{5.35a}$$

$$= \langle P(z)\, Q(1/z), 1\rangle \tag{5.35b}$$

$$= \langle Q(1/z), P(1/z)\rangle. \tag{5.35c}$$

In addition, from (5.33a), it is seen that if both $P(z)$ and $Q(z)$ are multiplied by the same power of z, z^{-k}, the inner product is unchanged since $z^k(1/z)^k=1$, i.e.,

$$\langle z^{-k} P(z), z^{-k} Q(z)\rangle = \langle P(z), Q(z)\rangle. \tag{5.36}$$

Furthermore, if

$$P(z) = aP_1(z) + bP_2(z)$$

where a and b are constants, then from (5.33),

$$\langle P(z), Q(z)\rangle = a\langle P_1(z), Q(z)\rangle + b\langle P_2(z), Q(z)\rangle. \tag{5.37}$$

The inner product expression as an integral in the z-plane will now be used to derive important relationships for generating speech synthesis structures. First the *orthogonality principle* is shown from (5.33 a). Direct substitution with (5.31) gives

$$\langle A(z), z^{-l}\rangle = \oint \frac{z^{l-1}}{A(1/z)} \frac{dz}{2\pi j}.$$

As $A(z)$ is a polynomial with only negative powers of z from (5.9), $A(1/z)$ is a polynomial with only positive powers of z. Since all roots of $A(z)$ are assumed to lie within the unit circle, roots of $1/A(z)$ will lie outside the unit circle. Therefore, since the above integrand has no singularities within the unit circle for $l \geq 1$, the integral must equal zero from Cauchy's theorem of complex variables, i.e.,

$$\langle A(z), z^{-l}\rangle = 0, \quad l = 1, 2, \ldots . \tag{5.38}$$

That is, $A(z)$ must be orthogonal to all negative powers of z, and thus orthogonal to the specific powers $z^{-1}, z^{-2}, \ldots, z^{-M}$. Rather than use a similar approach for $B(z)$, it is more convenient to consider the more general case of $B_m(z)$, using (5.6) rewritten here as

$$B_m(z) = z^{-(m+1)} A_m(1/z). \tag{5.39}$$

When $m = M$ this can be used for $B(z) = B_M(z)$ with $A(z) = A_M(z)$.

From the properties (5.34) through (5.36), it is seen that (5.39) leads to

$$\langle z^{-(m+1-l)}, B_m(z)\rangle = \langle z^{-l}, A_m(z)\rangle, \tag{5.40}$$

so that if it is known that $A_m(z)$ is orthogonal to the powers $z^{-1}, z^{-2}, \ldots, z^m$ (as is the case for $m = M$), then it follows that $B_m(z)$ is orthogonal ot the powers z^{-m}, $z^{-m+1}, \ldots, z^{-1}$, the same powers for which orthogonality with $A_m(z)$ exists.

From (5.11) and linearity of (5.37),

$$\begin{aligned} \langle z^{-l}, A_{m-1}(z)\rangle &= \frac{\langle z^{-l}, A_m(z)\rangle - k_m \langle z^{-l}, zB_m(z)\rangle}{1 - k_m^2} \\ &= \frac{\langle z^{-l}, A_m(z)\rangle - k_m \langle z^{-(l+1)}, B_m(z)\rangle}{1 - k_m^2}. \end{aligned} \tag{5.41}$$

If $A_m(z)$ is orthogonal to $z^{-1}, z^{-2}, \ldots, z^{-m}$, then (5.40) shows that $B_m(z)$ is orthogonal to the same powers. Using $l = 1, 2, \ldots, m-1$ in (5.41) leads to the result that $A_{m-1}(z)$ must be orthogonal to the powers $z^{-1}, z^{-2}, \ldots, z^{-(m-1)}$. Thus by starting with $A(z) = A_M(z)$, which is orthogonal to all of the negative powers of z as indicated by (5.38), we recursively find that for $m = M-1, M-2, \ldots, 1$,

$$\langle A_m(z), z^{-l} \rangle = \langle B_m(z), z^{-l} \rangle = 0 \quad \text{where} \quad l = 1, 2, \ldots, m. \tag{5.42}$$

These orthogonality results were simply stated and used in the PARCOR formulation of Chapter 2, where the inner product was introduced as an analysis tool. From these results several other properties are obtained. First,

$$\alpha_m = \langle A_m(z), A_m(z) \rangle = \langle 1, A_m(z) \rangle = \langle B_m(z), B_m(z) \rangle. \tag{5.43}$$

By applying the orthogonality property for $A_m(z)$,

$$\begin{aligned} \langle \sum_{i=0}^{M} a_i z^{-i}, A_m(z) \rangle &= \sum_{i=0}^{M} a_i \langle z^{-i}, A_m(z) \rangle \\ &= \langle 1, A_m(z) \rangle + 0 + \cdots + 0. \end{aligned}$$

By applying the orthogonality properties for $B_m(z)$,

$$\begin{aligned} \langle B_m(z), B_n(z) \rangle &= \sum_{i=1}^{n+1} b_{ni} \langle B_m(z), z^{-i} \rangle = 0 \qquad n < m \\ &= \sum_{i=1}^{m+1} b_{mi} \langle z^{-i}, B_n(z) \rangle = 0 \qquad m < n. \end{aligned} \tag{5.44}$$

Therefore, it is shown that the $B_m(z)$ polynomials are orthogonal to each other, i.e., $\langle B_m(z), B_n(z) \rangle = 0 \quad m \neq n$. In the same manner the following results are obtained,

$$\begin{aligned} \langle A_m(z), B_m(z) \rangle &= \langle 1, B_m(z) \rangle \\ &= \langle A_m(z), z^{-(m+1)} \rangle. \end{aligned} \tag{5.45}$$

From (5.4b),

$$z\, B_{m+1}(z) = k_{m+1} A_m(z) + B_m(z)$$

so that

$$\begin{aligned} \langle A_m(z), z B_{m+1}(z) \rangle &= \langle z^{-1} A_m(z), B_{m+1}(z) \rangle \\ &= k_{m+1} \langle A_m(z), A_m(z) \rangle + \langle A_m(z), B_m(z) \rangle. \end{aligned} \tag{5.46}$$

Application of the orthogonality property to $B_{m+1}(z)$ gives

$$\langle z^{-1} A_m(z), B_{m+1}(z) \rangle = \sum_{i=0}^{m} a_{mi} \langle z^{-(i+1)}, B_{m+1}(z) \rangle = 0 \tag{5.47}$$

so that

$$\langle A_m(z), B_m(z)\rangle = -k_{m+1}\,\alpha_m = \langle 1, B_m(z)\rangle. \tag{5.48}$$

Taking the inner product of (5.4a) with unity gives

$$\begin{aligned}\langle A_m(z), 1\rangle &= \langle A_{m-1}(z) + k_m B_{m-1}(z), 1\rangle \\ &= \langle A_{m-1}(z), 1\rangle + k_m \langle B_{m-1}(z), 1\rangle,\end{aligned}$$

or by using (5.43),

$$\begin{aligned}\alpha_m &= \alpha_{m-1} + k_m(-k_m\,\alpha_{m-1}) \\ &= (1-k_m^2)\,\alpha_{m-1}.\end{aligned} \tag{5.49}$$

Numerator Representation

A procedure is now presented for representing the numerator polynomial $P(z)$ as a linear combination of the orthogonal polynomials $z\,B_m(z)$. The polynomials $P(z) = P_M(z)$ and $z\,B_m(z)$ are given by

$$P_M(z) = \sum_{i=0}^{M} p_{Mi}\,z^{-i} = \sum_{i=0}^{M-1} p_{Mi}\,z^{-i} + p_{MM}\,z^{-M}$$

and

$$\begin{aligned}z\,B_M(z) &= z^{-M} A_M(1/z) = z^{-M} A(1/z) \\ &= z^{-M} \sum_{i=0}^{M} a_i z^i \qquad (a_0 = 1) \\ &= \sum_{i=0}^{M-1} a_{M-i} z^{-i} + z^{-M}.\end{aligned} \tag{5.50}$$

If $z\,B_M(z)$ is multiplied by p_{MM} and then subtracted from the Mth order polynomial $P_M(z)$, a new polynomial of order $M-1$ is obtained. Defining the multiplicative constant as v_M, the $M-1$ order polynomial $P_{M-1}(z)$ is obtained as

$$P_{M-1}(z) = P_M(z) - v_M\,z\,B_M(z).$$

Since each polynomial $zB_m(z)$ has a unity coefficient for the highest power of z^{-1}, the same procedure can be used to define $P_{M-2}(z)$, $P_{M-3}(z)$ down to $P_0(z)$. Define

$$P_m(z) = \sum_{i=0}^{m} p_{mi}\,z^{-i} \tag{5.51}$$

and

$$v_m = p_{mm}. \tag{5.52}$$

Then $P_{m-1}(z)$ is computed from

$$P_{m-1}(z) = P_m(z) - v_m\,z\,B_m(z) \tag{5.53}$$

for $m=M, M-1, \ldots, 1$. By calculating $P_m(z)$ from the above equation for $m=0, 1, \ldots, M$, with $P_{-1}(z)=0$, the numerator is equivalently described in terms of the orthogonal polynomials $z\,B_m(z)$ as

$$P(z)=P_M(z)=\sum_{m=0}^{M} v_m z\,B_m(z). \tag{5.54}$$

Synthesis Filter Relationships

The speech synthesis filter $G(z)$ can be represented in terms of the orthogonal polynomials $z\,B_m(z)$ and $1/A(z)$ as

$$\begin{aligned} G(z) &= P(z)/A(z) \\ &= \sum_{m=0}^{M} v_m [z\,B_m(z)/A(z)]. \end{aligned} \tag{5.55}$$

If $E(z)$ defines the synthesis filter input and $X(z)$ defines the output, the input-output relationship can then be described as

$$\begin{aligned} X(z) &= G(z)E(z) \\ &= \sum_{m=0}^{M} v_m [z\,B_m(z)E(z)/A(z)]. \end{aligned} \tag{5.56}$$

The inner product properties can now be used to efficiently evaluate the energy of the filter. From (5.29) with $G(z)=P(z)/A(z)$ the total energy of the filter $G(z)$ in response to a unit sample is computed as

$$\begin{aligned} \sum_{n=0}^{\infty} g^2(n) &= \int_{-\pi}^{\pi} |G(e^{j\theta})|^2 \frac{d\theta}{2\pi} \\ &= \int_{-\pi}^{\pi} \left| \frac{P(e^{j\theta})}{A(e^{j\theta})} \right|^2 \frac{d\theta}{2\pi} \\ &= \oint \frac{P(z)\,P(1/z)}{A(z)\,A(1/z)} \frac{dz}{2\pi jz} \\ &= \langle P(z), P(z) \rangle. \end{aligned}$$

But from (5.54) and (5.43), the inner products can be written as

$$\begin{aligned} \langle P(z), P(z) \rangle &= \sum_{n=0}^{M} \sum_{m=0}^{M} v_m v_n \langle z\,B_n(z), z\,B_m(z) \rangle \\ &= \sum_{m=0}^{M} v_m^2 \alpha_m \,. \end{aligned}$$

Therefore, the filter energy is directly computed in terms of v_m and α_m as

$$\boxed{\sum_{n=0}^{M} g^2(n) = \oint \frac{P(z)P(1/z)}{A(z)A(1/z)} \frac{dz}{2\pi jz} = \sum_{m=0}^{M} v_m^2 \alpha_m} \quad . \tag{5.57}$$

Closed form solutions for the integrals of low order have been tabulated [Jury, 1964, pp. 297–299]. Their evaluation becomes extremely tedious as the filter order increases. Since v_m is recursively evaluated from (5.52) and (5.53), and α_m is recursively evaluated from (5.49), these integrals can be recursively evaluated for arbitrary order. A similar recursive solution was first presented by Astrom, et al. [1970].

5.3.2 Equation Summary with Program

The various results to be used in the speech synthesis filter derivations are summarized below.

Polynomial Definitions

$$P(z) = P_M(z) \qquad \text{(numerator polynomial)}$$

$$A(z) = A_M(z) \qquad \text{(denominator polynomial)}$$

$$P_m(z) = \sum_{i=0}^{m} p_{mi} z^{-i}$$

$$A_m(z) = \sum_{i=0}^{m} a_{mi} z^{-i} \quad (a_{m0} = 1)$$

$$z B_m(z) = z^{-m} A(1/z) \qquad \text{(reciprocal polynomial)}$$

$$m = 0, 1, \ldots, M.$$

Step-up Procedure

$$\begin{aligned} A_m(z) &= A_{m-1}(z) + k_m B_{m-1}(z) \qquad \text{(recursion)} \\ &= A_{m-1}(z) + k_m z^{-m} A_{m-1}(1/z) \end{aligned}$$

$$\begin{aligned} z B_m(z) &= k_m A_{m-1}(z) + B_{m-1}(z) \\ &= k_m A_{m-1}(z) + z^{-m} A_{m-1}(1/z) \end{aligned}$$

$$m = 1, 2, \ldots, M$$

$$A_0(z) = z B_0(z) = 1 \qquad \text{(initial conditions)}$$

$$A(z) = A_M(z). \qquad \text{(final polynomial)}$$

Step-down Procedure

$$A_M(z) = A(z) \qquad \text{(initial condition)}$$

$$k_m = a_{mm} \qquad \text{(reflection coefficient)}$$

$$A_{m-1}(z) = \frac{A_m(z) - k_m z B_m(z)}{1 - k_m^2}$$

$$= \frac{A_m(z) - k_m z^{-m} A_m(1/z)}{1 - k_m^2}$$

$$B_{m-1}(z) = \frac{-k_m A_m(z) + z B_m(z)}{1 - k_m^2}$$

$$= \frac{-k_m A_m(z) + z^{-m} A_m(1/z)}{1 - k_m^2}$$

for $m = M, M-1, \ldots, 1$.

Stability Condition for $1/A(z)$

The necessary and sufficient condition for the stability of $1/A(z)$ is

$$|k_m| < 1 \qquad m = 1, 2, \ldots, M.$$

Orthogonality Relations

$$\langle A_m(z), z^{-l} \rangle = 0$$

$$\langle B_m(z), z^{-l} \rangle = 0$$

$$\langle B_m(z), B_l(z) \rangle = 0 \quad \text{for} \quad m \neq l$$

$$l = 1, 2, \ldots, m$$

Inner Product Results

$$\langle A_m(z), A_m(z) \rangle = \langle 1, A_m(z) \rangle = \alpha_m$$

$$\langle A_m(z), B_m(z) \rangle = \langle 1, B_m(z) \rangle$$

$$= \langle A_m(z), z^{-(m+1)} \rangle$$

Recursions

$$\alpha_{m-1} = \alpha_m / (1 - k_m^2)$$

$$P_{m-1}(z) = P_m(z) - v_m z^{-m} A_m(1/z)$$

$$v_m = p_{mm}$$

Synthesis Filter

$$g(n) \leftrightarrow G(z) = X(z)/E(z) = P(z)/A(z)$$
$$= \sum_{m=0}^{M} v_m \frac{z B_m(z)}{A(z)}$$

Input-Output Relationships

$$X(z) = \frac{P(z)}{A(z)} E(z)$$
$$= \sum_{m=0}^{M} v_m \frac{z B_m(z) E(z)}{A(z)}$$

Filter Energy Evaluation

$$\sum_{n=0}^{\infty} g^2(n) = \int_{-\pi}^{\pi} |G(e^{j\theta})|^2 \frac{d\theta}{2\pi}$$
$$= \oint \frac{P(z) P(1/z)}{A(z) A(1/z)} \frac{dz}{2\pi j z}$$
$$= \sum_{m=0}^{M} v_m^2 \alpha_m$$

A Fortran Program — EVAL

Figure 5.5 presents subroutine EVAL which 1) transforms the inverse filter $A(z)$ via the step-down recursion procedure into the acoustic tube reflection coefficients $\{k_m\} = \{k_1, \ldots, k_M\}$, 2) transforms the numerator polynomial coefficients into the coefficients $\{v_m\} = \{v_0, v_1, \ldots, v_M\}$, 3) computes the total squared error terms

```
C
      SUBROUTINE EVAL(A,P,RC,M,TAP,AL,TI)
      DIMENSION A(1),P(1),RC(1),TAP(1)
      DIMENSION AL(1),B(21)
      MP1=M+1
      TI=0.
      AL(MP1)=1.
      DO 20 J=1,M
      MR=M+1-J
      D=1.-A(MR+1)*A(MR+1)
      PS=P(MR+1)*P(MR+1)
      TI=TI+AL(MR+1)*PS
      AL(MR)=AL(MR+1)/D
      DO 10 K=1,MR
      MM=MR+2-K
   10 B(K)=A(MM)
      DO 20 K=1,MR
      P(K)=P(K)-P(MR+1)*B(K)
   20 A(K)=(A(K)-A(MR+1)*B(K))/D
      TI=TI+P(1)*P(1)*AL(1)
      DO 30 J=1,M
      RC(J)=A(J+1)
   30 TAP(J)=P(J)
      TAP(MP1)=P(MP1)
      RETURN
      END
```

Fig. 5.5 Fortran subroutine EVAL for evaluating speech synthesis filter parameters, total energy, and total squared error terms.

$\{\alpha_m\}=\{\alpha_0, \alpha_1, \ldots, \alpha_M\}$, and 4) computes the total energy of the filter $G(z)$, in response to a unit sample input. The first two outputs $\{k_m\}$ and $\{v_m\}$ define the parameters used to describe the general synthesis structure in the following section. The program variables are defined as follows:

INPUT VARIABLES:

$A(I)=a_{I-1} \quad I=1,2,\ldots,M+1$
$P(I)=P_{I-1} \quad I=1,2,\ldots,M+1$
$M=$order of the inverse filter

OUTPUT VARIABLES:

$RC(I)=k_I, \quad I=1,2,\ldots,M$
$TAP(I)=v_{I-1}, \quad I=1,2,\ldots,M+1$
$AL(I)=\alpha_{I-1}, \quad I=1,2,\ldots,M+1$

$$TI=\int_{-\pi}^{\pi}\left|\frac{P(e^{j\theta})}{A(e^{j\theta})}\right|^2\frac{d\theta}{2\pi}.$$

The program was written directly from the various equations in this section. Note that the step-down procedure presented earlier as the subroutine STEPDN is embedded in EVAL. The only feature left out is the stability test (which can be easily inserted if desired). This program assumes that $1/A(z)$ is a stable filter.

5.4 A General Synthesis Structure

In the digital signal processing literature [Rabiner and Rader, 1972; Oppenheim and Schafer, 1975; Rabiner and Gold, 1975], there are numerous filter structures that can be used to implement the general linear filter form $P(z)/A(z)$. The most common filters are referred to as the direct form, parallel form, and cascade form. Research in linear prediction of speech has uncovered a number of new structures for implementing $P(z)/A(z)$ [Itakura and Saito, 1971b; Gray and Markel, 1973, 1975a]. These structures are of considerable importance in speech synthesis for two reasons: 1) they allow the filter to be implemented directly from the reflection coefficients $\{k_m\}$ and 2) in an actual computer implementation, they allow one to trade off accuracy, the number of multiplications and additions per section, and complexity. Here, the general structure of the synthesis filter is developed in terms of the reflection coefficients, tap parameters, and a set of *pi-parameters* that will allow generation of the specific structures from a unified starting point.

In Section (5.3), the polynomial recursions for $A_m(z)$ and $z\,B_m(z)$ were summarized along with their initial conditions. The input-output relation from $E(z)$ to $X(z)$ was also presented. Multiplication of each of these relations by $E(z)/A_M(z)=E(z)/A(z)$ gives

$$\frac{A_m(z)\,E(z)}{A(z)}=\frac{A_{m-1}(z)\,E(z)}{A(z)}+\frac{k_m\,B_{m-1}(z)\,E(z)}{A(z)} \tag{5.58a}$$

$$\frac{z\,B_m(z)\,E(z)}{A(z)}=\frac{k_m\,A_{m-1}(z)\,E(z)}{A(z)}+\frac{B_{m-1}(z)\,E(z)}{A(z)} \tag{5.58b}$$

and

$$\frac{z\,B_0(z)\,E(z)}{A(z)}=\frac{A_0(z)\,E(z)}{A(z)}=\frac{E(z)}{A(z)}. \tag{5.59}$$

The purpose of this multiplication is to transform $A_0(z)=1$ into the desired output $X(z)=E(z)/A(z)$ as shown by (5.59). Note that for $m=M$, the input is defined by

$$\frac{A_M(z)\,E(z)}{A(z)}=E(z).$$

Therefore, the above equations evaluated from the input $E(z)$ at $m=M$ down to $m=0$ result in the output of the all-pole portion $E(z)/A(z)$.

Block diagrams of the above equations for the input $m=M$, an arbitrary point m, and the output at $m=0$ are shown in Fig. 5.6. The blocks are referred to as section M, section m, and section 1, respectively. The section number m is equal to the subscript number of the polynomial $A_m(z)$ or $B_m(z)$ at the left side. Note that in order to start from the input $m=M$ it is only necessary to rewrite (5.58a) as

$$\frac{A_{m-1}(z)\,E(z)}{A(z)}=\frac{A_m(z)\,E(z)}{A(z)}-k_m\frac{B_{m-1}(z)\,E(z)}{A(z)}. \tag{5.60}$$

These sections from $m=M$ down to $m=1$ define the all-pole portion of the filter $1/A(z)$. The implementation of the numerator polynomial $P(z)$ is trivial since from (5.56) it is necessary only to tap off of each lower node $zB_m(z)E(z)/A(z)$, multiply it by v_m, and combine all $m+1$ terms. The result is $X(z)=E(z)P(z)/A(z)$ directly in terms of the reflection coefficients $\{k_m\}$ and tap parameters $\{v_m\}$. By introducing a new parameter π_m, referred to as the *pi-parameter*, (5.58) and (5.59) will allow all of the specific speech synthesis structures to be derived from one unified set of equations. The definitions

$$\hat{A}_m(z)=A_m(z)\,\pi_m \tag{5.61a}$$

$$\hat{B}_m(z)=B_m(z)\,\pi_m \tag{5.61b}$$

$$\hat{v}_m=v_m/\pi_m, \tag{5.61c}$$

are introduced with

$$\pi_M=1, \tag{5.61d}$$

so that

$$\hat{A}_M(z)=A_M(z)=A(z). \tag{5.61e}$$

The terms $\hat{v}_m$ will be referred to as *modified tap parameters*. From (5.58) and (5.61),

$$\frac{\hat{A}_{m-1}(z)\,E(z)}{A(z)\,\pi_{m-1}}=\frac{\hat{A}_m(z)\,E(z)}{A(z)\,\pi_m}-k_m\frac{\hat{B}_{m-1}(z)\,E(z)}{A(z)\,\pi_{m-1}}$$

$$\frac{z\,\hat{B}_m(z)\,E(z)}{A(z)\,\pi_m}=k_m\frac{\hat{A}_{m-1}(z)\,E(z)}{A(z)\,\pi_{m-1}}+\frac{\hat{B}_{m-1}(z)\,Ez)}{A(z)\,\pi_{m-1}},$$

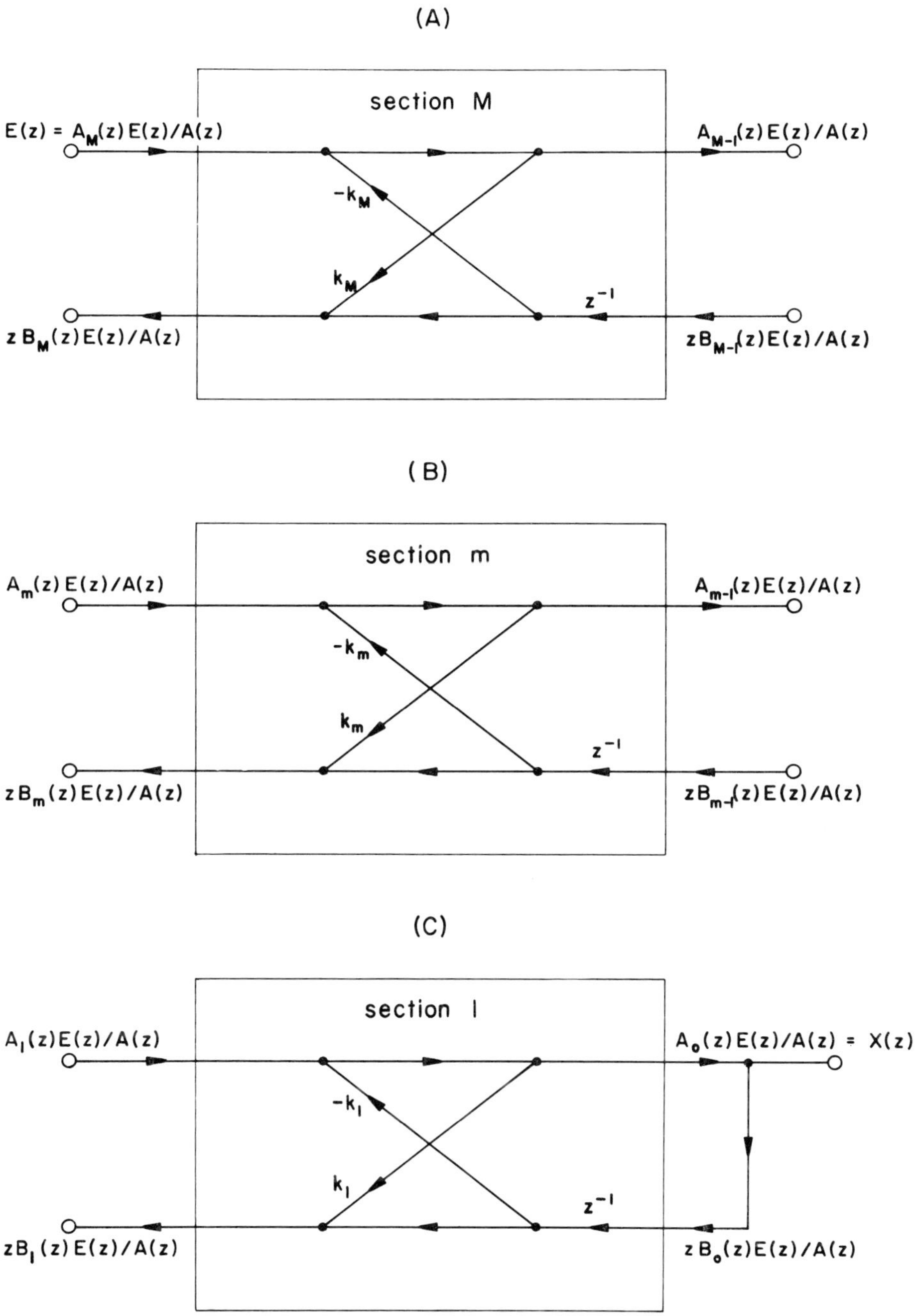

Fig. 5.6 Flow graphs of filter sections: A) section M, B) section m, C) section 1.

and from (5.59) and (5.61),

$$\frac{z\,\hat{B}_0(z)\,E(z)}{A(z)\,\pi_0}=\frac{\hat{A}_0(z)\,E(z)}{A(z)\,\pi_0}=\frac{E(z)}{A(z)}.$$

In order to simplify these equations, new symbols are introduced as follows:

$$E_m^+(z)=\frac{\hat{A}_m(z)\,E(z)}{A(z)}\leftrightarrow e_m^+(n) \tag{5.62a}$$

$$E_m^-(z)=\frac{\hat{B}_m(z)\,E(z)}{A(z)}\leftrightarrow e_m^-(n). \tag{5.62b}$$

The all-pole portion of the filter is then implemented from the equations

$$E_{m-1}^+(z)=\frac{\pi_{m-1}}{\pi_m}E_m^+(z)-k_m\,E_{m-1}^-(z) \tag{5.63a}$$

$$z\,E_m^-(z)\frac{\pi_{m-1}}{\pi_m}=k_m\,E_{m-1}^+(z)+E_{m-1}^-(z) \tag{5.63b}$$

for $m=M, M-1, \ldots, 1$.

The input is

$$E(z)=E_M^+(z)=\hat{A}_M(z)\,E(z)/A_M(z) \tag{5.64}$$

while the output from the $1/A(z)$ portion (the *termination*) is

$$z\,E_0^-(z)/\pi_0=E_0^+(z)/\pi_0. \tag{5.65}$$

The general output $X(z)$, in terms of these variables, is from (5.56), (5.61), and (5.62),

$$X(z)=G(z)\,E(z)$$

$$=\sum_{m=0}^{M}\hat{v}_m\,z\,\frac{\hat{B}_m(z)E(z)}{A(z)}$$

or

$$X(z)=\sum_{m=0}^{M}\hat{v}_m E_m^-(z) \quad . \tag{5.66}$$

Note that in the all-pole case,

$$X(z)=\hat{v}_0 z E_m^-(z)=z\,E_m^-(z)/\pi_0$$

or

$$X(z)=E_0^+(z)/\pi_0 \qquad (G(z)=1/A(z)) \quad . \tag{5.67}$$

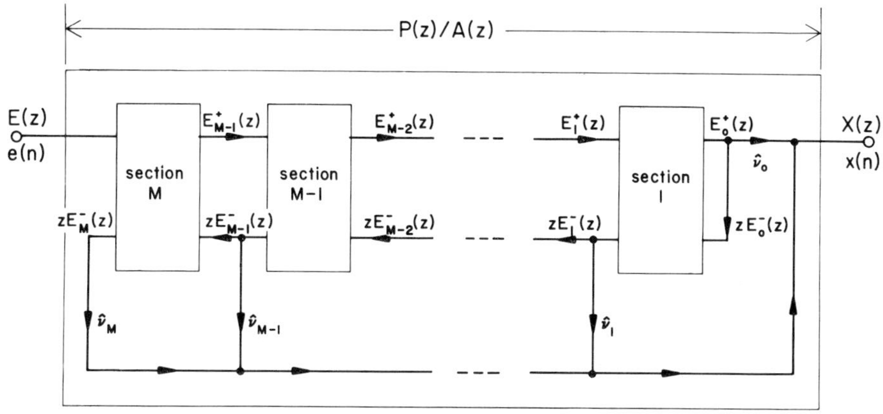

Fig. 5.7 General speech synthesis structure based upon the reflection coefficients (within each section) and the modified tap parameters.

A block diagram of the overall synthesis filter structure is shown in Fig. 5.7. The diagram is drawn directly from (5.63) and (5.66) in a way that easily shows the relation between the all-pole case and the more general pole and zero case. For example, in the all-pole case, $\hat{v}_1 = \hat{v}_2 = \ldots = \hat{v}_M = 0$, so that the only contribution to the output is from $\hat{v}_0 e_0^+(n)$, giving $x(n) = e_0^+(n)\hat{v}_0 = e_0^+(n)/\pi_0$ as the sampled data equivalences to the z-transform notation. This structure is completely general in the sense that it is necessary only to manipulate (5.63) to obtain the desired all-pole structure $1/A(z)$ in terms of the reflection coefficients $\{k_m\}$, and then compute $\{\hat{v}_m\}$ from (5.61c) to specify the overall filter. The input, termination, and output are specified by (5.64), (5.65), and (5.66), respectively, for all cases. Starting from $G(z) = P(z)/A(z)$, the step-down procedure is used to obtain the parameters $\{k_m\}$ and $\{v_m\}$. Definition of the structure via (5.63) and the modified tap parameters $\{\hat{v}_m\}$ via (5.61c) then uniquely describes the total filter.

5.5 Specific Speech Synthesis Structures

The following structures will be developed from (5.63) and (5.61c): 1) a two-multiplier lattice form, 2) a four-multiplier ladder form having the form of the Kelly-Lochbaum model (see Chapter 4), 3) a one-multiplier form, and 4) a four-multiplier normalized form. These filters, excluding the normalized form, were first derived for the all-pole transfer function $1/A(z)$ by Itakura and Saito [1971b]. Gray and Markel [1973] generalized the results to include poles and zeros, and developed the normalized form [1975a].

Of the standard digital filter structures, only the direct form and cascade form are widely used in speech synthesis. The cascade form is of use in formant synthesis where the roots corresponding to a polynomial $A(z)$ are known [Rabiner, 1968]. The direct form, as the name implies, directly implements the transfer function $P(z)/A(z)$, without further transformations. As this filter is also used in linear prediction work, it is presented first.

5.5.1 The Direct Form

The direct form is defined by an implementation of the difference equations corresponding to

$$X(z)/E(z)=P(z)[1/A(z)]=G(z)$$
$$=\sum_{i=0}^{M} p_i z^{-i} / \sum_{i=0}^{M} a_i z^{-i} \tag{5.68}$$

with $a_0=1$. Defining $D(z)=E(z)/A(z)$, the output of the all-pole section is

$$D(z)=E(z)-\sum_{i=1}^{M} a_i z^{-i} D(z). \tag{5.69}$$

The output is then

$$X(z)=P(z)D(z)=P(z)E(z)/A(z)$$
$$=\sum_{i=0}^{M} p_i z^{-i} D(z). \tag{5.70}$$

Eqs. (5.69) and (5.70) are shown as a signal flow graph in Fig. 5.8. A Fortran program for this filter is given in Fig. 5.9. The program input variables are $A(I)=a_{I-1}, I=1,2,\dots,M+1$, $P(I)=p_{I-1}, I=1,2,\dots,M+1$, the filter order M, a storage buffer $D(I), I=1,2,\dots,M+1$, and the input sample $XIN=e(n)\leftrightarrow E(z)$. The output variable is $XOUT=x(n)\leftrightarrow X(z)$.

5.5.2 Two-Multiplier Lattice Model

The two-multiplier lattice structure is defined by setting

$$\pi_m=1 \qquad m=0,1,\dots,M. \tag{5.71}$$

Direct substitution into (5.63) gives the structure of section m as

$$E^+_{m-1}(z)=E^+_m(z)-k_m E^-_{m-1}(z) \tag{5.72a}$$
$$zE^-_m(z)=k_m E^+_{m-1}(z)+E^-_{m-1}(z) \tag{5.72b}$$
$$\text{for} \quad m=M, M-1,\dots,1 \; .$$

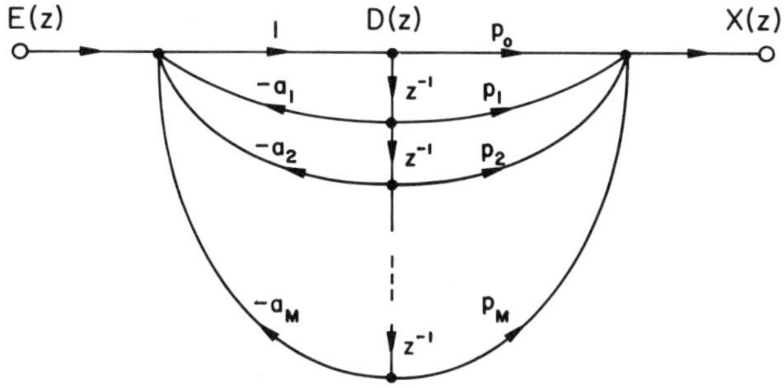

Fig. 5.8 Signal flow graph for direct form filter.

```
C
C        DIRECT FORM FILTER
C
         SUBROUTINE DIRECT(A,P,M,D,XIN,XOUT)
         DIMENSION A(1),P(1),D(1)
         XOUT=0.
         D(1)=XIN
         DO 10 J=1,M
         JJ=M+1-J
         XOUT=XOUT+D(JJ+1)*P(JJ+1)
         D(1)=D(1)-A(JJ)*D(JJ+1)
   10    D(JJ+1)=D(JJ)
         XOUT=XOUT+D(1)*P(1)
         RETURN
         END
```

Fig. 5.9 Fortran program DIRECT for implementing the direct form filter.

The tap parameters are simply

$$\boxed{\hat{v}_m = v_m/\pi_m = v_m} \quad . \tag{5.73}$$

Two equivalent signal flow graphs of section m are shown in Fig. 5.10. In Fig. 5.10A the flow graph is shown with paths crossing each other (as a lattice). In Fig. 5.10B, the flow graph is redrawn without crossover paths. In this section there are two multiplications (and thus the name *two-multiplier* lattice), two additions, and one delay element.

A Fortran implementation of the two-multiplier lattice form is shown in Fig. 5.11. The input parameters are $RC(I)=k_I, I=1,2,\ldots,M$; $TAP(I)=v_{I-1}, I=1,2,\ldots, M+1$; $EM(I)=e_{I-1}^-(n), I=1,2,\ldots,M+1$; and $EP=e_M^+(n)=e(n)$. As before, M defines the filter order. The output is $XOUT=x(n)$. It is necessary to store values only where delays are encountered. Therefore, EP at the input is continually updated until at $m=0$, EP equals the output of the all-pole portion.

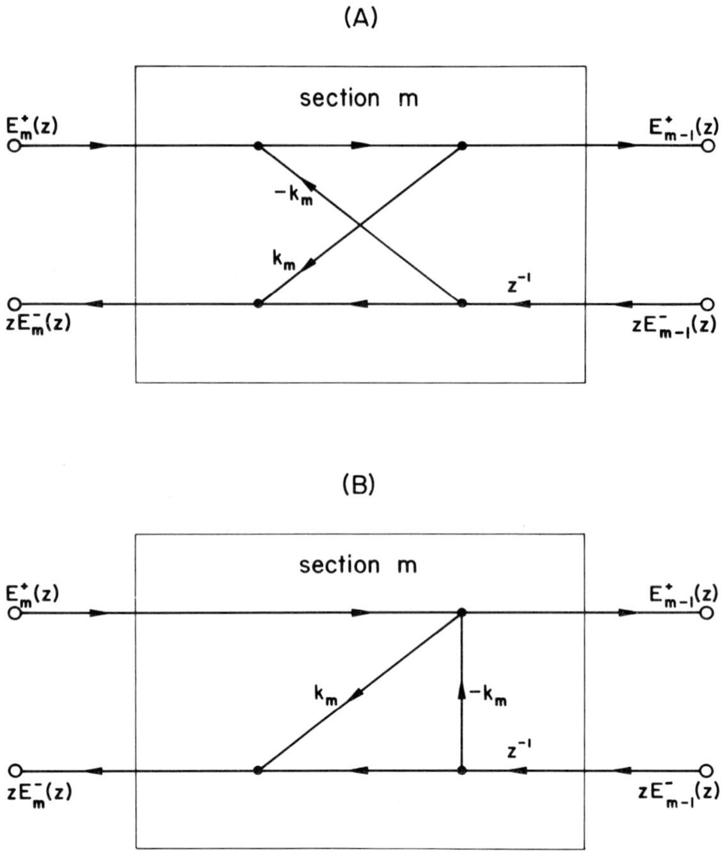

Fig. 5.10 Signal flow graph for the two-multiplier lattice filter: A) showing lattice form with crossover paths, B) without crossover paths.

```
C
C         TWO MULTIPLIER LATTICE SYNTHESIS MODEL
C
          SUBROUTINE TWOMUL(RC,TAP,M,EM,EP,XOUT)
          DIMENSION RC(1),TAP(1),EM(1)
          XOUT=0.
          DO 10 I=1,M
          II=M+1-I
          JJ=II+1
          EP=EP-RC(II)*EM(II)
          EM(II+1)=EM(II)+RC(II)*EP
   10     XOUT=XOUT+EM(II+1)*TAP(II+1)
          EM(1)=EP
          XOUT=XOUT+EM(1)*TAP(1)
          RETURN
          END
```

Fig. 5.11 Fortran program TWOMUL for implementing the two-multiplier lattice filter.

5.5.3 Kelly-Lochbaum Model

By a proper choice of the pi-parameters, it is possible to transform the lattice into the form of the Kelly-Lochbaum model [Kelly and Lochbaum, 1962]. The only differences are a sign change (due to the use of k_m instead of $-k_m$) and the combination of the split delays from one section into another.

From Fig. 5.10B it is seen that to have the form of the Kelly-Lochbaum model, $zE_m^-(z)$ must be obtainable in terms of $E_m^+(z)$ and $E_m^-(z)$. Leaving (5.63a) unchanged and then substituting $E_{m-1}^+(z)$ into (5.63b), gives the result

$$E_{m-1}^+(z)=(\pi_{m-1}/\pi_m)E_m^+(z)-k_m E_{m-1}^-(z) \tag{5.74a}$$

$$zE_m^-(z)=k_m E_m^+(z)+(\pi_m/\pi_{m-1})(1-k_m^2)E_{m-1}^-(z). \tag{5.74b}$$

Also to match the Kelly-Lochbaum form, the pi-parameters must be chosen so that $\pi_{m-1}/\pi_m=1\pm k_m$. Since $\pi_M=1$, π_{M-1}, π_{M-2}, etc., can be recursively computed to give the general term

$$\pi_m=\begin{cases}1 & m=M\\ \prod_{i=m+1}^{M}(1+\varepsilon_i k_i) & m=0,1,\dots,M-1\end{cases} \tag{5.75}$$

where $\{\varepsilon_i\}=\{\varepsilon_1,\varepsilon_2,\dots,\varepsilon_M\}$ defines the sequence of sign parameters having the values ± 1. From (5.75), the result

$$\begin{aligned}\pi_{m-1}/\pi_m&=1+\varepsilon_m k_m\\ &=\frac{1-k_m^2}{1-\varepsilon_m k_m}\end{aligned}$$

is easily verified for $\varepsilon_m=\pm 1$. Therefore, (5.74) can be expressed in the form

$$E_{m-1}^+(z)=(1+\varepsilon_m k_m)E_m^+(z)-k_m E_{m-1}^-(z) \tag{5.76a}$$

$$zE_m^-(z)=k_m E_m^+(z)+(1-\varepsilon_m k_m)E_{m-1}^-(z)\quad. \tag{5.76b}$$

The tap parameters $\hat{v}_m$ are computed from (5.75) and (5.61c) as

$$\hat{v}_m=\begin{cases}v_M & m=M\\ v_m/\prod_{i=m+1}^{M}(1+\varepsilon_i k_i) & m=M-1,M-2,\dots,0\end{cases}\quad. \tag{5.77}$$

Note that for the all-pole implementation $1/A(z)$, $\hat{v}_m = 0$ for $m \neq 0$, and

$$\hat{v}_0 = 1 / \prod_{i=1}^{M} (1 + \varepsilon_i k_i). \tag{5.78}$$

Eqs. (5.76) are shown as signal flow graphs for section m in Fig. 5.12. Fig. 5.12A shows the section for $\varepsilon_m = +1$, while Fig. 5.12B shows the section for $\varepsilon_m = -1$. These figures have the same form as the Kelly-Lochbaum acoustic tube model of Fig. 4.4 except for the sign change due to the fact that the variables in that figure are volume velocities. Each of these sections requires four multiplications, two additions, and one delay.

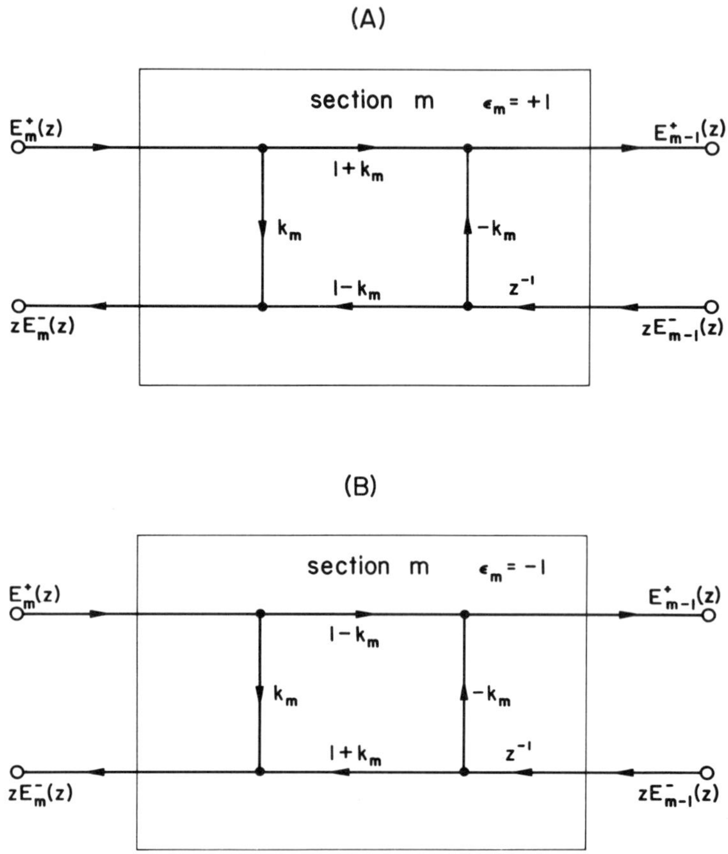

Fig. 5.12 Signal flow graph for the Kelly-Lochbaum Model. A) $\varepsilon_m = +1$. B) $\varepsilon_m = -1$.

A Fortran program for implementing the Kelly-Lochbaum model is shown in Fig. 5.13. The only variable not previously discussed is $TAPM(I) = \hat{v}_{I-1}$, $I = 1, 2, \ldots, M+1$. As long as $\{k_m\}$ and $\{v_m\}$ remain constant, $\{\hat{v}_m\}$ need be computed only once.

```
C
C       KELLY-LOCHBAUM
C       SYNTHESIS MODEL (EPS=+)
C
        SUBROUTINE KLOCH(RC,TAPM,M,EM,EP,XOUT)
        DIMENSION RC(1),TAPM(1),EM(1)
        XOUT=0.
        DO 10 I=1,M
        II=M+1-I
        EM(II+1)=(1.-RC(II))*EM(II)+EP*RC(II)
        EP=EP*(1.+RC(II))-EM(II)*RC(II)
 10     XOUT=XOUT+EM(II+1)*TAPM(II+1)
        EM(1)=EP
        XOUT=XOUT+EM(1)*TAPM(1)
        RETURN
        END
```

Fig. 5.13 Fortran program KLOCH for implementing the Kelly-Lochbaum model with $\varepsilon_m = +1$.

5.5.4 One-Multiplier Models

Itakura and Saito [1971b] showed that by properly combining terms, three of the four multipliers in the Kelly-Lochbaum model can be eliminated.

From (5.76), it is seen that both equations have the term

$$T_m(z) = [E_{m-1}^-(z) - \varepsilon_m E_m^+(z)]k_m \tag{5.79}$$

in common, so that

$$E_{m-1}^+(z) = E_m^+(z) - T_m(z) \tag{5.80a}$$

$$zE_m^-(z) = E_{m-1}^-(z) - \varepsilon_m T_m(z) \quad . \tag{5.80b}$$

The parameters $\hat{v}_m$ remain unchanged and are given by (5.77). These equations are shown as signal flow graphs for section m in Fig. 5.14. Figure 5.14A shows the section for $\varepsilon_m = +1$, while Fig. 5.14B shows the section for $\varepsilon_m = -1$. It is seen that each section requires one multiplication, three additions (or subtractions), and one delay.

A Fortran program for implementing the one-multiplier model with $\varepsilon_m = +1$ is shown in Fig. 5.15. The arguments are identical to those of the Kelly-Lochbaum model.

5.5.5 Normalized Filter Model

The normalized filter model was developed by Gray and Markel [1975a]. This filter is normalized in the sense that the polynomials $zB_m(z)$ are modified by proper

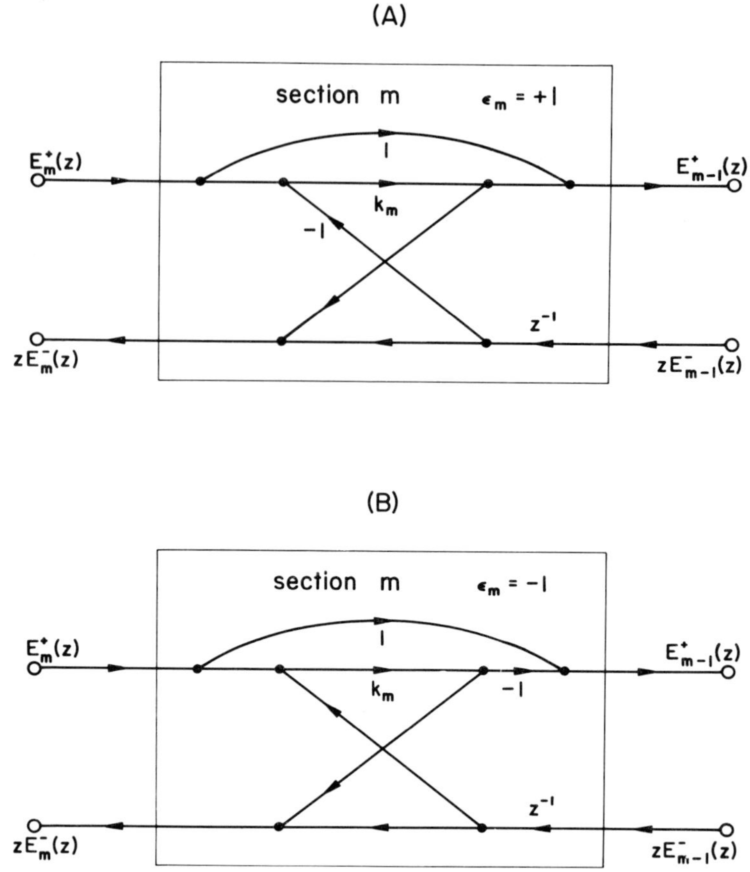

Fig. 5.14 Signal flow graph for the one-multiplier model. A) $\varepsilon_m = +1$. B) $\varepsilon_m = -1$.

```
C
C       ONE MULTIPLIER
C       SYNTHESIS MODEL (EPS=+)
C
        SUBROUTINE ONEMUL(RC,TAPM,M,EM,EP,XOUT)
        DIMENSION RC(1),TAPM(1),EM(1)
        XOUT=0.
        DO 10 I=1,M
        II=M+1-I
        TM=(EP-EM(II))*RC(II)
        EP=TM+EP
        EM(II+1)=TM+EM(II)
   10   XOUT=XOUT+EM(II+1)*TAPM(II+1)
        EM(1)=EP
        XOUT=XOUT+EM(1)*TAPM(1)
        RETURN
        END
```

Fig. 5.15 Fortran program ONEMUL for implementing the one-multiplier model with $\varepsilon_m = +1$.

choice of the pi-parameters to be not only orthogonal but orthonormal, that is, $z\hat{B}_m(z)$ satisfies

$$\langle z\hat{B}_m(z), z\hat{B}_n(z)\rangle = \delta_{mn}. \tag{5.81}$$

For this to occur, π_m, π_n must satisfy

$$\pi_m\pi_n\langle zB_m(z), zB_n(z)\rangle = \delta_{mn}.$$

From (5.43) and (5.36), π_m is then given by $\pi_m^2\alpha_m = 1$ or

$$\boxed{\pi_m = \pm(1/\alpha_m)^{1/2}}\ . \tag{5.82}$$

By choosing the plus sign, (5.49) gives the relationship

$$\frac{\pi_{m-1}}{\pi_m}\frac{\alpha_m^{1/2}}{\alpha_{m-1}^{1/2}} = (1-k_m^2)^{1/2}. \tag{5.83}$$

Since $|k_m| < 1$ for a stable filter, it is possible to define a change of variables

$$\theta_m = \sin^{-1}(k_m) \tag{5.84}$$

so that

$$k_m = \sin\theta_m \tag{5.85a}$$

and

$$(1-k_m^2)^{1/2} = \cos\theta_m. \tag{5.85b}$$

Direct substitution into (5.74) gives the normalized equations

$$E_{m-1}^+(z) = \cos\theta_m E_m^+(z) - \sin\theta_m E_{m-1}^-(z) \tag{5.86a}$$

and

$$zE_m^-(z) = \sin\theta_m E_m^+(z) + \cos\theta_m E_{m-1}^-(z)\ . \tag{5.86b}$$

The modified tap parameters are calculated from (5.61c) as

$$\hat{v}_m = v_m/\pi_m$$

or

$$\hat{v}_m = v_m\alpha_m^{1/2}.$$

But from (5.83) and (5.85b),

$$\alpha_{m-1}^{1/2} = \alpha_m^{1/2}/\cos\theta_m$$

with $\alpha_M = 1$. Therefore,

$$\boxed{\hat{v}_m = v_m \prod_{i=m+1}^{M} (\cos\theta_i)^{-1}} \quad . \tag{5.87}$$

The implementation of these equations as a signal flow graph is shown in Fig. 5.16. In this form there are four multiplications, two additions, and one delay just as in the Kelly-Lochbaum model.

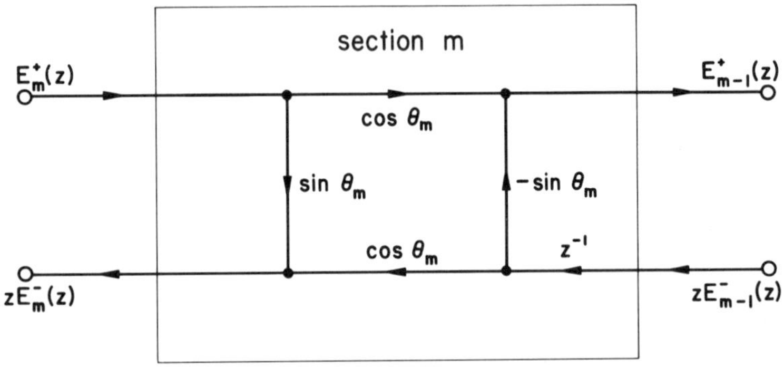

Fig. 5.16 Signal flow graph for the normalized model.

Since there is also the necessity of computing the $\sin\theta_m$ and $\cos\theta_m$ terms from (5.85), it might seem doubtful that anything has been gained from this new structure. From a practical point of view, if the arithmetic operations are not too costly with respect to the one-multiplier or two-multiplier models, a great deal can be gained in computational accuracy using fixed-point arithmetic [Markel and Gray, 1975b, c]. In addition, this structure has the unique property that each section involves precisely one complex multiply. By expanding the complex multiplication

$$E_{m-1}^{+}(z) + jzE_m^{-}(z) = [e^{j\theta_m}]\,[E_m^{+}(z) + jE_{m-1}^{-}(z)],$$

it is seen that the real and imaginary parts give precisely (5.86a) and (5.86b).

A Fortran program for implementing the normalized form with real arithmetic is shown in Fig. 5.17. The arguments not previously defined are $SN(I) = \mathrm{SIN}(\theta_I) = k_I$ and $CN(I) = \mathrm{COS}(\theta_I) = (1 - k_I^2)^{1/2}$, $I = 1, 2, \ldots, M$. The single necessary operation per section, if complex arithmetic is available, is shown as a COMMENT statement.

5.5.6 A Test Example

The application of the Fortran subroutine EVAL and one of the synthesizer filter structures ONEMUL is illustrated in Fig. 5.18. The numerator and denominator coefficients of the stable filter

$$G(z) = P(z)/A(z)$$

```
C
C       NORMALIZED SYNTHESIS FILTER STRUCTURE
C
        SUBROUTINE NORMAL(SN,CN,TAPM,M,EM,EP,XOUT)
        DIMENSION SN(1),CN(1),TAPM(1),EM(1)
        XOUT=0.
        DO 10 I=1,M
        II=M+1-I
        EM(II+1)=EP*SN(II)+EM(II)*CN(II)
        EP=EP*CN(II)-EM(II)*SN(II)
C
C       WITH COMPLEX ARITHMETIC EACH SECTION
C       REQUIRES PRECISELY 1 COMPLEX MULT.
C
C       (EP,EM(II+1))=CMUL( (CN(II),SN(II)),
C                 (EP,EM(II)) )
C
   10   XOUT=XOUT+EM(II+1)*TAPM(II+1)
        EM(1)=EP
        XOUT=XOUT+EM(1)*TAPM(1)
        RETURN
        END
```

Fig. 5.17 Fortran program NORMAL for implementing the normalized model.

are contained in a data statement. The filter order is $M=8$. The reflection coefficients and tap parameters denoted by $RC(\cdot)$ and $TAP(\cdot)$, respectively, are computed by EVAL along with the corresponding total squared error terms $AL(\cdot)$ and total energy in $G(z)$, TI. The modified tap parameters are computed according to (5.77) and then thirty samples of the one-multiplier filter in response to the scaled unit sample input $EP=256\,\delta_{n,0}$ for $n=0, 1, \ldots, 29$, are computed.

```
C
C        TEST PGM FOR EVAL
C        AND 1 MULT FILTER FORM
C
         DIMENSION A(9),RC(8),TAPM(9)
         DIMENSION P(9),TAP(9),AL(9),EM(9)
         DATA A/ 1.      ,-2.3464 , 1.657  ,
     1         -0.005988,  .32305,-1.4821 ,
     2          1.1546  ,-0.18966,-0.05899/
         DATA P/1.,.9,.8,.7,.6,.5,.4,.3,.2/
         DATA EM/9*0./
C
         M=8
         MP1=M+1
C
         CALL EVAL(A,P,RC,M,TAP,AL,TI)
         WRITE (5,10) TI
         WRITE (5,10) (RC(J),TAP(J),A(J),J=1,MP1)
   10    FORMAT(3E14.5)
C
C        MODIFIED TAP CALC
C
         T=1.
         TAPM(M+1)=TAP(M+1)
         DO 20 J=1,M
         JJ=M+1-J
         T=T*(1.+RC(JJ))
   20    TAPM(JJ)=TAP(JJ)/T
C
C        CALC UNIT SAMPLE RESPONSE
C
         EP=256.
         DO 40 J=1,30
         CALL ONEMUL(RC,TAPM,M,EM,EP,XOUT)
         WRITE (5,30) J,XOUT
   30    FORMAT(2X,I3,F12.3)
   40    EP=0.
         END
```

J	RK(J)	TAP(J)	AL(J)
1	-0.94207E 00	0.29074E 01	0.10000E 01
2	0.92394E 00	0.75296E 01	-0.94207E 00
3	-0.56202E 00	0.93966E 01	0.92394E 00
4	-0.94574E-01	0.78006E 01	-0.56202E 00
5	0.20216E 00	0.55704E 01	-0.94574E-01
6	0.53593E 00	0.33043E 01	0.20216E 00
7	-0.32922E 00	0.18886E 01	0.53593E 00
8	-0.58990E-01	0.76928E 00	-0.32922E 00
9	0.00000E 00	0.20000E 00	-0.58990E-01

J	XOUT
1	256.000
2	831.079
3	1730.649
4	2864.433
5	3929.294
6	4722.633
7	5066.975
8	4892.707
9	4314.281
10	3414.103
11	2362.507
12	1302.678
13	297.280
14	-565.112
15	-1268.357
16	-1792.315
17	-2098.274
18	-2192.563
19	-2063.141
20	-1735.560
21	-1270.564
22	-723.811
23	-178.602
24	303.412
25	684.202
26	935.556
27	1064.342
28	1080.885
29	1000.819
30	848.600

TI = 0.25759E 04

Fig. 5.18 An example program for testing EVAL and the one-multiplier synthesis model.

6. Spectral Analysis

6.1 Introduction

The applications, properties, and considerations of linear prediction in the spectral analysis of speech are presented in this chapter. Although numerous methods have been proposed for representing the spectral characteristics of speech, such as band-pass filter sampling and analysis-by-synthesis, there are a number of reasons why linear prediction techniques are becoming widely used.

- Parameter determination for the spectral model is non-iterative.
- Generally a very small number of parameters is necessary to accurately represent the spectral trend characteristics.
- A gain constant σ^2 is easily obtained to match spectral energies of the model and the data, using the autocorrelation method.
- The model spectrum represents a smoothed version of the data spectrum.

These reasons apply to wide classes of signals with speech being only one example. In spectral analysis of speech there are two additional compelling reasons for linear prediction.

- The spectral resonances (formant peaks) of voiced speech are weighted most heavily in the error criterion and thus represented most accurately.
- The all-pole model can be accurately fit to the log spectrum of a voiced sound with a sufficiently small number of resonances (complex pole pairs in the model) so that the problem of formant extraction in many cases reduces to simple peak picking.

The model used for representing the input data spectrum $|X[\exp(j\theta)]|^2$ is given by

$$\boxed{\frac{\sigma^2}{|A(e^{j\theta})|^2} = \left|\frac{\sigma}{A(z)}\right|^2_{z=e^{j\theta}}} \quad . \tag{6.1}$$

The spectral model in this chapter will be based mainly upon the autocorrelation method due to the complexity of a spectral interpretation in the covariance method. The covariance method does not have a one-dimensional frequency domain interpretation for the data sequence, however, Makhoul and Wolf [1972] have shown an interpretation of the covariance method based upon a frequency domain formulation using a two-dimensional all-pole spectrum to approximate a two-dimensional spectrum of a non-stationary signal.

6.2 Spectral Properties

There are several properties of the all-pole spectral model obtained from linear prediction which lend insight into the modeling process.

6.2.1 Zero Mean All-Pole Model

The first property of importance is that on a log magnitude scale, $A[\exp(j\theta)]$ or $1/A[\exp(j\theta)]$ for either the autocorrelation or covariance method has zero mean provided that $A(z)$ has all of its zeros within the unit circle [Itakura and Saito, 1970], i.e.,

$$\boxed{\pm \int_{-\pi}^{\pi} ln|A(e^{j\theta})|^2 \frac{d\theta}{2\pi} = 0} \tag{6.2}$$

If $A(z)$ has all of its zeros within the unit circle, then $A(1/z)$ will be analytic on and within the unit circle since all of its zeros are outside of the unit circle. Residue calculus can then be applied in the following manner to show (6.2) since

$$\begin{aligned}
&\int_{-\pi}^{\pi} ln\{|A[\exp(j\theta)]|^2\} \frac{d\theta}{2\pi} \\
&= \int_{-\pi}^{\pi} ln\{|A[\exp(-j\theta)]|^2\} \frac{d\theta}{2\pi} \\
&= 2\,\mathrm{Real}\left(\int_{-\pi}^{\pi} ln\{A[\exp(-j\theta)]\} \frac{d\theta}{2\pi} \right) \\
&= 2\,\mathrm{Real}\left(\oint_{\Gamma} ln[A(1/z)] \frac{dz}{2\pi j z} \right) \\
&= 2\,\mathrm{Real}\,\{ln[A(\infty)]\} = 2\,\mathrm{Real}\,[ln(1)] = 0.
\end{aligned}$$

The log magnitude spectra $LM(1/A) = 10\log_{10}|1/A[\exp(j\theta)]|^2$ from the analysis of a voiced and an unvoiced sound, based upon the autocorrelation method with twelve coefficients, are shown in Figs. 6.1A and B, respectively. Since the zeros of $A(z)$ are theoretically assured of being within the unit circle, as shown in Chapter 5, (6.2) must be satisfied. Therefore, $LM(1/A)$ has equal area above and below the 0 dB reference as shown in the figures.

6.2.2 Gain Factor for Spectral Matching

In the autocorrelation method of linear prediction, the gain coefficient σ^2 is equal to the total squared error α. This result was obtained from both the maximum likeli-

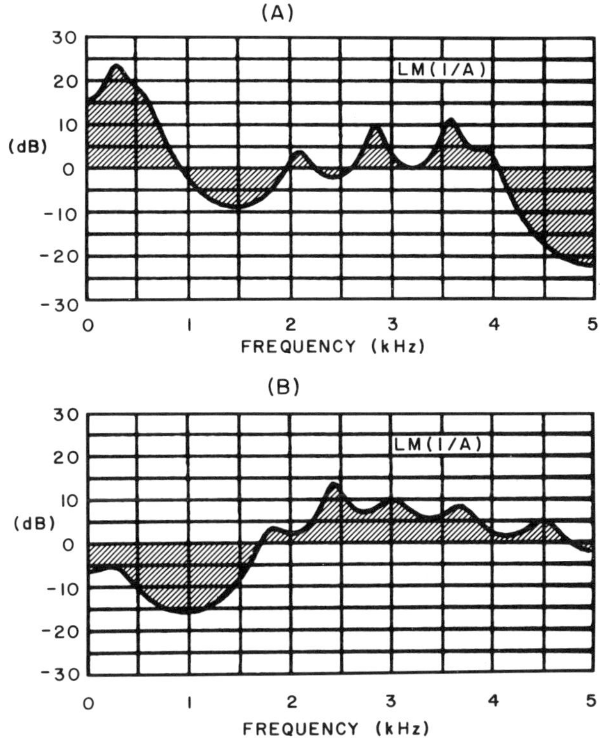

Fig. 6.1 Spectral model $1/A(z)$ based upon autocorrelation method. A) voiced speech. B) unvoiced speech.

hood approach and the correlation matching approach of Chapter 2, and is restated as

$$\sigma^2 = \alpha = r_e(0) \tag{6.3a}$$

where

$$r_e(0) = \int_{-\pi}^{\pi} |E(e^{j\theta})|^2 \frac{d\theta}{2\pi}. \tag{6.3b}$$

From the correlation matching formulation, it is seen that the gain factor matches the energy or average value of the input spectrum $|X[\exp(j\theta)]|^2$ to the model spectrum $\sigma^2/|A[\exp(j\theta)]|^2$, i.e.,

$$\boxed{\int_{-\pi}^{\pi} |X(e^{j\theta})|^2 \frac{d\theta}{2\pi} = r_x(0) = \int_{-\pi}^{\pi} \frac{\sigma^2}{|A(e^{j\theta})|^2} \frac{d\theta}{2\pi}} \quad . \tag{6.4}$$

The average value of the log spectrum of the model $\sigma^2/|A[\exp(j\theta)]|^2$ is from (6.2), simply

$$\int_{-\pi}^{\pi} ln \left| \frac{\sigma}{A(e^{j\theta})} \right|^2 \frac{d\theta}{2\pi} = ln(\sigma^2). \tag{6.5}$$

The log spectra for the voiced and unvoiced sounds analyzed in Fig. 6.1 are shown in Figs. 6.2A and B, respectively, along with the spectral model. The speech log spectrum $LM(X)$ is shown with a solid line while the model $LM(1/A)$ is shown with a dotted line. The averages of the model log spectra on a dB scale, $10\log_{10}\sigma^2$,

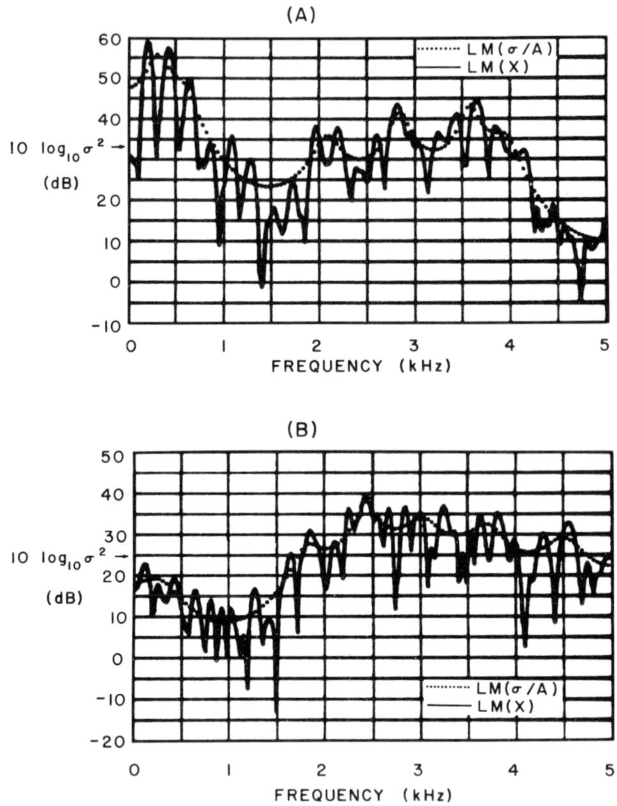

Fig. 6.2 Spectral model $\sigma/A(z)$ compared to input data. A) voiced speech. B) unvoiced speech.

are also shown on the vertical axis. It is seen that the model spectrum and speech spectrum are closely matched in terms of gain. The speech spectrum tends to lie below the model spectrum and also the model appears to more accurately represent peak behavior in the spectrum. Reasons for this behavior will be discussed in following sections.

6.2.3 Limiting Spectral Match

From the correlation matching formulation it can be noted that as the number of coefficients, M, approaches infinity, the autocorrelation sequences corresponding to $X(z)$ and to $\sigma/A(z)$ become equal. Since the spectrum equals the transform of the autocorrelation sequence,

$$|X(e^{j\theta})|^2 = \sigma^2/|A_\infty(e^{j\theta})|^2.$$

Substitution into (6.5) with $\sigma^2 = \alpha_\infty$, gives

$$ln(\alpha_\infty) = \int_{-\pi}^{\pi} ln|X(e^{j\theta})|^2 \frac{d\theta}{2\pi}$$

or (6.6)

$$\alpha_\infty = \exp \int_{-\pi}^{\pi} ln|X(e^{j\theta})|^2 \frac{d\theta}{2\pi} \quad .$$

The total squared error for $M = \infty$ might also be referred to as the minimum predictor total squared error since it describes the limiting value. It is also equivalent to the zero*th* value of the cepstrum [Makhoul and Wolf, 1972]. The computation of this expression will be considered in Section 6.2. The normalized squared error for a voiced speech sound obtained for $f_s = 6.5$ kHz with no pre-emphasis and a 19.6 ms window length is shown in Fig. 6.3 for selected values of M. In addition, the limiting value α_∞/α_0 is shown by a horizontal line on the figure. This graph shows that at some relatively small number such as $M=6$ to $M=10$, the total squared error in the representation is substantially reduced. It requires a much greater number of coefficients to closely approximate the limit α_∞/α_0.

As $ln(\alpha_M)$ represents the average of the log spectrum of the model and $ln(\alpha_\infty)$ represents the average of the log spectrum of the data sequence, the log spectrum of the model must always have an average larger than the log spectrum of the data sequence. It is for this reason that logarithmic plots of the two spectra such as

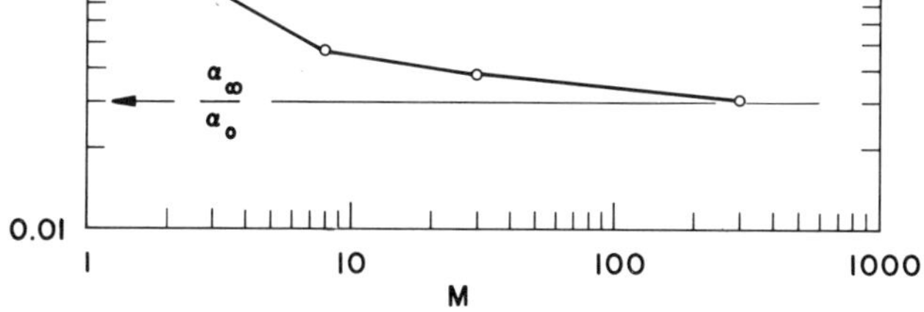

Fig. 6.3 Normalized total squared error for a voiced sound with the minimum predictor error.

shown in Fig. 6.2 always appear to show the data spectrum as on the average being slightly below the model spectrum, though the peaks of the model spectrum may be above those of the data spectrum. It should be noted that the behavior of the normalized squared error α_M/α_0 may be considerably different for the covariance method. As discussed in Chapter 2, if the signal being analyzed is composed of only M complex exponentials, it is possible to have $\alpha_M/\alpha_0 = 0$. In the autocorrelation method, this can occur only for $M = \infty$.

6.2.4 Non-uniform Spectral Weighting

Figure 6.2 illustrates the fact that the spectral model obtained through the autocorrelation method of linear prediction weights the local maxima more heavily than the local minima (of the log spectra) in the error criterion. This result has been discussed by Itakura and Saito [1968] using a maximum likelihood formulation, Makhoul and Wolf [1972], Makhoul [1973a, 1975b] using energy relations, and by Gray and Markel [1974b] using a spectral flatness measure. In this section the approach taken by Itakura and Saito will be discussed. Itakura and Saito showed that their maximum likelihood approach led to the minimization of an integral of the form

$$I = \int_{-\pi}^{\pi} [e^{V(\theta)} - V(\theta) - 1] \frac{d\theta}{2\pi} \tag{6.7}$$

where $V(\theta)$ defines the difference between the data log spectrum and the model log spectrum, i.e.,

$$\begin{aligned} V(\theta) &= ln|X(e^{j\theta})|^2 - ln\left|\frac{\sigma}{A(e^{j\theta})}\right|^2 \\ &= ln\left|\frac{X(e^{j\theta})}{\sigma/A(e^{j\theta})}\right|^2 = ln\left|\frac{E(e^{j\theta})}{\sigma}\right|^2. \end{aligned} \tag{6.8}$$

Therefore, $V(\theta)$ is also the log spectrum of the inverse filter output, normalized by the gain σ. First, minimization of (6.7) will be shown to result in the autocorrelation method, with the gain term $\sigma = \alpha_M$. The integrand of (6.7) will then be used to discuss the spectral weighting. By substituting (6.8), (6.7) takes on the form

$$\begin{aligned} I = &\frac{1}{\sigma^2}\int_{-\pi}^{\pi} |E(e^{j\theta})|^2 \frac{d\theta}{2\pi} - \int_{-\pi}^{\pi} ln[|X(e^{j\theta})|^2] \frac{d\theta}{2\pi} \\ &+ ln(\sigma^2) - \int_{-\pi}^{\pi} ln[|A(e^{j\theta})|^2] \frac{d\theta}{2\pi} - 1. \end{aligned} \tag{6.9}$$

By applying (6.2), (6.3b), and (6.6), this expression reduces to

$$I = \frac{1}{\sigma^2} r_e(0) + ln(\sigma^2) - ln(\alpha_\infty) - 1. \tag{6.10}$$

Since the last three terms are not a function of the filter coefficient values, minimization of I is equivalent to minimization of the inverse filter output energy $r_e(0)$. Direct minimization of (6.10) with respect to σ^2 gives the gain as

$$\sigma^2 = r_e(0) = \alpha_M = \alpha,$$

the same result as (6.3a). Substituting this result in (6.10) gives the minimum value of the integral as

$$I_{\min} = ln(\alpha_M/\alpha_\infty). \tag{6.11}$$

Now, since minimization of I has been shown to equivalently result in the autocorrelation method, it can be used to demonstrate the non-uniform spectral weighting in matching the model to the signal spectrum. This non-uniform weighting to be shown contrasts with that used in least squares minimization where

$$I_1 = \frac{1}{2} \int_{-\pi}^{\pi} V^2(\theta) \frac{d\theta}{2\pi}$$

is minimized. Minimization of I_1 with respect to the filter coefficients and gain term leads to the analysis-by-synthesis method [Bell, et al., 1961; Makhoul, 1974c]. The least squares approach equally weights positive and negative values of $V(\theta)$ since $V^2(\theta)$ is independent of the sign of $V(\theta)$. The least squares approach is also non-linear because of the logarithm in $V(\theta)$. Figure 6.4 illustrates a graph of the integrand of (6.7), shown in a solid curve, and $[V^2(\theta)]/2$ shown in a dashed line curve, both vs. $V(\theta)$. The autocorrelation method weights positive values of $V(\theta)$ more heavily than negative values.

For large positive $V(\theta)$, the integrand becomes approximately exponential

$$e^{V(\theta)} - V(\theta) - 1 \cong e^{V(\theta)} \quad \text{for} \quad V(\theta) \gg 1$$

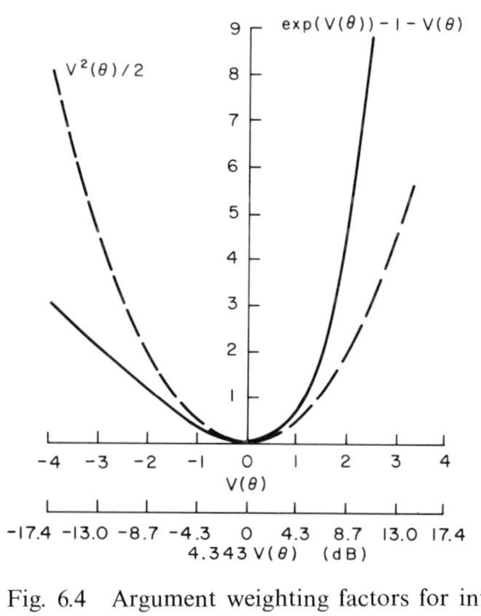

Fig. 6.4 Argument weighting factors for integral minimization in linear prediction and analysis by synthesis. [After Itakura and Saito, 1970]

and for large negative $V(\theta)$, the integrand becomes approximately linear

$$e^{V(\theta)} - V(\theta) - 1 \cong -V(\theta) \quad \text{for} \quad V(\theta) \ll -1.$$

For small values of $V(\theta)$ the two integrands are almost identical, for a Taylor series expansion of the integrand of (6.7) shows that

$$e^{V(\theta)} - V(\theta) - 1 \cong V^2(\theta)/2 \quad \text{for} \quad |V(\theta)| \ll 1.$$

In terms of the signal and model spectra, when the signal spectrum lies above the model spectrum ($V(\theta) \gg 1$), relatively large error contributions to the integral are generated. When the signal spectrum lies below the model spectrum ($V(\theta) \ll 1$), relatively small error contributions are made. Since the energy of the model is constrained to match the energy of the signal, small error contributions cannot be introduced by arbitrarily placing the model spectrum far below the signal spectrum. In addition, the model spectrum is constrained by its order M and its polynomial structure to cover the complete frequency range from 0 to $f_s/2$ with less than or equal to M poles at 0 or $f_s/2$, or less than or equal to $M/2$ resonances within 0 to $f_s/2$.

These results lead to the conclusion that *linear prediction spectral models tend to fit the peaks or regions of relatively high energy resonance behavior more accurately than the valleys or regions of low energy behavior*. If the order of the filter $A(z)$ is increased the spectral matching can be more accurate, but the weighting is still non-uniform.

In Fig. 6.2A, for example, this non-uniform spectral weighting can be easily seen. The dB difference from the signal and model spectrum is substantially smaller at the locations of the harmonic peaks than at the valleys between the harmonics. In this example, the first four peak locations of the model spectrum are estimates of the first four vocal tract resonances or formants (see Chapter 7). It is important to note that the peaks of the linear prediction model do not necessarily correspond to the location of a harmonic peak (a multiple of the fundamental frequency). This is a desirable property since the formant frequencies are not necessarily a multiple of the fundamental frequency in speech production. *The non-uniform weighting in the linear prediction method is preferred over uniform weighting (as in the analysis-by-synthesis method) since the peaks (or formants) play a dominant role in the perception of voiced speech* [Flanagan, 1972].

Although these same arguments cannot be directly applied to the covariance method, the same characteristics are nonetheless observed, as long as $A(z)$ has its zeros contained within the unit circle.

6.2.5 Minimax Spectral Matching

In spectral modeling with the autocorrelation method, each step in the model generation is an exact minimax approximation to the next step in terms of the log spectra [Gray and Markel, 1973]. We will show that if $\varepsilon_m(\theta)$ defines the difference in the log spectra of the model at steps $m-1$ and m, then

$$\varepsilon_m(\theta)=ln[\alpha_m/|A_m(e^{j\theta})|^2]-ln[\alpha_{m-1}/|A_{m-1}(e^{j\theta})|^2] \tag{6.12}$$

will have equal ripple in the sense that it will oscillate between the extreme values given by the log area function ratios of the acoustic tube model

$$\pm ln(\mathscr{A}_m/\mathscr{A}_{m-1})=\pm ln[(1+k_m)/(1-k_m)], \tag{6.13}$$

reaching these extreme values exactly $m+1$ times as θ increases from 0 to π. By a slight modification in the alternation theorem used in Tchebycheff approximation [Cheney, 1966, p. 75] it can be shown that demonstrating the equal ripple behavior of $\varepsilon_m(\theta)$ is equivalent to showing that $\sqrt{\alpha_{m-1}}/A_{m-1}(z)$ is the all-pole filter of order $m-1$ which minimizes the maximum log spectral deviation from $\sqrt{\alpha_m}/A_m(z)$. Eq. (6.12) can be rewritten in the form

$$\varepsilon_m(\theta)=ln[\alpha_m/\alpha_{m-1}]-2ln[|A_m(e^{j\theta})/A_{m-1}(e^{j\theta})|].$$

In (5.4), (5.6), and (5.49), the results

$$\alpha_m=(1-k_m^2)\alpha_{m-1} \tag{6.14a}$$

and

$$\begin{aligned} A_m(z)&=A_{m-1}(z)+k_m B_{m-1}(z)\\ &=A_{m-1}(z)+k_m z^{-m}A_{m-1}(1/z) \end{aligned} \tag{6.14b}$$

were obtained. Therefore,

$$\varepsilon_m(\theta)=ln[1-k_m^2]-2ln[|1+k_m Q_m(\theta)|], \tag{6.15}$$

where

$$Q_m(\theta)=e^{-jm\theta}[A_{m-1}(e^{-j\theta})/A_{m-1}(e^{j\theta})].$$

As the coefficients of $A_{m-1}(z)$ are real, $Q_m(\theta)$ can be expressed in the form

$$Q_m(\theta)=e^{-j\psi_m(\theta)},$$

where

$$\psi_m(\theta)=m\theta+2\arg[A_{m-1}(e^{j\theta})].$$

As the roots of $A_{m-1}(z)$ are within the unit circle, it is a minimum phase polynomial. This fact can be used to show that $\psi_m(\theta)$ is a monotonically increasing function of θ with

$$\frac{d}{d\theta}\psi_m(\theta)>1, \quad \psi_m(0)=0, \quad \text{and} \quad \psi(\pi)=m\pi. \tag{6.16}$$

The magnitude of $1+k_m Q_m(\theta)$ will thus oscillate between the values $1-k_m$ and $1+k_m$ as θ varies from $\theta=0$ to $\theta=\pi$. The extreme values will be realized when $\psi_m(\theta)$ equals 0, π, $2\pi, \ldots, m\pi$, a total of exactly $m+1$ times. From (6.15) this shows that $\varepsilon_m(\theta)$ will oscillate between the values $ln[(1-k_m)/(1+k_m)]$ and $ln[1+k_m)/(1-k_m)]$ and that it reaches those values exactly $m+1$ times as θ goes from $\theta=0$ to $\theta=\pi$. This shows the equiripple property, and gives the extreme values of (6.13).

The equiripple behavior has a useful application in interpreting the effect that a reflection coefficient has in the overall model spectrum. To see this we can sum (6.12) from $m=1$ through $m=M$ to obtain the result

$$\sum_{m=1}^{M} \varepsilon_m(\theta) = -ln[\alpha_0/|A_0(e^{j\theta})|^2] + ln[\alpha_M/|A_M(e^{j\theta})|^2].$$

Using the fact that $\alpha_0 = r(0)$ and $A_0(z)=1$, the relation between the model spectrum and the spectral differences is

$$ln[\alpha_M/|A_M(e^{j\theta})|^2] = ln[r(0)] + \sum_{m=1}^{M} \varepsilon_m(\theta). \tag{6.17}$$

Each $\varepsilon_m(\theta)$ oscillates in the range $\pm ln[(1+k_m)/(1-k_m)]$, and thus has little effect when k_m is small in magnitude. Major contributions to the spectrum arise when a value of k_m is near one in magnitude.

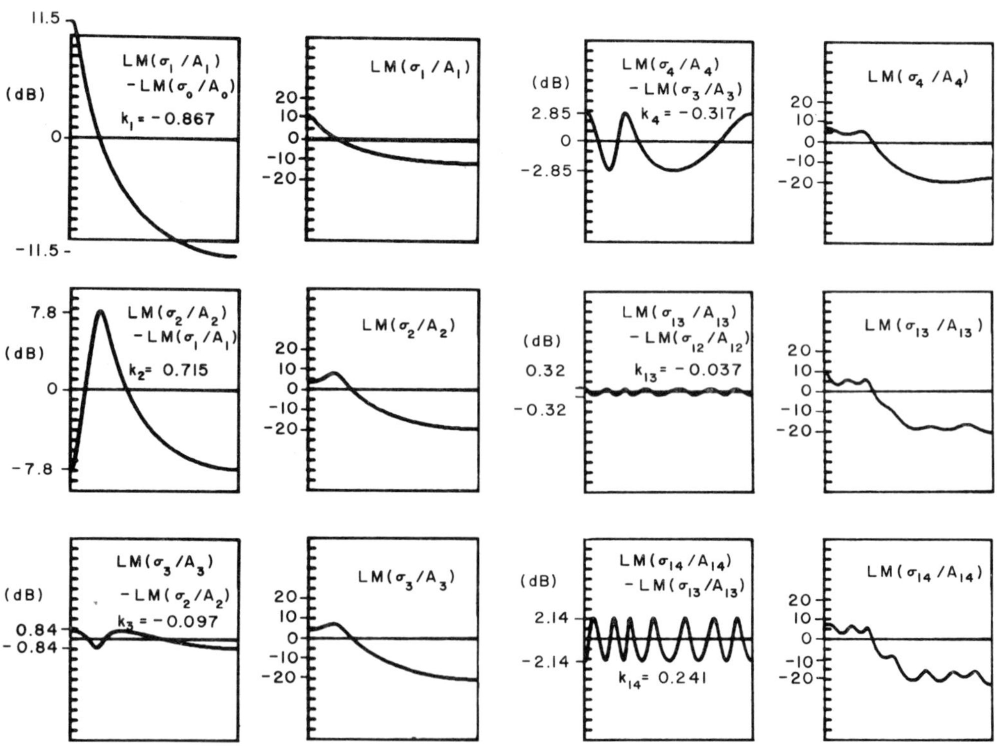

Fig. 6.5 Illustration of minimax error behavior in recursive buildup of spectral model.

Model spectra and the spectral differences are shown in Fig. 6.5 for a voiced sound sampled at 6.5 kHz without pre-emphasis. These spectra are shown with a dB scale so that the model spectra are labeled as $LM(\sigma_m/A_m)$ with $\sigma_m = \alpha_m^{1/2}$, and the spectral differences as $LM(\sigma_m/A_m) - LM(\sigma_{m-1}/A_{m-1})$. The spectra are normalized so that

$$\sigma_0^2 = \alpha_0 = r(0) = 1.$$

For $m=1$ the spectral difference ripple is identical to the model spectrum, and reaches the extreme values $\pm ln[(1+k_1)/(1-k_1)]$, or ± 11.5 dB since $k_1 = -.867$. The next reflection coefficient $k_2 = .715$ forces the model to contain a complex pole pair for representing the region of the first formant. The spectral difference here reaches the extreme values of $\pm$ 7.80 dB. Changes in the spectrum are seen to be small for the small reflection coefficients, as seen in the figure for $m=3$ and $m=13$. The figure illustrates the predictable nature of the model spectrum changes in the step-up procedure.

6.3 A Spectral Flatness Model

In the autocorrelation method, the error sequence $\{e(n)\}$, obtained by passing the original signal through the filter $A(z)$, is a whitened version of the original signal. In this section, the whitening process of the inverse filter $A(z)$ is studied in detail. First a spectral flatness measure is introduced for quantifying the flatness of a signal spectrum. This numerical measure is shown to be physically meaningful by presenting several representative analysis conditions for two important classes, voiced and unvoiced sounds.

It is shown that minimizing the spectral flatness measure of the error sequence is equivalent to the minimization of the error signal energy $r_e(0)$. Spectral flatness measures for idealized synthesis filter driving functions (idealized inverse filter outputs) are presented for voiced and unvoiced sounds and compared with experimental results.

In Chapter 9 it will be demonstrated that the flatness of the spectrum being analyzed is strongly correlated with potential numerical difficulties (ill-conditioning) in the computer implementation of the linear prediction algorithms. It will also be shown that the spectral flatness measure is a useful tool for lending insight into the effects of windowing and pre-emphasis of a speech signal.

6.3.1 A Spectral Flatness Measure

Let $\{f(n)\}$ represent a finite energy real time sequence, so that its z-transform $F(z)$ is analytic on the unit circle. If $r_f(0)$ denotes the energy of the time sequence, i.e.,

$$r_f(0) = \sum_{n=-\infty}^{\infty} f^2(n) = \int_{-\pi}^{\pi} |F(e^{j\theta})|^2 \frac{d\theta}{2\pi}, \tag{6.18}$$

then a *normalized log spectrum* of the time sequence can be defined by

$$\boxed{V(\theta)=ln[|F(e^{j\theta})|^2/r_f(0)]} \quad . \tag{6.19}$$

From (6.18), it can be seen that the energy of the time sequence equals the average of the spectrum, so that a perfectly flat or constant spectrum will yield a normalized log spectrum of zero, since if $|F[\exp(j\theta)]|^2$ equals a constant, then $r_f(0)$ is equal to the value of that constant.

One possible measure of deviation from a perfectly flat spectrum is the mean square of the normalized log spectrum, or any constant times that mean square. For example, a measure $\eta(F)$,

$$\eta(F)=\frac{1}{2}\int_{-\pi}^{\pi} V^2(\theta)\frac{d\theta}{2\pi} \tag{6.20}$$

such as applied in the analysis-by-synthesis method, can be used. Such a measure can equal zero only in the case of a flat spectrum. This measure weights positive and negative excursions of the normalized log spectrum equally, since the integrand, $V^2(\theta)/2$, is an even function of $V(\theta)$.

In the spectral analysis of speech it is preferable to use an integrand that is asymmetric, and which weights the positive excursions of $V(\theta)$ more heavily than the negative excursions. There is an infinite number of such integrands, but the discussion will be restricted to the form utilized by Itakura and Saito [1968] in their maximum likelihood approach to linear prediction as indicated in (6.7) and rewritten here as

$$\mu(F)=\int_{-\pi}^{\pi} [e^{V(\theta)}-V(\theta)-1]\frac{d\theta}{2\pi}. \tag{6.21}$$

In (6.7) $V(\theta)$ was based upon the error sequence whose z-transform was $E(z)$. In (6.21), $V(\theta)$ is based upon any finite energy sequence.

The expression (6.21) can be simplified by noting that as $V(\theta)$ represents the normalized log spectrum of the signal, the average value of $e^{V(\theta)}$ will be unity, giving the result

$$\mu(F)=-\int_{-\pi}^{\pi} V(\theta)\frac{d\theta}{2\pi}. \tag{6.22}$$

Viewed in this manner, $-\mu(F)$ is simply the average of the normalized log spectrum.

For purposes of normalization a spectral flatness measure $\Xi(F)$ is defined as

$$\boxed{\Xi(F)=e^{-\mu(F)}=\exp\left[\int_{-\pi}^{\pi} V(\theta)\frac{d\theta}{2\pi}\right]} \quad . \tag{6.23}$$

With this normalization, the spectral flatness measure will lie between zero and one, and equal one only for a perfectly flat spectrum. Substitution of (6.19) into (6.23) gives as an alternate expression for $\Xi(F)$ the result

$$\Xi(F)=\frac{\exp\left[\int_{-\pi}^{\pi} ln|F(e^{j\theta})|^2 \frac{d\theta}{2\pi}\right]}{\int_{-\pi}^{\pi} |F(e^{j\theta})|^2 \frac{d\theta}{2\pi}}. \tag{6.24}$$

In this form, $\Xi(F)$ appears as a ratio of a geometric to an arithmetic mean of $|F[\exp(j\theta)]|^2$ [Makhoul and Wolf, 1972]. To see this relationship more clearly, the integrals can be approximated by summations using N distinct points and rectangular integration with the result,

$$\boxed{\Xi(F)\simeq\frac{\exp\left[1/N \sum_{k=1}^{N} ln|F_k|^2\right]}{1/N \sum_{k=1}^{N} |F_k|^2}=\frac{\left[\prod_{k=1}^{N} |F_k|^2\right]^{1/N}}{1/N \sum_{k=1}^{N} |F_k|^2}}, \tag{6.25}$$

where N can be arbitrarily large, and $F_k=F\{\exp[j2\pi(k-1)/N]\}$.

6.3.2 Spectral Flatness Transformations

If the input sequence $\{x(n)\}$ with z-transform $X(z)$ is passed through the inverse filter

$$A(z)=1+\sum_{i=1}^{M} a_i z^{-i} \tag{6.26}$$

then the output sequence will have a z-transform of the form

$$E(z)=X(z)A(z). \tag{6.27}$$

If $A(z)$ is restricted to have its zeros within the unit circle, then Section 6.2.1 shows that its log spectrum has zero average value, and thus,

$$\int_{-\pi}^{\pi} ln|E(e^{j\theta})|^2 \frac{d\theta}{2\pi} = \int_{-\pi}^{\pi} ln|X(e^{j\theta})|^2 \frac{d\theta}{2\pi}.$$

Application of the spectral flatness measure definition, with $r_x(0)$ and $r_e(0)$ representing the energies of the filter output and input, results in the transformation

$$\Xi(E)=\Xi(X)r_x(0)/r_e(0). \tag{6.28}$$

If the input to the filter is fixed, the only portion of (6.28) that can produce a change in the output spectral flatness is $r_e(0)$, the energy of the output. $\Xi(E)$ will thus be maximum when $r_e(0)$ is minimum. Therefore, *the criterion of minimizing the energy of the inverse filter is equivalent to choosing the inverse filter that maximizes the spectral flatness at its output.*

Since $r_x(0)=\alpha_0$ and $r_e(0)=\alpha_M$, (6.28) can be written in the form

$$\boxed{10\log_{10}\Xi(E)=10\log_{10}\Xi(X)+10\log_{10}(\alpha_0/\alpha_M)} \quad . \tag{6.29}$$

Thus not only does the inverse filter maximize the spectral flatness at the output, but it adds to the flatness, on a dB scale, the amount $10\log_{10}(\alpha_0/\alpha_M)$. This amount plays an important role in describing the decomposition or deconvolution of the input log spectrum as a function of the filter length M, as will now be shown.

To see this, (6.27) can be rewritten as

$$X(z)=E(z)\frac{1}{A(z)}$$

so that the log spectrum of $X(z)$ is the summation of the log spectra of $E(z)$ and $1/A(z)$. It was shown in the correlation matching approach of Chapter 2 that the energy of the time sequence associated with $\sigma/A(z)$ is equal to $r_x(0)$ or α_0. Thus a direct application of the spectral flatness definition along with the zero mean property of Section 6.2.1 leads to the result

$$\Xi(1/A)=\alpha_M/\alpha_0.$$

Using this with (6.29) we have

$$\boxed{10\log_{10}\Xi(X)=10\log_{10}\Xi(E)+10\log_{10}\Xi(1/A)} \quad . \tag{6.30}$$

Thus the log spectrum of $X(z)$ can be decomposed into the sum of the log spectra of $E(z)$ and $1/A(z)$, and the spectral flatness measures similarly decomposed, as indicated by (6.30). Examples will be presented after discussing the numerical evaluation of spectral flatness measures.

6.3.3 Numerical Evaluation

In order to evaluate spectral flatness, it is necessary to evaluate an integral, as in (6.23), where the integral gives the average of the normalized log spectrum. The analytic nature of the z-transform can be used to show that the zeros of the spectrum must be isolated and the resultant effect will still lead to a finite average for the normalized log spectrum, even though the spectrum may go to zero at isolated points. For a data sequence with a finite number of non-zero points it has been shown that there is a theoretical lower bound on the resulting flatness measure [Gray and Markel, 1974b]. An analogous theoretical lower bound has been obtained for data having a specific dynamic range [Makhoul and Wolf, 1972].

A rough estimate of the spectral flatness measure, in dB, can be found by visually estimating the average dB value from a graph of the normalized log spectrum from (6.22). A closer estimate can be obtained by using a discrete summation to approximate the integral average. In particular, one can use the discrete Fourier transform to obtain discrete samples of the log spectrum, which can be normalized by using a numerically calculated energy for the signal. If the frequency samples are spaced closely enough and there are no zeros in the spectrum, a simple numerical discrete average of the normalized log spectrum can yield very good results. To estimate the accuracy, the frequency resolution can be increased by doubling the number of points in the discrete Fourier transform being used. As an experimental example, a typical voiced sound was truncated to 128 points, zeros were appended, and the spectral flatness measure was estimated using 256-point, 512-point, and 1024-point FFTs. The resulting spectral flatness measures differed by only 0.1 dB.

If there are any zeros in the spectrum, it becomes necessary to truncate the log spectrum from below to avoid a result of minus infinity in the numerical discrete average. More sophisticated numerical quadrature approaches could be utilized for greater accuracy, but we have not found them necessary for speech data, based upon numerous experiments with actual speech data using different-sized discrete Fourier transforms, different truncation limits, and discrete summations to approximate the integrals.

An acceptable rule of thumb is to apply a discrete Fourier transform (or FFT) having approximately twice as many points as the input data sequence. The spectra should be truncated from below, so that any value falling below 10^{-5} of the peak value, for example, is replaced by 10^{-5} of the peak, giving a 50 dB dynamic range.

Modifications to this numerical procedure may be necessary for non-speech signals, sampling rates greater than about 10 kHz, or different filtering before A/D conversion. In particular, if the filtering before A/D conversion has a cutoff frequency significantly below the half sampling frequency, the sampled signal will have a very low spectrum for a portion of the range, resulting in large negative excursions of the normalized log spectrum. The potential numerical difficulties can be handled by increasing the number of points used in the discrete Fourier transform and increasing the allowable dynamic range by lowering the truncation value. In general, however, the autocorrelation method should start with an analog signal sharply prefiltered at (not below) the discrete folding frequency, and low frequencies should not be filtered out of the signal. The very fact that the linear prediction approach is an attempt to model all-pole signals should alone be enough to discourage the artificial introduction of zeros into a data spectrum.

6.3.4 Experimental Results

To illustrate the spectral flattening effects of the inverse filter, results from an unvoiced and a voiced sound are shown in Figs. 6.6 and 6.7, respectively [Gray and Markel, 1974b]. The sampling rate was 6.5 kHz, and a 128-point time window was applied. A Hamming window was introduced before the analysis without any pre-emphasis. Each part of the figures is labeled with the filter order M, and a spectral flatness measure $SF(\cdot)$ defined by

$$SF(\cdot)=10\log_{10}\Xi(\cdot) \quad \text{(dB)}.$$

The separate figures show normalized log spectra, with the inverse filter output (or error) spectrum on the left and the reciprocal inverse filter, $1/A_M(z)$, on the right. The grid marks on the left of each figure indicate 5 dB differentials.

For $M=0$ the normalized log spectrum of the error signal equals the normalized log spectrum of the input, and the log spectrum of $1/A_M(z)$ equals zero. As M increases, the flatness of the error spectrum increases, both qualitatively and quantitatively. For the largest M used, $M=300$, it can be noted that almost all of the spectral flatness measure has been transferred from $E(z)$ to $1/A_M(z)$. The inverse filter output $E(z)$ is then an almost perfectly flat spectrum and $1/A_M(z)$ has a log spectrum that is almost identical to the input log spectrum. It should again be emphasized that each of the input spectra in Figs. 6.6 and 6.7 is normalized, so that the average of each spectrum is unity, but the average of their log spectra is necessarily below zero. It is the average of the normalized log spectrum that yields the spectral flatness measures.

It is interesting to compare Figs. 6.3 and 6.7 since they are both obtained from the same data using $10\log_{10}(\alpha_M/\alpha_0)=SF(1/A_M)$. It is not surprising that $\alpha_\infty/\alpha_0 \simeq \alpha_{300}/\alpha_0$ since $A_{300}(z)$ has essentially extracted the inverse of the input spectrum, leaving an error spectrum with very close to unity value.

6.3.5 Driving Function Models

Linear predictive analysis attempts to decompose the smoothed spectral structure, in the form of an all-pole filter model, from the speech signal, leaving the driving function or error signal information. The spectral flatness measure in (6.30) mathematically defines this decomposition, since the spectral flatness of the data sequence in dB can be expressed as the sum of the spectral flatness measures of the error signal and the model filter.

Ideally, a proper choice of filter order can lead to an error signal which is approximately an uncorrelated sequence for unvoiced sounds or a sequence of equally spaced samples (spaced by the pitch period) for voiced sounds. While the linear predictor maximizes the spectral flatness of the error signal, it does not seem desirable to use a filter order so high that this maximum is much greater than that predicted for ideal error signals.

For unvoiced sounds, Gray [1974a] has shown that the log spectrum of a windowed uncorrelated Gaussian sequence will have an expected value that is below the logarithm of the expected value of the spectrum by an amount equal to γ, where γ is Euler's constant 0.5772..., for all frequencies other than those adjacent to zero and the folding frequency. Thus a numerically evaluated spectral flatness measure for uncorrelated sequences should have an expected value of

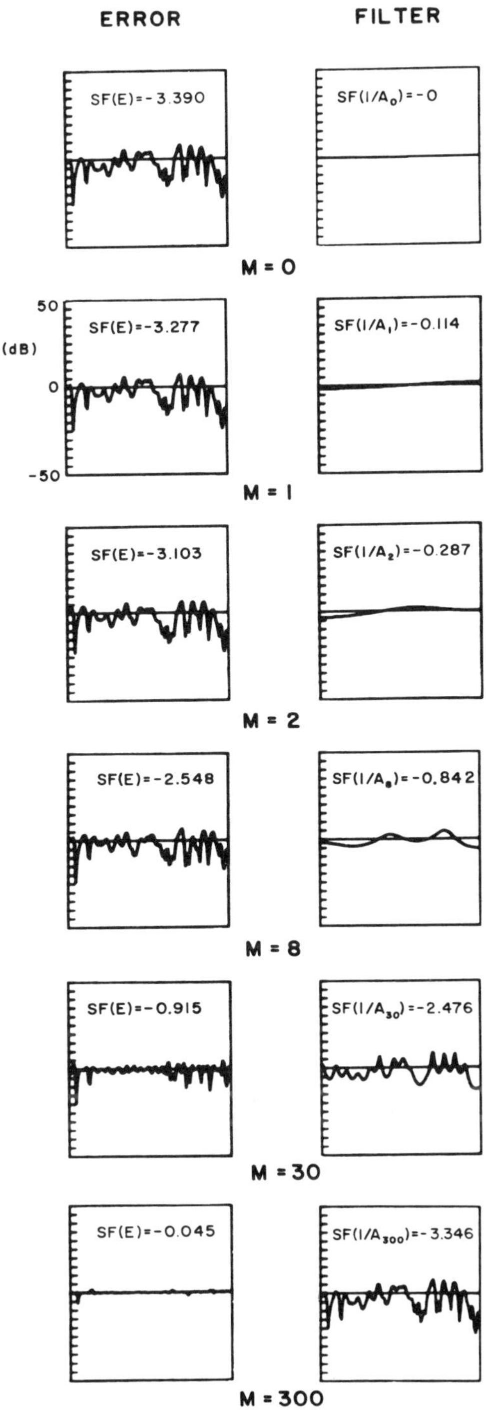

Fig. 6.6 Spectral flatness example for an unvoiced speech sound.

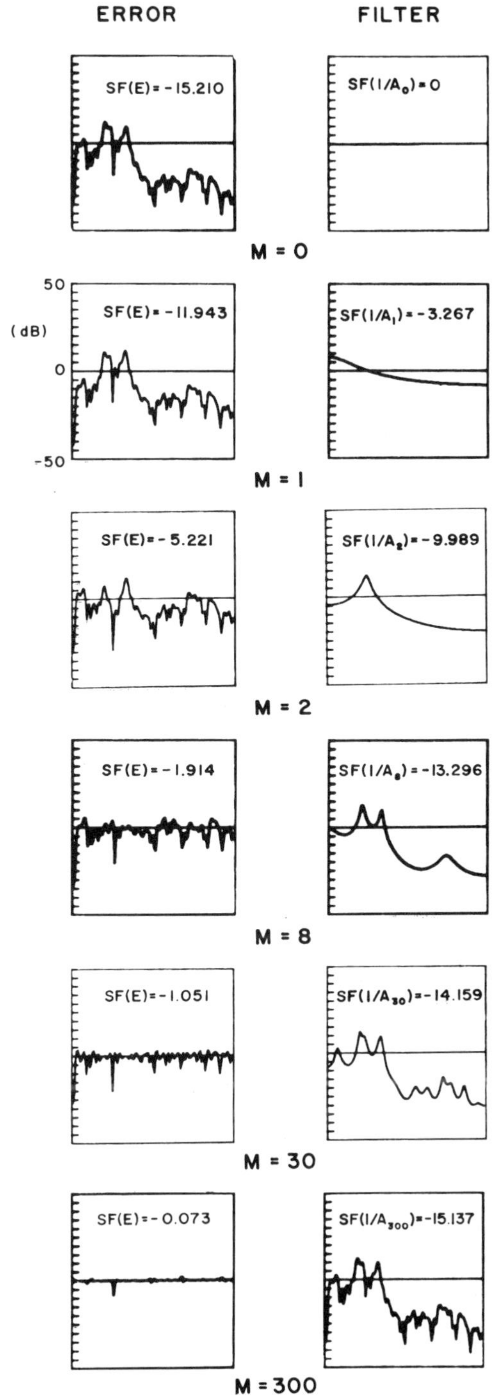

Fig. 6.7 Spectral flatness example for a voiced speech sound.

roughly $e^{-\gamma}$ or -2.5 dB on the average. Figure 6.6 shows an unvoiced example, where for $M=8$ the data sequence flatness has been decomposed into an error sequence flatness of -2.548 dB and a model flatness of $-.842$ dB. The close agreement in this case is much closer than one would expect for a single example, for the standard deviation of the log spectrum for Gaussian uncorrelated noise is of the order of 5.7 dB [Gray, 1974a], and thus measurements from a single frame of data should not be interpreted as reliable to test the model.

For voiced speech, the ideal error signal can be modeled as a set of equally spaced samples. For the sampling frequency and window size used in Fig. 6.6, there are roughly two pitch periods in each analysis frame, so the ideal result would be two equally spaced samples. It has been shown [Gray and Markel, 1974b] that a spectral flatness measure in such a case will be greater than 1/2 (-3 dB), with that value being obtained if the samples are of equal size. The spectral flatness measure for the error signal when $M=8$ in Fig. 6.7G was -1.914 dB. In terms of the spectral flatness measure, $M=8$ can be considered in Fig. 6.7 to carry out a reasonable decomposition of the original data sequence. When the number of equally spaced samples is increased, the spectral flatness measure for those samples decreases.

6.4 Selective Linear Prediction

In certain applications of linear prediction where the detailed spectral behavior of both voiced and unvoiced sounds is of interest, it is necessary to include a region from close to zero frequency to around 10 kHz. To assure a reasonable smoothed spectral match one should sample the signal at a rate of at least 20 kHz and then use an inverse filter whose order is greater than 20. Considerations in the choice of filter length are presented in Section 6.5.

Figure 6.8 shows the log magnitude spectrum of a speech signal sampled at 20 kHz and a smoothed spectral fit based upon the autocorrelation method with

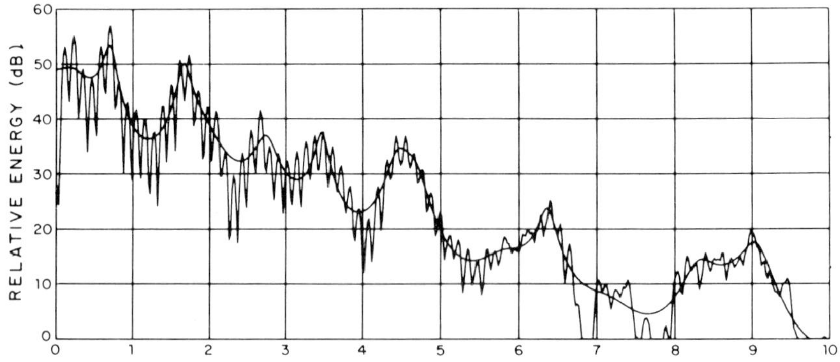

Fig. 6.8 A 28-pole linear prediction model of a voiced speech input spectrum. [From Makhoul, 1975b]

28 coefficients for the inverse filter. It is seen that a very reasonable spectral match is obtained throughout the total range from zero to 10 kHz. For voiced speech, the range from zero to 4 or 5 kHz is of major importance, as this is the region containing the first four or five formants and most of the signal energy. For unvoiced speech, however, considerable energy generally exists outside this range. In fact, for /s/ as in *s*it, nearly all the energy may reside in the interval from 5 to 8 kHz.

Since accurate spectral representation of unvoiced sounds does not appear to require as many coefficients as representation of voiced speech, it is worthwhile to consider how a low frequency and a high frequency region might be analyzed separately with fewer total coefficients while retaining accurate spectral representation.

If a signal is sampled at 20 kHz, and only the range from zero to 5 kHz is of interest, one might be tempted to sharply filter the signal to remove information above 5 kHz. This unfortunately increases the dynamic range of the log spectrum, and as discussed earlier, results in greatly decreased spectral flatness. In addition, the inverse filter will use many of its coefficients to represent the filtered portion of the spectrum.

The formulation for analyzing only portions of the spectrum and its solution is due to Makhoul [1975b] and is termed *selective linear prediction.* The essential idea is to translate the selected portion of the signal spectrum to the angular range from zero to π on the unit circle in the z-plane. By computing the autocorrelation coefficients from this new spectrum, the solution to the autocorrelation equations from Chapter 3 can then be applied. Since the selected frequency range will span the total range from zero to a new folding frequency (which is less than the original), fewer inverse filter coefficients are needed.

An illustrative example, based upon the same data as used in Fig. 6.8, is shown in Fig. 6.9. The two ranges from zero to 5 kHz and from 5 to 10 kHz were selected and separately analyzed, using 14 coefficients for the inverse filter in the lower frequency range and 5 coefficients for the inverse filter in the higher frequency range. There is an obvious discontinuity at the joining of the selected frequency regions, due to the fact that the 5 kHz point is a folding frequency in the translated spectrum in one case and zero frequency in the translated spectrum in the other case.

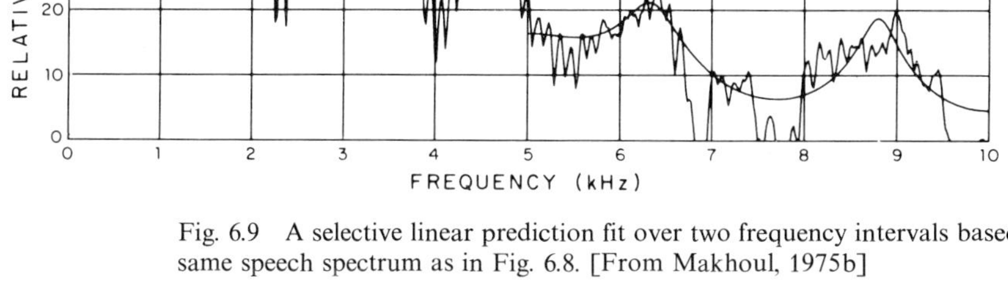

Fig. 6.9 A selective linear prediction fit over two frequency intervals based upon the same speech spectrum as in Fig. 6.8. [From Makhoul, 1975b]

6.4.1 Selective Linear Prediction (SLP) Algorithm

Assuming that selective linear prediction is desired on uniformly spaced data samples, the following steps are applied.

Step 1: Compute the data sequence spectrum.

Given an N-point data sequence $\{x(0), x(1), \ldots, x(N-1)\}$, zeros are appended so that $x(n)=0$ for $n=N, N+1, \ldots, I-1$, where I is a power of two if a radix-2 FFT is desired for computation. I should be larger than or equal to $N+M$, where M is the filter order to be used, as discussed in Section 6.4.3. The resultant DFT (or FFT if I is a power of 2) is given by

$$X(e^{j2\pi k/I}) = \sum_{n=0}^{I-1} x(n)\, e^{-jn2\pi k/I}, \tag{6.31}$$

and its magnitude squared is given by

$$R(e^{j2\pi k/I}) = |X(e^{j2\pi k/I})|^2, \tag{6.32}$$

for $k=0, 1, \ldots, I-1$.

Step 2: Create a translated spectrum.

Define a new spectrum based upon the selected frequency range. In normalized frequency terms, an angular range (π represents the folding frequency, one half of the original sampling frequency) from the angle $k_1 2\pi/I$ through $(k_1+l)2\pi/I$ is desired. If k_1 is zero and l is $I/2$, the range becomes the full range from zero to the folding frequency so that the entire spectrum is selected, and results become identical to ordinary linear prediction using the autocorrelation method provided that $I \geq N+M$ as discussed in Section 6.4.3.

A translated spectrum is defined by

$$R_t(e^{j2\pi k/L}) = R[e^{j2\pi(k+k_1)/I}] \tag{6.33a}$$

for $k=0, 1, \ldots, l$, where L can be chosen as either $2l$ or $2l+1$. In order to have an even translated spectrum, the reflected portion is defined by

$$R_t(e^{j2\pi k/L}) = R_t[e^{j2\pi(L-k)/L}] \tag{6.33b}$$

for $k=l+1, l+2, \ldots, L-1$. This spectrum can be thought of as the result of analyzing some data sequence with an L-point DFT.

Step 3: Compute the autocorrelation sequence.

The autocorrelation sequence is obtained from a discrete spectrum by computing the inverse DFT, but since the spectrum is real and symmetric, application of the DFT scaled by $1/L$ gives the same result,

$$r_t(n)=\frac{1}{L}\sum_{k=0}^{L-1} R_t(e^{j2\pi k/L})\,e^{-j2\pi kn/L}. \tag{6.34}$$

If $L$is a power of 2, this computation can be easily carried out by using an FFT. If L is not a power of 2, the summation must be directly evaluated for $n=0, 1, \ldots, M$, where M is the order of the inverse filter to be used.

From the real symmetric property of the spectrum, $r_t(n)$ can be expressed in terms of real variables by

$$\begin{aligned} r_t(n)=&\frac{2}{L}\sum_{K=1}^{l-1} R_t(e^{j2\pi k/L})\cos(2\pi kn/L) \\ &+\frac{1}{L}[R_t(1)+(-1)^n R_t(-1)] \end{aligned} \tag{6.35a}$$

when $L=2l$, and

$$\begin{aligned} r_t(n)=&\frac{2}{L}\sum_{k=1}^{l} R_t(e^{j2\pi k/L})\cos(2\pi kn/L) \\ &+\frac{1}{L}R_t(1) \end{aligned} \tag{6.35b}$$

when $L=2l+1$. These equations can be directly evaluated to obtain the translated autocorrelation coefficients $\{r_t(0), r_t(1), \ldots, r_t(M)\}$ for the translated spectrum.

Step 4: Computation of SLP model parameters.

The parameters $\{a_1, a_2, \ldots, a_M, \sigma^2\}$ are obtained in the same manner as discussed in Chapter 3, based upon the translated autocorrelation coefficients. The resultant model $\sigma^2/|A[\exp(j\theta)]|^2$ then represents the smoothed estimate of the selected spectral portion of the original data.

6.4.2 A Selective Linear Prediction Program

A Fortran subroutine SLP for performing selective linear prediction on the data sequence $\{x(n)\}$ is shown in Fig. 6.10 where $L=2l$. This program makes use of the FFT subroutine given in the next section. Inputs to the subroutine are as follows: N defines the number of input data samples in the X array $\{x(0), x(1), \ldots, x(N-1)\}$. The selective linear prediction filter order is M and the power of two for the input spectral computation is IP which must satisfy $M+N\leq 2^{IP}\leq 512$. The upper limit is simply a function of the dimension size on the X and Yarray in the FFT. The arguments $K1$ and $K2$ define the respective lower and upper index limits of the selected frequency range. Actual frequency limits are given by $f_1=(K1)(f_s)2^{-IP}$ and $f_2=(K2)(f_s)2^{-IP}$ where f_s is the sampling frequency of the input data array. Note that index limits are $0\leq K1<I/2$ and $K1<K2\leq I/2$ where $I=2^{IP}$. The output arguments are A, the array of filter coefficients $\{a_1, a_2, \ldots, a_M\}$, $ALPHA$, the gain term σ^2, and RC, the array of reflection coefficients $\{k_1, k_2, \ldots, k_M\}$.

```
C
      SUBROUTINE SLP(N,X,M,IP,K1,K2,A,ALPHA,RC)
      DIMENSION X(1),Y(512),A(1)
      DIMENSION RC(1),B(21),R(21)
      DATA PI/3.141592653/
      I=2**IP
      DO 10 J=1,I
10    Y(J)=0.
      NP=N+1
      DO 20 J=NP,I
20    X(J)=0.
      CALL FFT(X,Y,IP)
      L=K2-K1
      LP=L+1
      DO 30 J=1,LP
      JK=J+K1
30    X(J)=X(JK)*X(JK)+Y(JK)*Y(JK)
      MP=M+1
      RL=X(LP)
      DO 50 J=1,MP
      R(J)=X(1)+RL
      RL=-RL
      LM=L-1
      DO 40 K=1,LM
      ARG=(PI*K*(J-1))/L
40    R(J)=2.*X(K+1)*COS(ARG)+R(J)
50    R(J)=.5*R(J)/L
      ALPHA=R(1)
      RC(1)=-R(2)/R(1)
      A(1)=1.
      A(2)=RC(1)
      ALPHA=ALPHA-RC(1)*RC(1)*ALPHA
      MF=M
      DO 90 MINC=2,MF
      M=MINC-1
      DO 60 J=1,MINC
      JB=MINC-J+1
60    B(J)=A(JB)
      M=M+1
      S=0.
      DO 70 IP=1,M
      MIP=M-IP+2
70    S=S+R(MIP)*A(IP)
      RC(M)=-S/ALPHA
      DO 80 IP=2,M
80    A(IP)=A(IP)+RC(M)*B(IP-1)
      A(M+1)=RC(M)
      ALPHA=ALPHA-RC(M)*RC(M)*ALPHA
90    CONTINUE
      RETURN
      END
```

Fig. 6.10 A Fortran subroutine SLP for performing selective linear prediction.

6.4.3 Computational Considerations

As the input data sequence is of length N, the error sequence resulting from the autocorrelation method is of length $N+M$, for the inverse filter has a memory of length M. If a DFT of length $N+M$ or more is used with a sufficient number of zeros appended to the N-point data sequence, the energy of the error sequence can be found from either the sampled data or from the discrete frequency domain, since Parseval's relation gives

$$\sum_{n=0}^{N+M-1} e^2(n) = \frac{1}{L} \sum_{k=0}^{L-1} |E(e^{j2\pi k/L})|^2, \tag{6.36}$$

if $L \geq N+M$.

A second reason for appending zeros lies in the evaluation of autocorrelation coefficients, since

$$r_x(n) = \frac{1}{L} \sum_{k=0}^{L-1} |X(e^{j2\pi k/L})|^2 \, e^{j2\pi kn/L}, \tag{6.37}$$

for $n=0, 1, \ldots, L-N$. An inverse DFT can yield the correct autocorrelation coefficients, but only for arguments in a limited range. In order to use an Mth order inverse filter, the sequence $\{r_x(0), r_x(1), \ldots, r_x(M)\}$ is necessary, and thus $L \geq N+M$.

In using selective linear prediction, a result equivalent to the autocorrelation method of linear prediction is obtained if the full frequency range is selected, $k_1=0, l=I/2$, and $L=2l$ with $I \geq N+M$. These conditions can be used as a test for the selective linear prediction program. In general, smaller frequency ranges are utilized for selective linear prediction. The frequency resolution obtained is a function of the order of the original FFT used to obtain the data spectrum (resolution between frequency samples is $f_s/2^I$). If the entire spectrum is selected, additional frequency resolution does not affect the result. If only a portion of the spectrum is selected, then the choice of I can strongly influence the results. Thus it is necessary to append a minimum of M zeros to the data. Larger values may be beneficial for the spectral model at the expense of computation time.

6.5 Considerations in Choice of Analysis Conditions

To obtain useful results with linear prediction it is important to understand the relationships and effects of the choice of method, sampling rates, order of the model, time windows, and pre-emphasis. These choices will be discussed as they relate to speech analysis.

6.5.1 Choice of Method

The choice of the autocorrelation or the covariance method for spectral analysis depends upon somewhat different factors from those for speech transmission or for estimation of vocal tract area functions.

As a general statement, the autocorrelation method for spectral estimation is global in that several pitch periods must be contained within the analysis window for meaningful results with voiced speech. The covariance method, on the other hand, can be used as either a local method (applied to intervals less than a pitch period) or as a global one. If several pitch periods are analyzed, the spectral results for both methods will tend to be quite similar. For unvoiced speech, both methods will give similar results for intervals greater than 5 – 10 ms. The similarity of the results lies in the fact that when the number of samples is sufficiently large, the covariance coefficients c_{ij} will be approximately equal to the autocorrelation coefficients $r(i-j)$. The autocorrelation method requires somewhat less calculation, is assured of being stable, and allows a meaningful spectral matching gain to be computed. As a result, it would seem to be the preferable choice for the circumstances described in the preceding paragraph.

Based upon a linear all-pole model of speech that assumes glottal closure, only complex exponential behavior is possible until glottal opening occurs [Pinson, 1963]. Theoretically, the measured center frequencies and bandwidths would then correspond to the formants of the vocal tract. When analysis is performed over several pitch periods it is known that the estimated bandwidths will be larger due to the fact that energy is dissipated into the subglottal region during glottal opening. Since it was demonstrated in Chapter 2 that the covariance method can be used to automatically extract the parameters of a linear combination of complex exponentials that combine into a real signal, it might seem that the problem of automatic formant measurement had been solved. For example, if voiced speech is sampled at 10 kHz and it is assumed that precisely five formants exist in the vocal tract impulse response (10 complex exponentials), then to obtain the formant bandwidths and center frequencies it should be necessary to analyze only 20 samples from a 2 ms interval during glottal closure.

Unfortunately, if this procedure is attempted with actual speech using only a 2 ms analysis interval, poor results are likely to be obtained. The reason is that speech is not precisely composed of a small number of complex exponentials. Furthermore, there is the physical problem of finding the interval of glottal closure when in fact closure does not even occur as a general rule [Rosenberg, 1971]. Increasing the number of speech samples used in the analysis and using a least squares fit tends to overcome this difficulty at the expense of some accuracy.

The success of formant estimation for a given value of N (the number of speech data points used in the analysis) will depend strongly upon the particular speaker, utterance, and speaking conditions. Best results can be predicted for a male speaker with low fundamental frequency producing a vowel such as /ɑ/ in f*a*ther rather forcefully. These conditions tend to ensure glottal closure over a relatively long interval of time. Conditions leading to possibly unreliable results are high fundamental frequency, insufficient interval of glottal closure, and utterances of low intensity since low subglottal pressure will cause the volume velocity waveform to appear almost sinusoidal [Rosenberg, 1971].

In spectral modeling with linear prediction, it is desirable to include the gain term σ so that the energy of the model spectrum matches that of the signal spectrum. In the autocorrelation method this is included in the solution process, since in the autocorrelation matching formulation it was shown that $\sigma/A(z)$ will have exactly the same energy as the original data sequence used in the analysis if σ^2 is chosen equal to the prediction error, α.

In the covariance approach certain problems can arise. For example, if the inverse filter, $A(z)$, has a root on or outside the unit circle, the unit sample response of $1/A(z)$ will have infinite energy. Also, the prediction error can theoretically be zero for the case where the model may exactly represent the signal, as was seen in the discussion of Prony's method.

In general, for both the autocorrelation and covariance method, if $A(z)$ has no roots on or outside of the unit circle, the energy of the unit sample response of $1/A(z)$ can be obtained by the step-down procedure of Chapter 5, giving α_0/α_M, as

$$\alpha_0/\alpha_M = \prod_{m=1}^{M} [1-k_m^2]^{-1},$$

where k_m, $m=1,2,\ldots,M$ defines the reflection coefficients determined by the step-down procedure. Then, if $\sigma/A(z)$ is to be used for a smoothed version of a data sequence spectrum, it is necessary only to choose σ so that $\sigma^2 \alpha_0/\alpha_M$ matches the energy of the signal $\{x(n)\}$ used to compute the spectrum. In the covariance method, the number of samples used in the linear prediction analysis may cover less than a pitch period.

To compute the signal spectrum several pitch periods are usually used (for voiced speech) with the assumption that the signal is quasi-periodic. Thus, in the covariance method, the energy calculation may utilize a larger set of data points than that used in obtaining the inverse filter, $A(z)$. In the autocorrelation method, the same set of data is generally used for calculating both the inverse filter and the energy. More discussion of these methods within the context of formant estimation is presented in the next chapter.

6.5.2 Sampling Rates

As the sampling rate is increased, the representation of a continuous speech signal becomes more accurate. However, more samples imply larger storage requirements and greater computation. A general rule of thumb is to sample as slowly as possible without destroying the significant features of the signal.

In Chapter 9, it will also be seen that increasing the sampling rate causes computational accuracy problems. Based upon an average vocal tract length of 17 cm, the first three formant frequencies will lie in the frequency range of about 250–2800 Hz. Shorter vocal tract lengths, characteristic of women and children, will have somewhat higher formant frequencies in the range 300–3500 Hz. The telephone supports a voice bandwidth of about 300–3200 Hz.

To ensure against spectral aliasing, it is necessary to sample at twice the desired bandwidth. Representative sampling frequencies for use in low bit rate vocoder systems cover the range $f_s = 6.5-10$ kHz (see Chapter 10). For accurate estimation of voiced speech, the sampling frequency should exceed 6 kHz to include a speech bandwidth of at least 3 kHz. For accurate representation of fricative sounds, it is necessary to have very high signal-to-noise ratio (due to these being lower intensity sounds) and high sampling frequencies, such as 20 kHz (due to the high frequency content, often out to 8–10 kHz).

Before sampling, the data should be sharply prefiltered at or near the folding frequency $f_s/2$. If prefiltering occurs below $f_s/2$, the effect is to decrease the spectral flatness of the signal and to possibly cause some coefficients to be wasted in representing the artificially introduced non-flat behavior. Analog filters having minimum attenuation of 40 dB at frequencies beyond 1.06 of the cutoff frequency are commercially available and seem to work quite satisfactorily.

6.5.3 Order of Filter

As a practical matter, it is generally desirable to use the minimum number of parameters necessary to accurately model the significant features of the signal. In spectral modeling of speech, these features are the vocal tract resonances or formants, and to a lesser extent, the regions between the resonances. In Chapter 4, it was shown that to adequately represent the vocal tract under ideal circumstances, the memory of the model $A(z)$ must be equal to twice the time required for sound waves to travel from the glottis to the lips, that is, $2L/c$, where L is the length of the vocal tract and c is the speed of sound. For example, the representative values $c \simeq 34$ cm/ms and $L \simeq 17$ cm result in a necessary memory of 1 ms. When the sampling rate is 10 kHz, the filter order, M, must be at least 10, and for 6.5 kHz at least 7 (by rounding to the next highest integer).

As the glottal and lip radiation characteristics have not been accounted for in the above model, these numbers must be taken as lower limits. The glottal spectral slope characteristics can vary from -10 to -18 dB/octave while the lip radiation characteristic is usually assumed to have an approximate $+6$ dB/octave slope. Coupled with the fact that digitized speech waveforms are not exactly all pole waveforms (consider the fact that the signal being analyzed also has the prefilter characteristics superimposed upon it), it is generally necessary to add several more coefficients for these factors. Markel [1971b] suggested a reasonable value for formant trajectory estimation as f_s (the sampling frequency in kHz) plus 4 or 5. This value was based upon experimental results, and amounts to adding 4 or 5 poles for shaping to account for an average 17 cm vocal tract length and a velocity of sound of 34 cm/ms.

Spectral models for $M=10$ and $M=15$ are shown on an $LM(\sigma/A)$ vs. frequency and time scale in Figs. 6.11A and B. The utterance analyzed was the /iɚ/ portion of lin*ear*. The sampling frequency was 10 kHz and analysis was performed using 19.2 ms windows ($N=192$); each spectral slice corresponds to an increment of 5 ms or 50 samples. From Fig. 6.11A with $M=10$, it would appear that there are three formants in the region from $0-5$ kHz. However, inspection of the spectrogram of the same portion in Chapter 1 shows four formants in this interval. Ten coefficients are insufficient to accurately represent the spectral structure of voiced speech sampled at 10 kHz. By increasing M to 15 the four formants are clearly represented as shown in Fig. 6.11B. Initial and final frames corresponding to /i/ and /ɚ/, respectively, now show formant structures representative of these sounds uttered in isolation, e.g., very low second and third formants for the /ɚ/.

In many sounds analyzed using the autocorrelation method, the last few reflec-

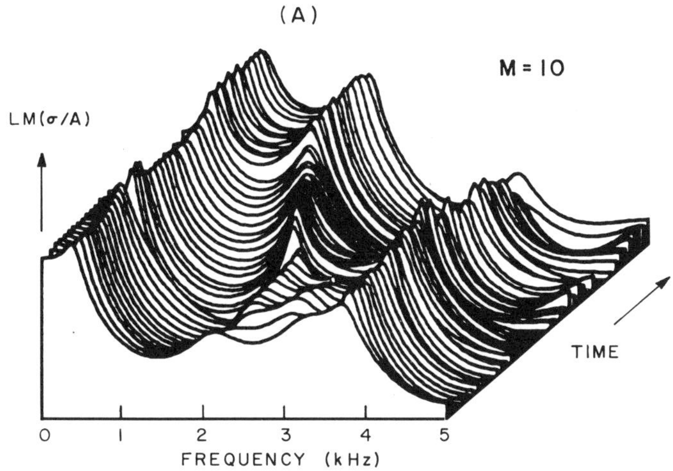

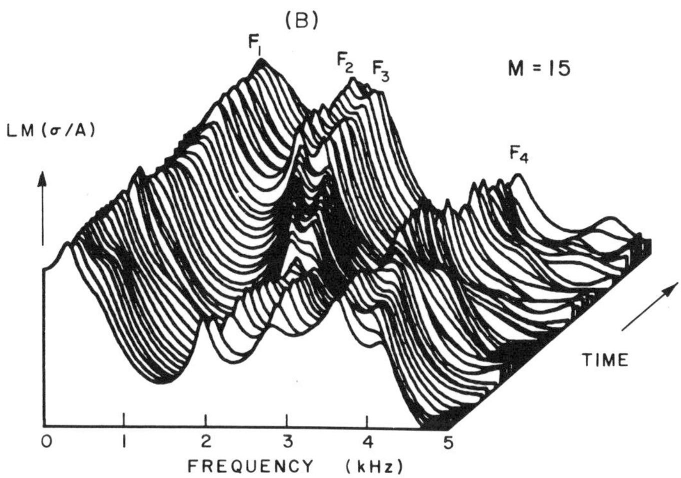

Fig. 6.11 Model spectra from $1/A(z)$ based upon analysis of the word "linear". A) $M=10$ coefficients. B) $M=15$ coefficients.

tion coefficient values are very small, and thus have little effect. It was noted in Section 6.3 that the log spectra $\alpha_m/|A_m[\exp(j\theta)]|^2$ and $\alpha_{m-1}/|A_{m-1}[\exp(j\theta)]|^2$ differ by an amount that oscillates between plus and minus $ln[1+k_m)/(1-k_m)]$. In decibels this factor results in a variation of ± 0.87 dB for $k_m = \pm .1$ and ± 1.76 dB for $k_m = \pm .2$. Thus, if the last value k_M is small, its effect is negligible and it can be ignored. For example, if $M=10$ and $k_{10}=0.1$, then the only difference between using $M=9$ and $M=10$ as far as the spectrum is concerned is the fact that the two resulting spectra differ by a maximum of 0.87 dB.

Although the order of the analysis filter is generally chosen to have a constant value, many fewer coefficients will be necessary for accurate smoothed spectral representation of unvoiced sounds. For example, fifteen coefficients may be necessary to clearly represent the formant structure of a voiced sound while four coefficients very adequately represent the trend characteristic of a particular fricative sound having at most one rather broad spectral peak. Figure 6.12 shows the same unvoiced sound as in Fig. 6.2B, only with a fourth-order spectral model instead of a twelfth-order model. The fit is quite reasonable and in fact would be more than adequate for synthesizing a realistic sounding /ʃ/ as in predic*ti*on.

In summary, it is reasonable to choose M for voiced speech analysis as the sampling frequency in kHz (since from (4.53) the sound velocity divided into twice the length of an average vocal tract is about 1 ms) with the addition of several more terms (from two to five depending upon the desired results). This choice of M places an upper bound on the number necessary for the analysis of unvoiced speech.

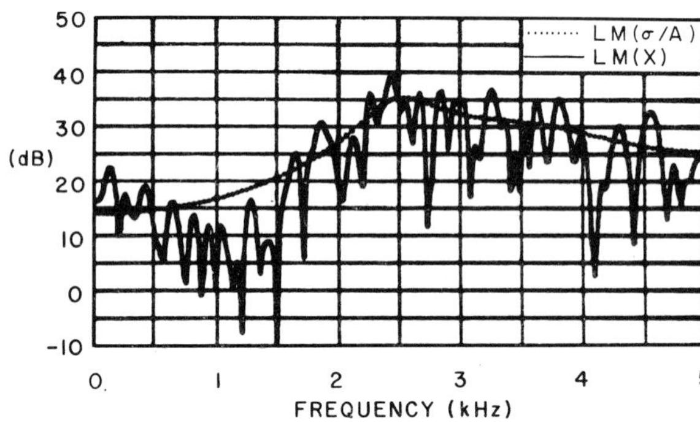

Fig. 6.12 Analysis of an unvoiced sound with $M=4$.

6.5.4 Choice of Analysis Interval

The choice of the analysis interval includes two factors, the location of the interval (placement with respect to the pitch period) and the length of the interval. A good choice of the interval length depends upon whether one is analyzing transient data or quasi-periodic data. It is desirable to perform spectral analysis within an interval where the vocal tract movement is negligible, on the order of 15 – 20 ms for most vowels. Although there can be considerable movement for glides (such as /w/ in *we*) within an interval of this length, it represents a reasonable range that results in reliable spectral estimates for both methods.

Thus a reasonable number of samples N to use in the analysis interval is given by f_s (kHz) multiplied by 15–20 (ms). This value is a compromise between the desire to have stable spectral estimates while minimizing averaging of the time-varying signal. Absolute placement of a 15 – 20 ms interval will not substantially affect the results of either the covariance or the autocorrelation method in most instances. However, if the covariance method is applied to voiced speech with a much smaller time interval, then the placement of the interval is critical as discussed in Chapter 2. Localized placement within a single pitch period is referred to as *pitch synchronous analysis.* Arbitrary placement of the time interval is referred to as *pitch asynchronous analysis.*

For unvoiced speech, the length of the interval can be substantially smaller than 15–20 ms. In fact, the burst associated with the release of an unvoiced stop consonant in the initial position such as /t/ in *t*ick may exist for only a few ms. Asynchronous analysis will often cause averaging in to the voiced portion following the /t/ or into the silence preceding the /t/ release. It would appear that for accurate analysis of transient sounds a smaller interval on the order of 10 ms is desirable.

6.5.5 Windowing

Briefly stated, windowing in both methods is advisable when they are used globally, but windowing in the covariance method used locally is inadvisable. The covariance method attempts to approximate a given data sequence by a linear combination of complex exponentials, and if the original data approximately correspond to such a linear combination, the use of a window destroys this effect.

When the analysis interval is long enough to include a number of pitch periods of voiced sounds or a sufficiently long interval for unvoiced sounds, the two methods give similar results. Even if no window is explicitly introduced, there is a rectangular window implicit in the treatment of the data sequence, for only a given sequence of N samples $\{x(0), x(1), \ldots, x(N-1)\}$ is utilized in the analysis. Though the samples are utilized in a different manner for the covariance and the autocorrelation methods, when N is large, the results are similar.

In the autocorrelation method a model spectrum $\sigma^2/|A[\exp(j\theta)]|^2$ is being used to represent a data spectrum $|X[\exp(j\theta)]|^2$. If no explicit windowing is carried out, discontinuities between the values of x_0 and or x_{N-1} and the numerical values of zero (outside of the implicit rectangular window) can cause spectral distortion.

The general properties of windows have been covered in great detail in Blackman and Tukey [1958]. Markel [1971c] and Makhoul and Wolf [1972] have discussed specific properties of windows as they apply to linear prediction, with examples. One extreme example of the effects of rectangular versus tapered windows is shown in Fig. 6.13. Figures 6.13 A–C show the input signal, input spectrum, and model spectrum for a 32 ms portion of a voiced sound. Note the deviation from zero at the end points in the input signal.

If the input signal is directly transformed, the resultant log spectrum shown in Fig. 6.13B is obtained. This log spectrum is normalized so that its peak is at 0 dB. Figure 6.13C shows the log spectrum of the model (again normalized so that the peak is at 0 dB) for the case where $M=14$, $N=320$, and $f_s=10$ kHz. At most, one resonance can be estimated. However, by observing a spectrogram of the utterance from which this segment was taken, it is quite clear that there are two closely spaced low formants and two additional formants, one at about 2.2 kHz and the other at about 3.1 kHz.

Figures 6.13 D–F show the input signal, input spectrum, and model spectrum of the same data, but with a 320-point Hamming window. Figure 6.13E shows the spectrum of the windowed data, which now shows a substantial reduc-

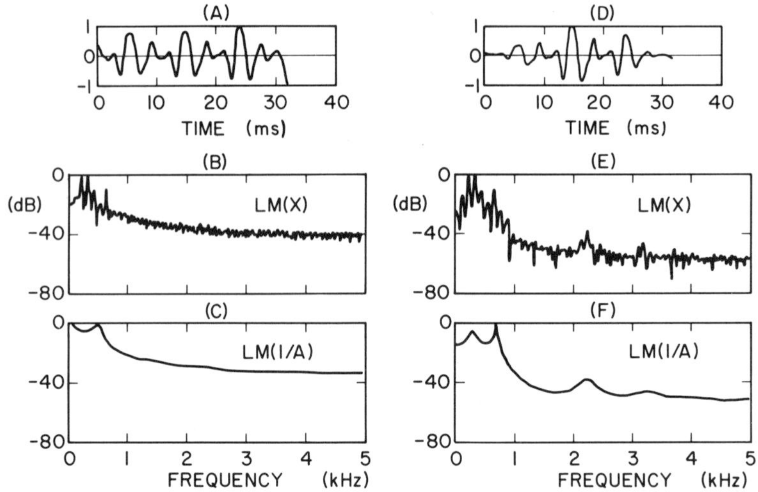

Fig. 6.13 Illustration of rectangular and non-rectangular window effects.

tion of the spectral distortion. From Fig. 6.13F it is evident that there are two formants at low frequencies, and two formants at 2.2 kHz and 3.1 kHz.

A general rule for linear prediction methods used over intervals greater than 15 ms or several pitch periods is that windowing should be used on the data sequence.

6.5.6 Pre-emphasis

If spectral properties of the vocal tract, without the effects of glottal waveform and lip radiation characteristics, are to be estimated, then the speech should be pre-emphasized before analysis. In Chapter 4 it was stated that a simple way to effect this pre-emphasis is to pass the signal through a simple one-zero filter of the form $1-\mu z^{-1}$, where μ is near or equal to one. For speech analysis the value of μ is not critical, and values in the range from 0.9 to 1.0 yield roughly equivalent results.

Thus, if vocal tract spectral characteristics are desired, pre-emphasis by a filter $1-\mu z^{-1}$ with μ in the range of 0.9 through 1 should be applied. There appears to be little difference in the results when pre-emphasis is applied before windowing and when pre-emphasis is applied after windowing.

The preceding comments are with reference to voiced speech. In the case of unvoiced speech there seems to be little reason to pre-emphasize the data. Indeed, it may only grossly exaggerate the high frequencies present in most unvoiced utterances. From this point of view, it appears that a pre-emphasis factor based upon the signal characteristics might be of value. Both Gray and Markel [1974b] and Makhoul and Vishwanathan [1974a] have considered an optimal or adaptive pre-emphasis factor given by $\mu=r_x(1)/r_x(0)$ where $\{r_x(n)\}$ is the autocorrelation sequence of the data $\{x(n)\}$. For unvoiced sounds, μ will usually be small, whereas for voiced sounds μ will be near unity.

6.6 Spectral Evaluation Techniques

If $\{x(n)\}=\{x(0), x(1), \ldots, x(N-1)\}$ represents an N-length sequence of data, the spectrum (where it is understood to be a *discrete* spectrum) is obtained from the discrete Fourier transform (DFT) by

$$\begin{aligned} X\left(e^{\frac{j2\pi k}{N}}\right) &= \sum_{n=0}^{N-1} x(n) e^{-\frac{j2\pi nk}{N}} \\ &= \mathrm{DFT}\{x(n)\} \\ k &= 0, 1, \ldots, N-1. \end{aligned} \tag{6.38}$$

The inverse discrete Fourier transform (IDFT) allows $x(n)$ to be computed from $X[\exp(j2\pi k/N)]$ by

$$x(n)=1/N\{\mathrm{DFT}\{X^*\left(e^{\frac{j2\pi k}{N}}\right)\}\}^* \tag{6.39}$$

where * defines the complex conjugate. This expression can be used to obtain the DFT and IDFT from a single DFT program. The fast Fourier transform [Brigham, 1974] can be used to efficiently implement the DFT of any sequence whose length is a power of two. If the sequence length is not a power of two, zeros can be appended to the sequence, effecting an interpolation in the spectrum.

A Fortran FFT subroutine implementation of a radix-2 algorithm is shown in Fig. 6.14. The real and imaginary parts of the input are placed in the X and Y array, respectively. The transform size N' is determined by L where $N'=2^L \geq N$. If $N' > N$, then zeros must be appended in the input sequence out to N'. At the subroutine completion, the real and imaginary parts of the transform are contained in the X and Y arrays, respectively. Assuming that the input data are based upon a sampling frequency $f_s=1/T$ and that the FFT input sequence length is $N'=2^L$, the relationship between real time (t) and frequency (f) in terms of the Fortran index K is

$$t=(K-1)T=(K-1)/f_s$$

for $\quad K=1, 2, \ldots, N',\quad$ and

$$f=(K-1)f_s/N'=2^{-L}(K-1)f_s,$$

for $K=1, 2, \ldots, N'/2+1$. If the input sequence is real, then the values for $K=N'/2+2$ out to N' contain no new information. More specifically,

$$X\left(e^{\frac{j2\pi k}{N'}}\right)=X^*\left(e^{\frac{j2\pi(N'-k)}{N'}}\right) \quad k=0, 1, \ldots, N'/2.$$

```
C
      SUBROUTINE FFT(X,Y,L)
      DIMENSION X(1),Y(1)
C
C     RADIX-2 FFT
C
      NP=2**L
      LMX=NP
      SCL=6.283185303/NP
      DO 20 LO=1,L
      LIX=LMX
      LMX=LMX/2
      ARG=0.
      DO 10 LM=1,LMX
      C=COS(ARG)
      S=SIN(ARG)
      ARG=ARG+SCL
      DO 10 LI=LIX,NP,LIX
      J1=LI-LIX+LM
      J2=J1+LMX
      T1=X(J1)-X(J2)
      T2=Y(J1)-Y(J2)
      X(J1)=X(J1)+X(J2)
      Y(J1)=Y(J1)+Y(J2)
      X(J2)=C*T1+S*T2
   10 Y(J2)=C*T2-S*T1
   20 SCL=2.*SCL
C
C     BIT REVERSAL
C
      J=1
      NV2=NP/2
      NPM1=NP-1
      DO 50 I=1,NPM1
      IF (I.GE.J) GO TO 30
      T1=X(J)
      T2=Y(J)
      X(J)=X(I)
      Y(J)=Y(I)
      X(I)=T1
      Y(I)=T2
   30 K=NV2
   40 IF (K.GE.J) GO TO 50
      J=J-K
      K=K/2
      GO TO 40
   50 J=J+K
      RETURN
      END
```

Fig. 6.14 Radix-2 fast Fourier transform (FFT) subroutine.

As an example of the FFT usage, assume that a 15th order spectral model with parameters $\{a_1, a_2, \ldots, a_{15}, \sigma\}$ has been calculated, based upon input data sampled at 10 kHz. To have a frequency resolution of 30 Hz or less in the spectrum, N' must satisfy $f_s(\text{kHz})/N' < .03$ or $N' > 333$. Choosing the closest power of two gives $L=9$ and $N'=512$ and a frequency resolution (distance between discrete sample) of 19.53 Hz. To compute $LM(\sigma/A)$ at the discrete frequencies $f_k = 19.53k$, $k=0, 1, \ldots, 256$, the Y array is filled with zeros since the input sequence is real, and the X array is filled as follows:

$$X = \{1, a_1, a_2, \ldots, a_{15}, 0, 0, \ldots, 0\}.$$

$$\begin{array}{lccccc} & \uparrow & \uparrow & \uparrow & \uparrow & \uparrow \\ \text{index} & 1 & 2 & 16 & 17 & 512 \end{array}$$

After calling the subroutine with $L=9, LM(\sigma/A)$ is computed from the Fortran variables $X(J)$ and $Y(J)$ as

$$10\log_{10}\left\{\frac{\sigma^2}{\left|A\left[e^{\frac{j2\pi(J-1)}{N'}}\right]\right|^2}\right\}=20\log_{10}\sigma-10\log_{10}[X^2(J)+Y^2(J)] \quad (6.40)$$

for

$$J=1,2,\ldots,257.$$

As closing notes on the FFT, it is important to note that further efficiency in spectral computation can be obtained at a slight cost in programming complexity and storage requirements. For example,

- The trigonometric functions can be calculated in advance and stored.
- N'-point real sequences can be transformed using $N'/2$-point FFTs [Bergland, 1969].
- N'-point real and even sequences (such as a sequence representing the magnitude square of a spectrum) can be transformed using $N/4$-point FFTs [Cooley, et al., 1970].
- When a significant number of input sequence values is zero, FFT pruning can be used to decrease the number of calculations [Markel, 1971c].

6.7 Pole Enhancement

In the next chapter, it will be seen that there is an occasional need to enhance the peaks of a spectral model so that poles of a model can be more easily located without resorting to root solving routines.

One of the simplest approaches for accomplishing this goal is that of evaluating the z-transform on a circle closer to the poles of the spectral model. Makhoul and Wolf [1972] have used the name *off-axis spectrum* computation since if the s-plane is mapped into the z-plane using the mapping $z=\exp(-sT)$, a circle of radius r in the z-plane becomes a line of constant real part in the s-plane, to the left of the $j\omega$ axis if r is less than one.

If $A(z)$ represents the z-transform of the polynomial

$$A(z)=\sum_{i=0}^{M} a_i z^{-i} \quad (6.41)$$

then an off-axis spectrum is defined by

$$A(re^{j\theta})=\sum_{i=0}^{M}(a_i r^i)e^{-ji\theta}. \quad (6.42)$$

Thus, it is necessary only to replace a_i by $a_i r^i$ to effect the off-axis spectrum.

It has become customary to describe r in terms of a bandwidth. This follows from an s-plane result that bandwidth of a pole in radians per second is equal to twice the distance of the pole from the $j\omega$ axis when the pole is isolated from other poles and zeros. If the mapping between the s and z domain is given by $z = \exp(-sT)$, then a pole with radius ϱ in the z-plane will be related to its bandwidth in Hz by the relation $\varrho = \exp(-\pi BT)$ where B is the bandwidth. The bandwidth reduction of a resonance in the off-axis spectrum is denoted by B_0 where $r = \exp(-\pi B_0 T)$.

The method of off-axis spectrum evaluation can be used effectively, within limits, for formant extraction. The potential value and problems of this approach are illustrated in Fig. 6.15. The log spectrum of the inverse filter $A(z)$ for a voiced sound is shown by the solid line. The sampling frequency was 17.5 kHz. The first, second, and fourth formants are easily obtained by finding the local

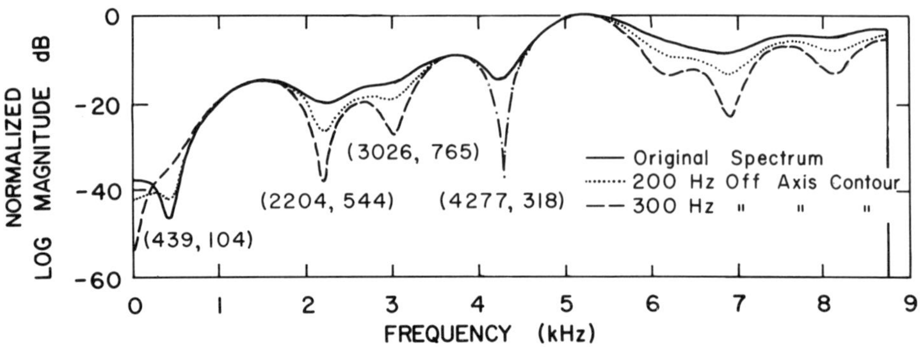

Fig. 6.15 Comparison of model and off-axis spectra.

minima of the spectrum. By applying a spectral enhancement (bandwidth reduction) of 200 Hz, the third formant can be seen, as indicated by the dotted curve in the figure. Note that the first formant becomes less dominant, for it appears as though the zero of $A(z)$ which produced it is now further from the circle on which the z-transform is evaluated (its original bandwidth was less than 100 Hz).

Increasing the bandwidth reduction to 300 Hz leads to the curve plotted with dashed lines. The first formant is lost completely, whereas the previously undetected minimum representing a fifth formant now appears. The bandwidth and behavior near the fourth formant appear unchanged, indicating that its original bandwidth was about halfway between 200 and 300 Hz. This example clearly illustrates the compromise necessary in choosing B_0. For spectral operations such as peak picking, it is not necessary that the off-axis contour enclose all the roots (in the z-plane). However, as a rule of thumb, B_0 must be sufficiently small that it does not greatly exceed twice the smallest bandwidth of the original spectrum.

An alternate and more general contour is obtainable from the chirp z-transform [Rabiner, 1969]. The chirp z-transform is conceptually desirable for it allows for a contour which is not a unit circle in the z-plane (or constant real part in the s-plane under the mapping $z=\exp(-sT)$). The chirp z-transform allows for the evaluation of the z-transforms on spirals of decreasing radius, but this is at the expense of a substantial increase in computation time.

The pole enhancement methods are not needed if computation time is not a consideration, since widely available polynomial root-solving programs will produce all of the roots of $A(z)$. If computation time is an important consideration, then pole enhancement methods can be of value in formant extraction.

7. Automatic Formant Trajectory Estimation

7.1 Introduction

A primary purpose of speech analysis is to extract features or parameters that represent important characteristics of the waveform. The two most basic acoustical speech parameters (in the sense that they are relatable to the physical production of speech and are parameters of the analog vocal tract model) are the formant and fundamental frequency parameters.

The formant is generally defined as a damped sinusoidal component of the vocal tract acoustic impulse response. In the classical model of speech, a formant is equivalently defined as a complex pole-pair of the vocal tract transfer function. For an average vocal tract length of approximately 17 cm there will generally be three or four formants within 3 kHz and four or five formants within 5 kHz. Speech synthesis studies have demonstrated that the first three formants are of primary importance in the representation of voiced speech. In fact, quite intelligible synthetic voiced speech is possible by applying only the first three estimated time-varying formant frequencies (formant trajectories) to a formant model of speech [Fant, 1960; Schafer and Rabiner, 1970].

The formant parameters play a dominant role in many areas such as acoustic phonetics, automatic speech recognition [Broad, 1972] and, low-bit rate speech transmission [Flanagan, 1972]. Automatic formant analysis is a major problem due to the fact that the vocal tract impulse response is not a directly observable quantity. Parameters of an all-pole model are desired where the signal to be processed is the model convolved with a quasi-periodic glottal driving function. For accurate estimation, it is therefore necessary to perform a deconvolution to separate the impulse response and the driving function. The two techniques being widely used at present for formant estimation are based upon cepstral analysis and linear prediction.

Schafer and Rabiner [1970] presented the first detailed approach for automatically estimating formant structure from voiced speech using cepstral analysis. Itakura and Saito [1970], Atal and Hanauer [1971b], and Markel [1971b, 1972b] discussed the application of linear prediction to formant estimation. Markel [1972a, 1973b] has presented a simplified procedure for automatically estimating formant trajectories using linear prediction techniques. McCandless [1974a] has presented a comprehensive algorithm based upon linear prediction spectra for dealing with several problem areas.

The purpose of this chapter is to demonstrate that linear prediction techniques can be used to obtain accurate formant trajectory estimation for voiced speech. With respect to other techniques, linear prediction techniques offer the advantages of minimal complexity, minimal computation time, and maximal accuracy in formant estimation. The estimates are shown to be quite reasonable when compared to standard spectrographic (sonograph) representations and other accepted techniques such as cepstral analysis. The representation is of nearly minimal complexity in the sense that the raw data (corresponding to the null or minima locations in the inverse filter spectrum for each frame of data) generally define the first three formant frequencies in order of frequency occurrence, approximately 85–90 percent of the time, without further processing [Markel, 1971b, 1972b]. The raw data from cepstral analysis, for example, will generally have much more non-conformant information which must be removed by decision algorithms.

The desirability of linear prediction for formant estimation can be qualitatively justified by the fact that in signal representation, the most efficient basis set (or set of approximating functions) is one which most closely matches, in some sense, the desired characteristics of the signal. The impulse response of the vocal tract model and the analog counterpart of the unit sample response of $1/A(z)$ are made up of only complex exponential terms as shown in Chapter 3. In this chapter, a general structure for use in formant estimation is now presented with examples of formant estimation from raw data. Then two algorithms for automatic formant trajectory estimation are presented.

7.2 Formant Trajectory Estimation Procedure

7.2.1 Introduction

A general procedure for formant trajectory estimation based upon linear prediction analysis is shown in Fig. 7.1. Each frame of speech to be analyzed is denoted by the N-length sequence $\{s(n)\}$. The speech is first preprocessed by pre-emphasis and possibly windowing. Based upon the linear speech production model and the discussion in Chapter 4 on estimation of vocal tract area functions from speech,

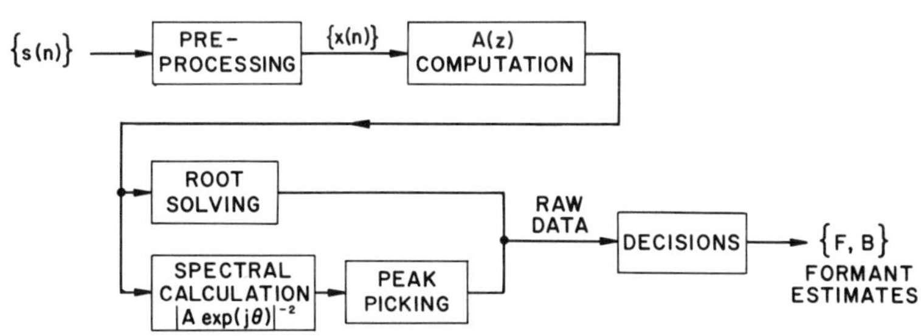

Fig. 7.1 A general procedure for formant trajectory estimation.

it should be clear that some amount of pre-emphasis is necessary to eliminate the spectral slope characteristic introduced by the effect of the glottal volume velocity waveform (modeled by $\simeq$ −12 dB/octave slope) and the lip radiation characteristic (modeled by $\simeq$ +6dB/octave slope). Wakita's experiments [1972, 1973b] have shown that a reasonable pre-emphasis factor is $1-\mu z^{-1}$ where μ is near unity. For $\mu=1$, the result is an approximate +6dB/octave slope with a maximum error of about 3.9 dB at the half sampling frequency. Pre-emphasis will generally result in a slight upward shift for the estimated formant frequency locations with respect to no pre-emphasis ($\mu=0$). These shifted values are preferred, based upon the linear speech production model, since the pre-emphasized spectrum is the estimate of the periodic pulse train spectrum times the vocal tract spectral model. Any technique that results in an inverse filter $A(z)$ can be applied for formant estimation; the most commonly used ones are either the covariance or the autocorrelation method of linear prediction. There has also been some application of the Kalman filter for estimating $A(z)$, but as yet, advantages over the other more easily implemented procedures have not been shown.

In the covariance method, pitch-synchronous analysis over single pitch periods does not theoretically require pre-emphasis. (See Section 2.4). Under certain conditions this procedure can result in very accurate bandwidth estimates. Windowing and pre-emphasis are not beneficial unless $N \gg M$ (where M equals the number of analysis coefficients), in which case the covariance method will behave similarly to the autocorrelation method. In the autocorrelation method it has generally been found to be beneficial to both pre-emphasize and window the data before processing further (Chapter 6).

Given an input sequence $\{x(n)\}$, algorithms for designing an inverse filter $A(z)=1+\sum_{i=1}^{M} a_i z^{-i}$ have been presented in Chapters 2 and 3. If the sampling frequency, pre-emphasis conditions, number of coefficients, etc., are reasonably chosen as discussed in Chapters 4 and 6, then $1/|A(\exp[j\theta])|^2$ will represent the estimated spectrum of the vocal tract impulse response.

Initial estimates of the formant frequencies and bandwidths are defined by either solving for the roots of the polynomial $A(z)$ or computing the interpolated spectrum of the inverse filter with an FFT, and searching for the peaks in the spectrum $1/|A[\exp(j\theta)]|^2$. Solving for the roots guarantees that all possible formant frequency and bandwidth candidates will be extracted. Due to the computational expense required in polynomial root-solving, however, peak picking procedures are of considerable interest and practical utility. The major disadvantage of the spectral calculation and peak picking is that closely spaced resonances (complex pole pairs) may not be extracted from the spectrum. When this condition has been detected, the off-axis spectral enhancement procedure of Chapter 6 can be beneficial. Parabolic interpolation about the peaks results in initial estimates for the formant frequencies and bandwidths $\hat{F}_i$ and $\hat{B}_i$, respectively. These initial estimates are also referred to as *raw data* since $\hat{F}_i$, $\hat{B}_i$ may or may not correspond to the *lth* formant value for frame k, $F_l(k)$ and $B_l(k)$. For example, if only the second formant were missed in the estimation (an omission), then $\hat{F}_2(k)$ would be assigned to $F_3(k)$ and not to $F_2(k)$.

A plot of just the raw data (frequency vs. k) over only voiced speech intervals is generally sufficient to estimate the formant trajectories $F_l(k)$, $l=1,2,3$ for frame k by visual inspection. However, for automatic formant trajectory estimation, it is necessary to perform decisions to ensure continuity. For example, if there is reason to believe that a formant should exist at some location where one was not extracted, then one needs to be inserted. It is important to note that this kind of judgment makes sense only when at least several frames of raw data are available. In addition, a voiced-unvoiced decision is important in defining formant regions from continuous speech.

7.2.2 Raw Data from $A(z)$

If the raw data are computed from a root-solving procedure, the bandwidth $\hat{B}$ and frequency $\hat{F}$ for any complex root z are obtained from the s-plane to z-plane transformation $z=\exp(sT)$ where $s=-\pi\hat{B}\pm j2\pi\hat{F}$. If $z=R_e(z)+jI_m(z)$ defines the real and imaginary terms of a complex root, then

$$\hat{B}=-(f_s/\pi)\,ln|z| \quad \text{(Hz)} \tag{7.1a}$$

and

$$\hat{F}=(f_s/2\pi)\tan^{-1}[I_m(z)/R_e(z)] \quad \text{(Hz)} \tag{7.1b}$$

where $f_s=1/T$ defines the sampling frequency. Root-solving programs are computationally expensive but are generally standard items in scientific subroutine libraries. An alternative procedure is to calculate $1/|A[\exp(j\theta)]|^2$ at a discrete number of samples using an N-point FFT as discussed in Chapter 6 and then apply peak picking and parabolic interpolation. A parabola has the form

$$y(\lambda)=a\lambda^2+\text{b}\lambda+c. \tag{7.2}$$

If $y(0)$ defines a discrete peak value, and $y(-1)$ and $y(1)$ define the samples to the left and right of $y(0)$, then the parabola that passes through these points has coefficients

$$\begin{aligned} c&=y(0)\\ b&=[y(1)-y(-1)]/2\\ a&=[y(-1)+y(1)]/2-y(0). \end{aligned} \tag{7.3}$$

By solving $dy(\lambda)/d\lambda=0$, the peak location with respect to a zero index is $\lambda_p=-b/2a$. Thus if the discrete spectral peak is located at n_p, then the interpolated raw data estimate is

$$\hat{F}=(\eta_p+\lambda_p)f_s/N \tag{7.4}$$

The 3 dB bandwidth B, based upon passing a parabola through spectrum values, is satisfied by $y(\lambda)/y(\lambda p)=.5$ or

$$\hat{B}=\frac{-\{b^2-4a[c-0.5y(\lambda_p)]\}^{1/2}f_s}{aN}. \tag{7.5}$$

A Fortran program FINDPK for finding the peaks of $1/|A[\exp(j\theta)]|^2$ is given in Fig. 7.2. The F array contains the discrete spectral values $|A[\exp(2\pi k/2\,NPTS)]|^2$,

```
C
      SUBROUTINE FINDPK(F,NPTS,FSK,NPEAK,IX)
      DIMENSION F(1),IX(1)
      J00=0
      NPEAK=0
      NPTS2=NPTS*2
      NPTM1=NPTS-1
      TS=FSK/NPTS2
      TSS=TS*TS
      NF=0
      YM=F(2)
      YP=F(3)
      IF (YP .GT. YM) GO TO 10
      KCD=1
      YM=YP
      GO TO 20
   10 KCD=2
      YM=YP
   20 DO 60 K=4,NPTM1
      YP=F(K)
      GO TO (30,50),KCD
   30 IF (YP .LE. YM) GO TO 40
      KCD=2
      NF=NF+1
      J=K-1
      FN=(J-1)*TS
      FJ=1./F(J)
      FJM1=1./F(J-1)
      FJP1=1./F(J+1)
      AA=FJM1-2.*FJ+FJP1
      AA=AA/(2.*TSS)
      BB=(FJP1-FJM1)/(2.*TS)
      CC=FJ
      PK=CC-BB*BB/(4.*AA)
      DD=CC-PK/2.
      GG=BB*BB-4.*AA*DD
      BW=-SQRT(GG)/AA
      FI=-BB/(2.*AA+FN)
      IX(J00+1)=FI
      IX(J00+2)=BW
      IX(J00+3)=434.*ALOG(PK)
      J00=J00+3
      NPEAK=NPEAK+1
   40 YM=YP
      GO TO 60
   50 IF (YP .LT. YM)  KCD=1
      YM=YP
   60 CONTINUE
      RETURN
      END
```

Fig. 7.2 Subroutine FINDPK for performing peak picking and interpolation on the spectrum $1/|A[\exp(j\theta)]|^2$.

for $k=0,1,\ldots,NPTS=N/2$. The sampling frequncy is FSK. The output is $NPEAK$, the number of peaks found for the analysis frame, and the array IX. This array stores in sequential order $\hat{F}_1, \hat{B}_1, \hat{P}_1, \hat{F}_2, \hat{B}_2, \hat{P}_2, \ldots, \hat{F}_{NPEAK}, \hat{B}_{NPEAK}, \hat{P}_{NPEAK}$, where $\hat{P}_i$ defines the interpolated peak amplitude in units of dB times 100.

It is important to note that interpolation of the spectrum $|A[\exp(j\theta)]|^2$ using minima picking will not produce equivalent results with interpolation on $1/|A[\exp(j\theta)]|^2$ using peak picking. In fact, interpolation on $|A[\exp(j\theta)]|^2$ can result in negative valued extrema leading to imaginary bandwidths. Although parabolic interpolation of the spectrum generally leads to wider bandwidth estimates than would be obtained by root-solving, results appear to be consistent and in addition, they are more efficiently computed if maximum accuracy is not necessary.

7.2.3 Examples of Raw Data

An example of raw data estimation using peak picking from the voiced utterance "We were away" from Markel [1972b] is shown in Fig. 7.3. The parameter values

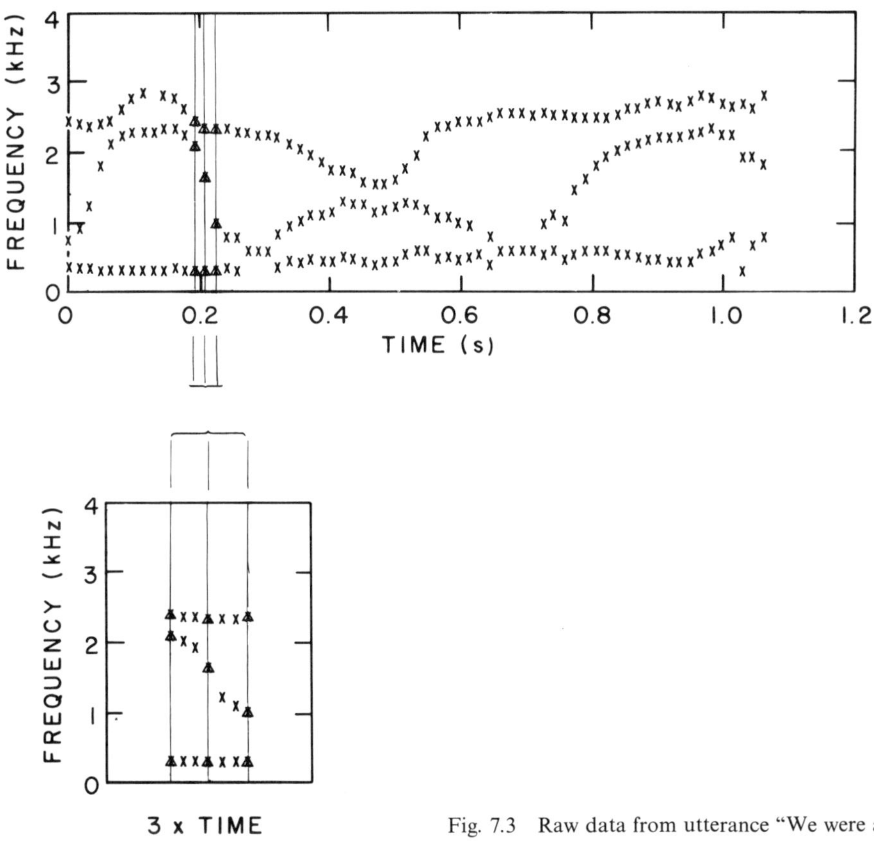

Fig. 7.3 Raw data from utterance "We were away".

are $M=14$, $N=256$, and $f_s=10$ kHz. A Hamming window and pre-emphasis factor of $\mu=1.0$ were also used, with the autocorrelation methods.

The raw data are plotted on a scale of peak frequency location versus time. With the knowledge that formant trajectories must be continuous and cannot intersect, it is quite easy to estimate the first three formant trajectories by inspection for voiced non-nasalized speech. It can be seen that an occasional omission of a peak occurs. There were no extraneous insertions (i.e., non-formant peaks) obtained for this example.

How easily the formant trajectories can be estimated is a strong function of the frame rate f_r. As f_r is increased, proportionally more data per unit time are obtained so that the effect of occasional omissions is minimized. In addition, more detailed behavior of the formant trajectories can be observed. To illustrate this point, an expanded resolution analysis was performed over the 32 ms time period shown in Fig. 7.3 by tripling f_r in this region. The results in the lower portion of the figure show the fast transitional second formant behavior in considerably more detail. The fastest rate of change of the trajectory is measured as -400 Hz/5 ms $= -80$ Hz/ms.

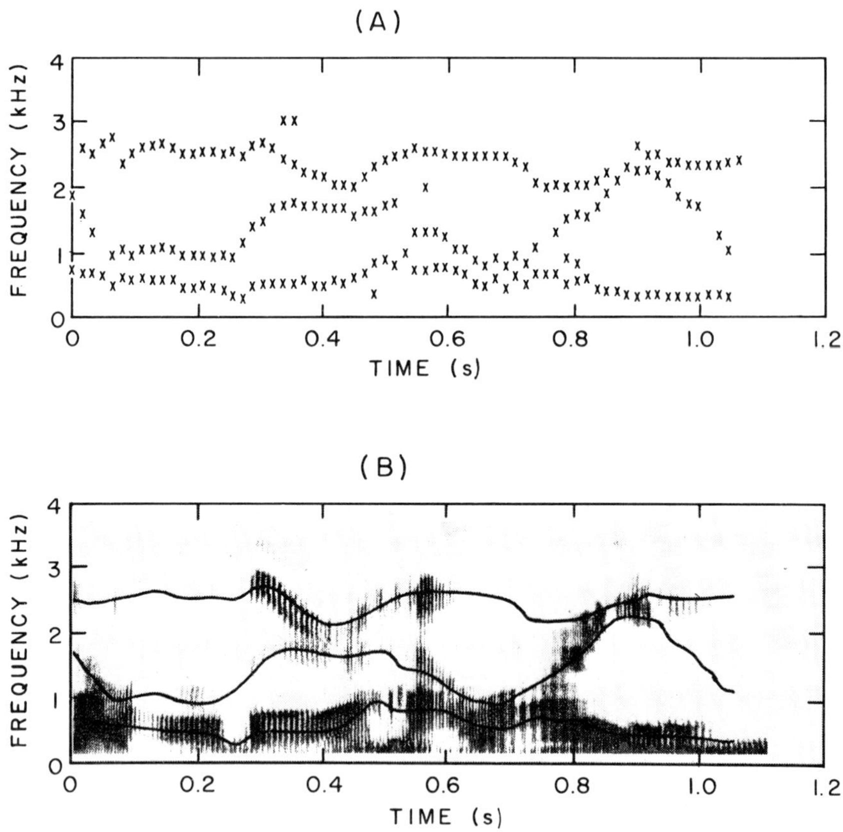

Fig. 7.4 Analysis of utterance "Hello there, how are you?" A) raw data. B) spectrogram with overlay of estimated formant trajectories from raw data.

It should be emphasized that simple minima picking of the inverse filter spectra (or peak picking of the reciprocal of the inverse filter spectra) will uniquely define the formant trajectories roughly 85–90 percent of the time. For the remaining frames, some kind of decision criteria will be necessary for automatic formant trajectory extraction.

Although the analysis model $A(z)$ results in an all-pole representation, formant frequencies can be reasonably well extracted with nasal and unvoiced sounds that contain antiresonance or zero behavior along with formant structure.

The phrase "Hello there, how are you?" from Markel [1971b] is shown with parameter values chosen as $N=256$, $M=15$, and $f_s=10$ kHz. The raw data are shown in Fig. 7.4A. An estimate of three non-intersecting continuous lines obtained by manual estimation of the formant trajectories from the raw data is superimposed on a sonogram of the utterance for comparison in Fig. 7.4B. Although there is a sharp transition in the interval (.24, .28) s, the raw data show the formants to be continuously tracked. In addition, very reasonable third formant estimates are obtained even though they are not apparent at several places on the sonogram.

In addition to the theoretical desirability of pre-emphasis, it also has practical value in extracting raw data from the spectrum. This is shown in Fig. 7.5A where the acoustic waveform for "linear" is shown along with the raw data using FINDPK. The analysis conditions were $N=192$, $M=15$, $f_r=0.156$ kHz, and $f_s=10$ kHz. Fig. 7.5B shows the results for no pre-emphasis ($\mu=0$), while Fig. 7.5C shows the results with $\mu=1$. Several observations can be made. First, with pre-emphasis, the F_1 is increased slightly and the peak locations appear less ragged. Secondly, the closely spaced F_2 and F_3 regions are more easily resolved by using the pre-emphasis. There is a rather sudden drop in the F_2 locations during the nasal /n/. With pre-emphasis, the transitional values on either side of the lowered F_2 trajectory are recovered. These same data were analyzed using the covariance method also. For the same analysis conditions excluding the Hamming window, nearly identical results were obtained for the peak locations. By lowering the window length to $N=50$ (5 ms) without performing pitch-synchronous analysis, somewhat erratic results were obtained with the covariance method. The formant estimation obtained by connecting the dots resulted in considerably rougher curves.

It is also possible to obtain reasonable formant bandwidth estimates, at least in terms of consistency, from both the autocorrelation and covariance methods. Results from analyzing the utterance "linear" with the autocorrelation method with $\mu=0$ and $\mu=1$ are shown in Figs. 7.6A and B, respectively. In these figures, a vertical line centered at the peak frequency is drawn which corresponds to the bandwidth value.

The most important observation is that the bandwidth estimates for B_1 are much more consistent when pre-emphasis is used and, in addition, are of smaller value. Even though the bandwidths are relatively large, pre-emphasis does result in estimates at the frames corresponding to the very rapid transition into and out of the nasal portion. Results obtained using the covariance method with similar analysis conditions, $f_s=10$ kHz, $N=192$, and $M=15$, but without a Hamming window were essentially equivalent to those just presented for both $\mu=0$ and $\mu=1$, and are therefore not shown.

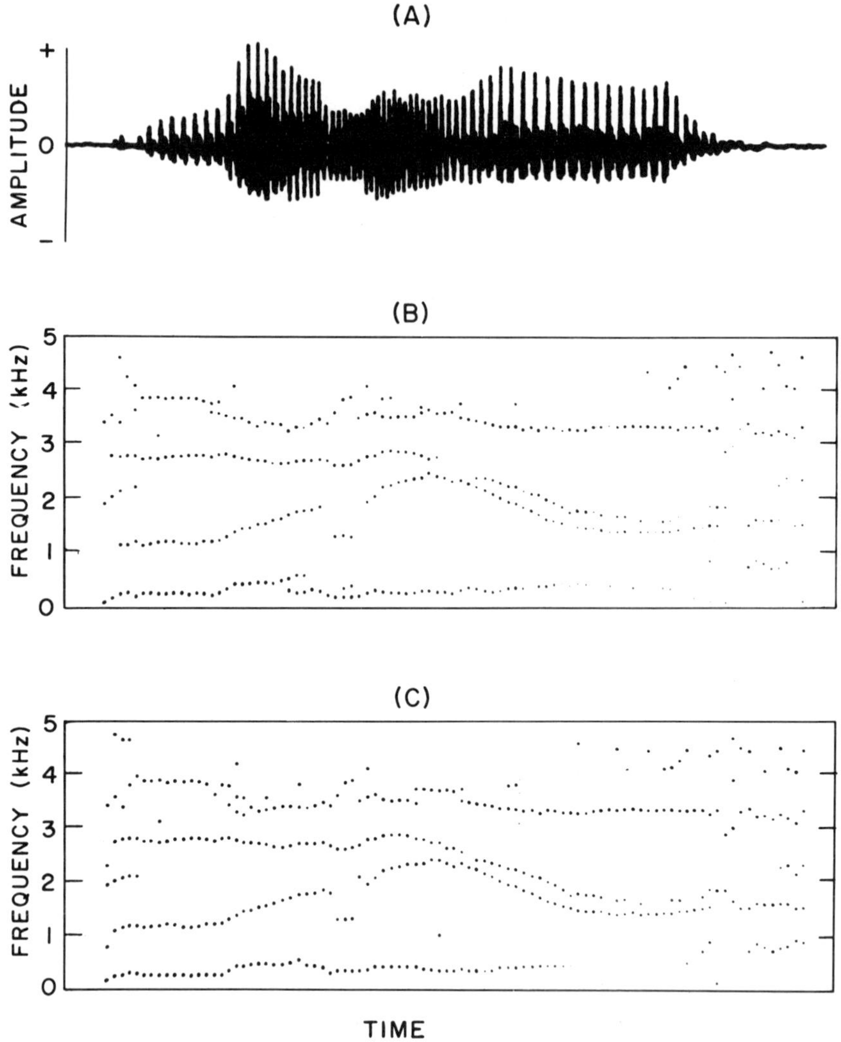

Fig. 7.5 Raw data from analysis of "linear". A) speech waveform. B) raw data with $\mu=0$. C) raw data with $\mu=1$.

7.3 Comparison of Raw Data from Linear Prediction and Cepstral Smoothing

In several respects formant estimation using linear prediction is quite similar to formant estimation using cepstral smoothing [Schafer and Rabiner, 1970]. In both approaches, frames of windowed speech data are transformed into

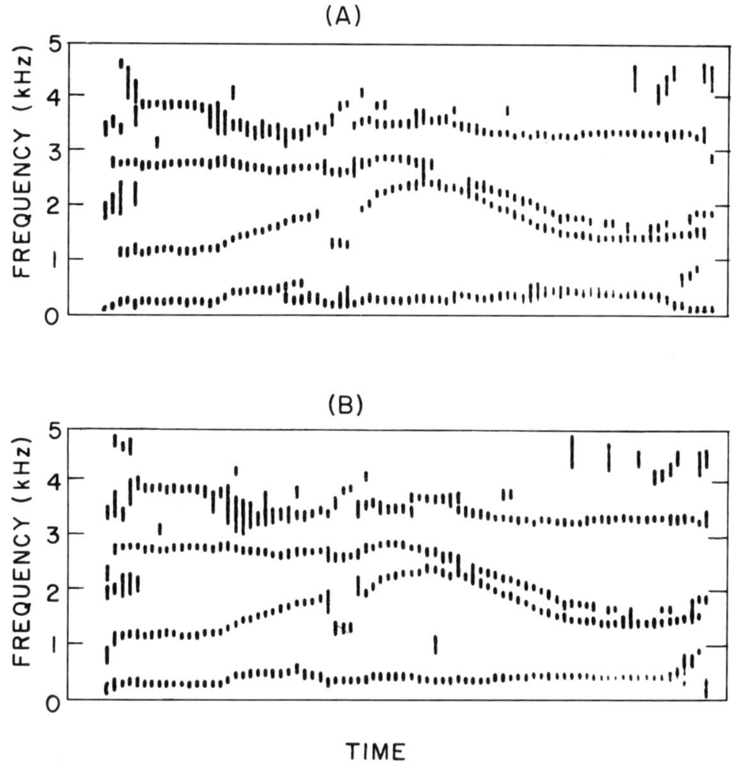

Fig. 7.6 Raw data from analysis of "linear" showing bandwidths A) $\mu=0$, B) $\mu=1$.

smoothed spectral representations which hopefully have the fundamental frequency associated with the fine-grain structure removed. From the smoothed spectra all peaks are picked as potential candidates for formant indicators. This set of peaks, in both algorithms, is referred to as the raw data. Since the cepstral smoothing algorithm has been demonstrated to produce accurate formant data both from a visual and perceptual point of view (by applying the data to a digital synthesizer) [Schafer and Rabiner, 1970], it is of interest to compare linear prediction results with cepstral smoothing algorithm results. In particular, the raw data are compared since the peak frequency locations from the smoothed spectra in both algorithms define the entire space of peak frequencies from which formant trajectories are estimated.

The significant result of this comparison is that, in general, the raw data from linear prediction will consist of a subset of the raw data from the cepstral analysis, namely the available formant trajectory markers plus a small percentage (10–15 percent) of extraneous, non-formant information. Although there are exceptions, such as missing data points, the exceptions are infrequent enough that additional decision criteria such as amplitude information and allowable ranges of the formant frequencies required in cepstral analysis can be omitted.

Another way of illustrating this point is to state that, by inspection of the raw data from linear prediction analysis, one can generally "connect the dots" to define the first three formant trajectories during voiced non-nasalized segments without reference to a spectrogram of the utterance, peak amplitude, or allowable formant range information. Formant trajectory estimation from the cepstral raw data without further decision criteria is generally not possible.

The above comments were illustrated by Markel [1971b] in the analysis of a voiced phrase taken from Schafer and Rabiner [1970] as an example of difficult situations (closely spaced formants and fast transitions). Instead of varying the Hamming window length as a function of the previous two pitch period estimates, it was fixed at 32 ms, which is approximately four times the average pitch period. A cepstral window of 4 ms duration was applied and the cepstrally smoothed spectra were obtained with 512-point FFTs. A frame rate of 62.5 Hz was used. The sampling frequency f_s was chosen as 10 kHz.

The raw data from analyzing the utterance "We were away" are shown in Fig. 7.7 from zero frequency to 3 kHz. The cepstral analysis data shown by the x's

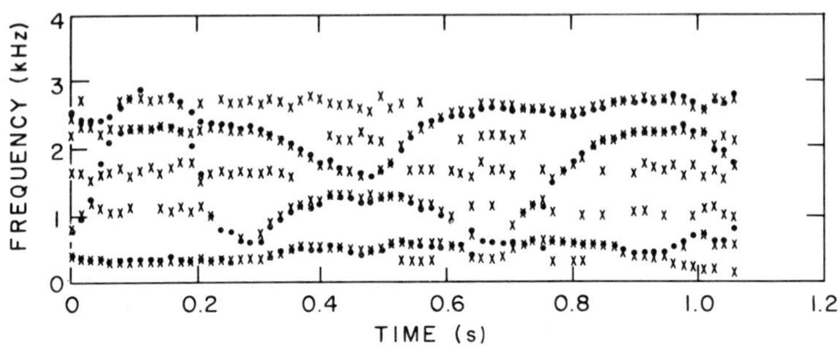

Fig. 7.7 Analysis of "We were away". Raw data from linear prediction analysis (dots) are superimposed upon cepstral smoothing results (*x's*).

are overlaid on the linear prediction data shown by the dots. By connecting the dots to estimate the three formant trajectories it is seen that, in essence, the linear prediction data are a subset of the cepstral analysis data — namely, they are very nearly the subset of cepstral data that correspond to the true formant frequency markers.

It is also interesting to compare the spectra from which the raw data are picked for the two cases. The left-hand side of Fig. 7.8 shows the results from cepstral analysis and the right-hand side shows the results from linear prediction analysis. Frames of data corresponding to a total time interval of 512 ms are presented.

A cepstrally smoothed spectrum is precisely a weighted least-squares approximation to the discrete log magnitude spectrum. Therefore, independent of whether or not a perturbation in the spectrum corresponds to a formant, the cepstral representation will follow the log magnitude spectrum (including all the fine grain structure) as best it can, based upon the number of terms in the series. For reasonable formant extraction, a cepstral window width of approximately 4 ms has been suggested [Schafer and Rabiner, 1970]. Therefore, it is possible to have, at

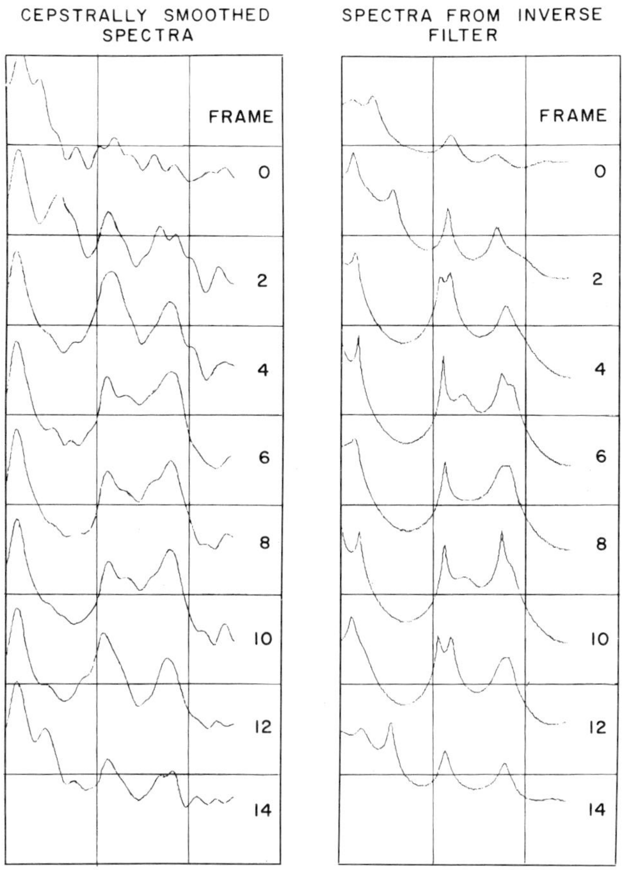

Fig. 7.8 Comparison of cepstrally smoothed spectra and inverse filter spectra. Vertical scale = 40 dB per division. Horizontal scale = 2 kHz per division.

most, oscillatory components spaced every 1/4 ms = .250 kHz in the frequency domain. In the region $(0, f_s/2)$ it is possible to find at most $Wf_s/2$ peaks where W is the cepstral window width. For $W=4$ ms and $f_s = 10$ kHz, 20 peaks are possible. Note, however, that the number of peaks is strongly dependent upon the spectral character of the speech wave since a least-squares fit is applied to the log magnitude spectrum. Furthermore, the number of peaks is strongly dependent upon the type of cepstral window used. The main reason for applying a smoothly decreasing low-time window function to the cepstrum is in fact to reduce the number of undesired local maxima. For Fig. 7.8 an average of 7.3 peaks per frame was obtained out to the half sampling frequency. For comparison purposes, a better figure of merit is the average number of peaks per frame within (0,3000) Hz. For the cepstral smoothing procedure, an average of 4.15 peaks was measured, while for the linear prediction procedure an average of 3.02 peaks was obtained. Based upon physical data it is expected that three true formant peaks would exist in this interval.

The ease with which it is possible to manually define the formant trajectories by inspection is due to the fact that the raw data (in the range 0 to 3 kHz for male speakers) specify the formant data with relatively few exceptions. In addition, one can observe the data globally before deciding on the best three trajectories. The only constraints are that the curves cannot intersect, and that each curve must be relatively smooth due to the physiological relationships between the vocal tract movement and formant frequencies. To simulate on the computer what is trivial for the human observer requires, in general, pattern recognition with embedded constraints. A much simpler approach to automatic analysis is to separate the problem into two parts. First the trajectories are estimated using only local information to satisfy continuity, then smoothing is applied to ensure that a maximum rate of change of the trajectories is not exceeded. One additional constraint of importance is a maximum range for the first three formant trajectories.

7.4 Algorithm 1

A relatively simple scheme for automatically extracting formant trajectories from the raw spectral data has been presented by Markel [1972a]. It is computationally efficient and uses only the previous frame information. The simplicity is based upon the restriction that F_3 is assumed to satisfy $F_3 \leq 3$ kHz. Although this is a reasonable assumption for most male voices, it is not valid for most female or children's voices. Let N_p define the number of peaks obtained for the present frame k being analyzed. A nearest neighbor criterion is defined by

$$v_{ij} = |\hat{F}_i(k) - F_j(k-1)| \tag{7.6}$$

where $F_j(k-1)$ is the jth formant frequency defined in the previous frame $k-1$ and $\hat{F}_i(k)$ is the ith raw data frequency estimate for frame k. The term $L \leq N_p$ defines the number of peaks which satisfy a bandwidth threshold such as $\hat{B}_i(k) < 500$ Hz. Depending upon L, different strategies are used. If $L=3$ (approximately 85–90 percent of the time) a direct assignment of the raw data to formant data via $F_i(k) = \hat{F}_i(k)$ is made. If $L=1$ (less than 1 percent of the time) $\hat{F}_i(k)$ is assigned to the nearest value $F_j(k-1)$. The remaining two formant slots are filled by the corresponding values from the previous frame. The case $L>4$ will generally not occur if M has been reasonably chosen using the criteria set forth in Chapters 4 and 6. The cases $L=2$ and $L=4$ (corresponding to one omission and one insertion of extra peaks, respectively) occur 10–15 percent of the time and therefore require the major attention.

The general procedure for $L=2$ is to define the two formant slots from the nearest neighbor criterion and then fill the third slot with the previous frame's corresponding formant value. For example, with $L=2$, if $v_{11} < v_{12}$, $\hat{F}_1(k)$ is a nearest neighbor to $F_1(k-1)$. Otherwise, it is a nearest neighbor to $F_2(k-1)$. If $|\hat{F}_2(k) - \hat{F}_1(k)| < 500$ Hz, then an assignment $F_i(k) = \hat{F}_i(k)$, $i=1,2$ is made. Then $F_3(k)$ is assigned to $F_3(k-1)$. For $L=4$ the general procedure is to assign three out of

```
C
      SUBROUTINE FORMNT(IX,IFF,IF1,IUV,IUV1,NPEAK)
C
C     THIS ROUTINE ESTIMATES THE FORMANTS
C     FROM THE RAW DATA
C
      DIMENSION IP(8),IX(1),IFF(1),IF1(1)
C
      IV(I,J)=IABS(IP(I)-IF1(J))
      IF (NPEAK.EQ.0) GO TO 1
      IHLF=500
C
C     IS FRAME VOICED?
C
      IF (IUV.EQ.1) GO TO 10
    1 IFF(1)=0
      IFF(2)=0
      IFF(3)=0
      GO TO 200
   10 L=0
C
C     OBTAIN PEAKS FOR THIS FRAME
C
      DO 20 J=1,NPEAK
      I=3*J-2
      IF (IX(I)  .GT.3000)  GO TO 20
      IF (IX(I)  .LE.   0)  GO TO 20
      IF (IX(I+1).GT. 500)  GO TO 20
      L=L+1
      IP(L)=IX(I)
   20 CONTINUE
C
C     DETERMINE F1-F3 FOR PRESENT FRAME
C
      IF (L.EQ.3) GO TO 30
      IF (L.EQ.1) GO TO 50
      GO TO 90
C
C     WHEN L=3
C
   30 DO 40 J=1,3
   40 IFF(J)=IP(J)
      GO TO 200
C
C     WHEN L=1
C
   50 DO 60 J=1,3
   60 IFF(J)=IF1(J)
      IF (IV(1,1).GT.IV(2,1)) GO TO 70
      IFF(1)=IP(1)
      GO TO 200
   70 IF (IV(3,1).GT.IV(2,1)) GO TO 80
      IFF(2)=IP(1)
      GO TO 200
   80 IFF(3)=IP(1)
      GO TO 200
C
C     WHEN L=2 OR L>3
C
   90 IF (IV(1,1).LT.IV(1,2)) GO TO 110
      IF (IP(2)-IP(1).LE.IHLF) GO TO 100
      ICD=2
      IFF(2)=IP(1)
      IFF(1)=IF1(1)
      GO TO 160
  100 IFF(1)=IP(1)
      IFF(2)=IP(2)
      ICD=3
      GO TO 160
  110 IFF(1)=IP(1)
      IF (L.EQ.2) GO TO 140
      IF (IV(2,2).GT.IV(3,2)) GO TO 120
      IF (IV(2,2).LT.IV(2,3)) GO TO 130
  120 IFF(2)=IP(3)
      IFF(3)=IP(4)
      GO TO 200
  130 IFF(2)=IP(2)
      ICD=3
      GO TO 170
  140 IF (IV(2,2).LT.IV(2,3)) GO TO 150
      IFF(3)=IP(2)
      IFF(2)=IF1(2)
      GO TO 200
  150 IFF(2)=IP(2)
      IFF(3)=IF1(3)
      GO TO 200
  160 IF (ICD.GT.L) GO TO 190
      IF (ICD.EQ.L) GO TO 180
  170 IF (IV(ICD,3).LT.IV(ICD+1,3)) GO TO 180
      IFF(3)=IP(ICD+1)
      GO TO 200
  180 IFF(3)=IP(ICD)
      GO TO 200
  190 IFF(3)=IF1(3)
  200 RETURN
      END
```

Fig. 7.9 Subroutine FORMNT for performing formant estimation on raw data.

four of the raw data locations based upon a nearest neighbor criterion. A Fortran subroutine FORMNT for implementing the formant estimation procedure is given in Fig. 7.9. For each set of raw data FORMNT is called to make the explicit formant trajectory estimates and assignments. The input array IX holds the raw data frequencies, bandwidths, and amplitudes in the same order as described for FINDPK. A reasonable bandwidth threshold value for $IBWT$ is 500 Hz since the first three formant bandwidths are expected to be physically less than 200 Hz, while linear prediction bandwidth estimates may be in error by as much as a factor of 2.5. At completion, IFF contains the formant frequency estimate for frame k, while $IF1$ contains the formant frequency estimates for the previous frame. IUV and $IUV1$ contain the voicing information about the present and previous frame, respectively. As initial conditions, if $L=3, F_i(0)\equiv\hat{F}_i(0)$, $i=1, 2, 3$. If $L\neq 3$, neutral vowel positions $\{F_i\}=\{.5, 1.5, 2.5\}$ kHz can be used.

An example of this algorithm is shown in Fig. 7.10. The acoustic waveform of the utterance "linear prediction" is shown in Fig. 7.10A. The raw data over this corresponding time interval are shown in Fig. 7.10B. The autocorrelation method

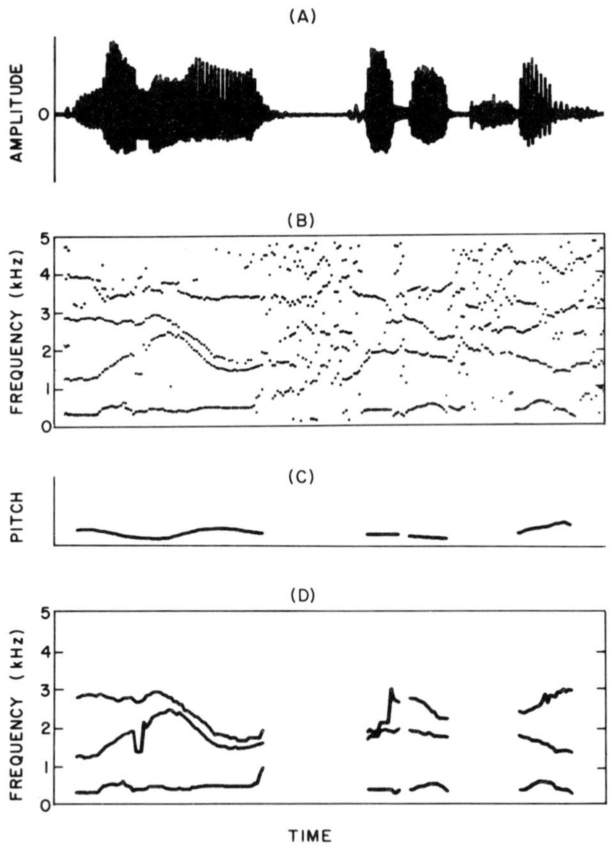

Fig. 7.10 Automatic formant trajectory estimation for the utterance "linear prediction". A) acoustic waveform. B) raw data. C) pitch and voicing function. D) formant trajectories using subroutine FORMNT.

was used with $N=192$, $M=15$, $\mu=0.9$, $f_r=0.156$ kHz, and $f_s=10$ kHz. During periods of silence (between "linear" and "prediction" and at the end of the utterance), rather random peak behavior is observed. In Fig. 7.10C, the pitch contour automatically extracted from this utterance (using methods described in the following chapter) is shown. Where pitch equals zero, unvoiced portions are defined. The automatically defined formant trajectories obtained by applying subroutine FORMNT to the raw data are shown in Fig. 7.10D.

It is seen that the relatively few problem areas such as insertions of extraneous peaks and the omissions due to close second and third formants have, with few exceptions, been correctly resolved. The major error would appear to be in the estimation of the third formant frequency of the second voiced segment. A rather sharp discontinuity is obtained due to the lack of raw data in the third formant region.

The rather sharp F_2 dip in "linear" during the nasal /n/ portion exists within the data as can be verified by referring to the spectrogram of the utterance in Chapter 1. In fact, these results very nicely illustrate the ability of linear prediction techniques to complement what is observable from spectrograms. During the /n/ portion, for example, the spectrogram shows that F_2 has been lowered, but it is not possible to accurately estimate its value. During the voiced plosive /d/, a lowering of F_1 is clearly seen in the linear prediction raw data estimates, whereas in the spectrogram it is possible to see only that nearly all spectral energy resides in the first formant region. The lowering is not seen in Fig. 7.10D since the /d/ was defined as unvoiced in the pitch extraction algorithm.

Several additional points should be emphasized. The algorithm as stated is for the analysis of adult male speech only, where the first three formant frequencies are assumed to lie within 3 kHz. For about 85–90 percent of the time $L=3$; thus no decisions are necessary to automatically obtain the formants for frame k. In addition, a forced choice is made in the algorithm so that three and only three formants will be defined at every frame where voicing occurs. For speakers having a third formant value greater than 3 kHz, this algorithm will perform poorly due to the direct assignment of peaks to formants when $L=3$.

The voicing parameter is also very important for accurate formant estimation, since any frame being defined as unvoiced will leave a gap in the trajectory. Without accurate voicing information it is preferable to set $IUV=1$ for all frames.

The procedure is similar to that presented by Schafer and Rabiner [1970] with three important exceptions: 1) minimum and maximum ranges for each formant are unnecessary, 2) amplitude information is unnecessary, and 3) corrections for $L\neq 3$ are applied only 10 or 15 percent of the time. These simplifications are possible because of the small percentage of non-formant peaks that exists in the reciprocal inverse filter spectra.

After the trajectories within a voiced region have been defined, they can be individually smoothed. There are three constraints which must be simultaneously satisfied to correctly smooth or filter the formant trajectories. The bandwidth must be adequate to pass the maximum rate of change expected by articulator movement but small enough to filter out extraneous perturbations due to occasional gross errors in the formant tracking algorithm. The filter should have either zero or known linear phase characteristics so that the trajectories are not distorted with respect to a fixed time reference. Finally, the filter should be critically damped with respect to a unit step input. Otherwise, perceptually significant overshoot can be obtained when fast transitions are analyzed.

In performing pitch-asynchronous analysis, it is important to filter the trajectories, as they do indicate some amount of random behavior about a smooth curvature as would be expected from physiological constraints. The cause of the slight irregularities is the fixed frame analysis rate and window conditions. Several periods may be placed at varying locations within any window. Pitch-synchronous analysis may resolve this problem. Unfortunately, correct pitch-synchronous analysis is much more difficult to perform automatically. A smoothing algorithm such as described in the next section will successfully smooth the irregularities from the data.

One of the most important advantages of this approach to formant extraction is that very few decisions are generally necessary to determine $F_1 - F_3$ for a particular frame of data. For approximately 85–90 percent of the time, formant frequency assignment is trivial. For the remainder of the time, the procedure is quite simple, requires memory of only the previous frame, and is thus applicable to on-line computation. An important feature of this procedure is that for each frame a forced choice decision of $F_1 - F_3$ is obtained.

The formant tracking procedure described has been used successfully in most analyses (in the sense that synthetic speech generated from the analyzed formant trajectories compares favorably with the original speech). There are, however, several problems which occasionally occur that can be improved upon by more sophisticated decision rules. One infrequent problem occurs when one non-formant peak is measured and one formant peak is suppressed, thus resulting in $L=3$. This condition can occur in a sound such as /i/ when considerable separation exists between F_1 and F_2, and F_3 happens to be slightly above the maximum frequency range searched. Since the procedure presented here is to assign $F_i(k) = \hat{F}_i(k)$, a gross error would be made. A second obvious problem exists for the analysis of voices in which F_3 often exceeds the maximum frequency range. Even though this situation is rather infrequent for male speech, it has been encountered. It is also certainly reasonable to consider the removal of the somewhat artificial maximum frequency range limit. If this is done, however, a more complex decision algorithm is necessary.

7.5 Algorithm 2

In this section, a somewhat more sophisticated formant tracking algorithm developed by McCandless [1974a] is presented. This algorithm also estimates the frequencies of the first three formants based upon linear prediction spectra. Formant slots are filled with the available spectral extrema, based upon relative frequency locations to a reference. Processing begins by estimating an *anchor point* (the center of a vowel segment where formant estimates are most likely to be correct), and then branching outward in both directions, using the most recent formant frequency estimates as the next reference. Frequency evaluation within the unit circle is then used where necessary to enhance potential formant peaks.

7.5.1 Definition of Anchor Points

To define the anchor points, voiced intervals must be separated from unvoiced intervals. This task can be accomplished by using any reasonably accurate pitch extraction algorithm (see Chapter 8). Since it is possible to have rather long voiced intervals with considerable formant frequency variation (e.g., "We were away" spoken in a normal conversational manner), each voiced interval is further subdivided into high and low energy categories on the basis of the total energy E_T and the energy E_R in the region 640 – 2880 Hz. The energies E_T and E_R are computed for each frame in the voiced interval and then the peaks and valleys of the functions E_T and E_R versus the frame index or time of occurrence are computed. If the peak to valley ratio is sufficiently high for either waveform, then subdivision boundary marks are placed between the valley and two surrounding peaks at the locations where the slope of the corresponding energy function is greatest.

7.5.2 Processing of Each Voiced Segment

Figure 7.11 is a flow chart of the anchor point scheme for processing each voiced segment. Processing of the backward branch is begun at the next anchor point and

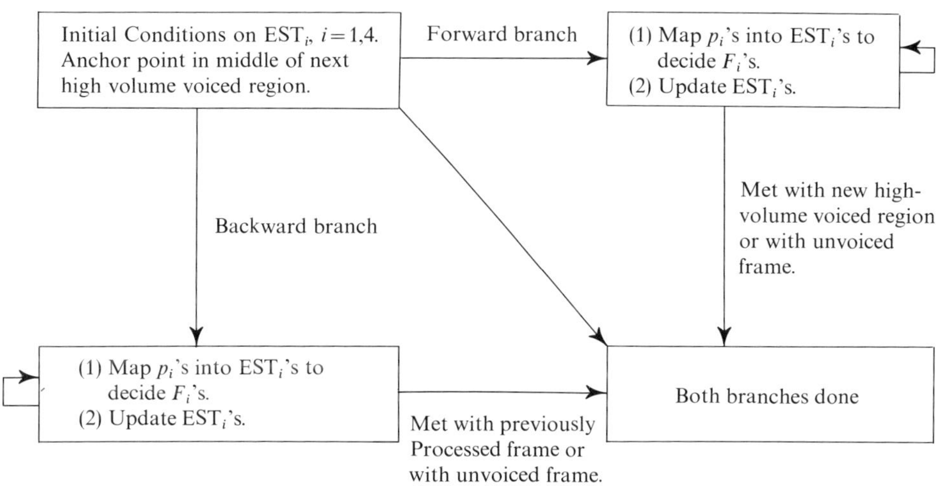

EST_i = Frequency estimates for four formants in current frame
p_i = Frequency locations of peaks in current frame
F_i = Formant frequencies decided in current frame

Fig. 7.11 Flow chart of anchor-point scheme. [After McCandless, 1974a]

continued until an unvoiced frame is encountered, or until a frame is encountered which has already been processed by the previous forward branch. Then the forward branch from the same anchor point is begun and continued until an unvoiced frame is encountered, or until a new subdivision boundary is reached. At this point, processing jumps to the next anchor point, begins again with a backward branch, and so forth, until the processing is complete.

Anchor points are initially defined to be the middle of each vowel segment. However, if the original anchor was found to be a poor choice, a new anchor can be selected according to the following scheme [McCandless, 1975]. If, within 25 ms of a given anchor point, the algorithm is unable to fill any of the first three formant slots, reset the anchor at the frame where the problem appeared, and proceed from there. The assumption being made is that a formant track which has disappeared is probably not really a formant. Anchors are only rarely reset in this manner, but this procedure alleviates the problem of false nasal formants most of the time.

The raw data are obtained from peaks of the spectrum $|1/A[\exp(j\theta)]|^2$. The six steps for estimating formant frequencies from the raw data in each voiced frame are as follows.

Step 1: Fetch Peaks. Find the frequencies and amplitudes of up to four peaks in the region from 150 to 3400 Hz.

Step 2: Fill Slots. Fill each formant slot S_i, $i=1$ to 4, with the best candidate F_j by placing the peak F_j closest in frequency to the estimates EST_i into slot S_i.

Step 3: Remove Duplicates. If the same peak F_j fills more than one slot S_i, keep it only in the slot S_k which corresponds to the estimate EST_k closest in frequency, and remove it from any other slots.

Step 4: Deal with Unassigned Peaks. If all frequencies F_j have been assigned to formant slots, go to Step 5. Otherwise, try to fill empty slots with values not assigned in Step 2 as follows.

a) If there is an unassigned value p_k, and an unfilled slot S_k, fill the slot with the peak and go to Step 5. If p_k is unassigned, but slot S_k is already filled, check the amplitude (amp) corresponding to p_k as follows: if amp $(p_k) < 1/2$ amp (peak already assigned to S_k) throw p_k away and go to Step 5. Otherwise, go to (b).

b) If p_k is still unassigned, but S_{k+1} is unfilled, move the peak in S_k to S_{k+1}, and put p_k in S_k. Go to Step 5.

c) If p_k is still unassigned, but S_{k-1} is unfilled $(k>1)$, move the peak in S_k to S_{k-1}, and put p_k in S_k. Go to Step 5. If a), b), and c) all fail, ignore p_k.

Step 5: Deal with Unfilled Slots. If S_1, S_2, and S_3 are all filled, go to Step 6. (F_4 may or may not be filled). Otherwise, recompute the spectrum on a circle with radius less than unity to hopefully separate merged peaks. Go to Step 1 to recompute the spectrum. Otherwise, go to Step 6.

Step 6: Update Estimate. Accept formant slot contents as formant estimates for this frame, i.e., $F_i = S_i$, $i=1, 2, 3$. Also, use formant slot contents as estimates for next frame, i.e., $EST_i = S_i$, $i=1, 2, 3, 4$. (If a slot is empty, keep the original formant estimate for that formant).

At the anchor point, Steps 1–6 cannot be applied unless formant estimates are available. In the original algorithm, the EST_i were set at fixed initial conditions, determined empirically. A modified version, with improved performance, uses the actual peak locations of the enhanced spectrum (r = .95) at the anchor point to determine initial conditions [McCandless, 1975]. The four EST_i slots are filled with peak values p_j by choosing the best candidate in an acceptable region for each formant (as shown in the flow chart of Fig. 7.11).

The only reason for including S_4 and EST_4 is to prevent F_4 when it does exist from competing with F_3 for the F_3 slot. The system sampling frequency is 10 kHz and 6 dB/octave pre-emphasis is used before sampling. (If all digital operation is desired the sampled data can be passed through a prefilter $1 - z^{-1}$). The autocorrelation method is applied to the sampled data after applying a Hamming window. The analysis frame length is defined as 256 samples (25.6 ms). A new analysis frame is defined every 50 samples (5.0 ms).

If spectral enhancement is required in the algorithm, the initial radius r is defined as .98. If the spectrum fails to yield a peak to fill the empty slot, then Steps 1 – 5 are repeated in this manner until a peak is finally found or until r = 0.88, at which point it is assumed that no peak exists to fill the empty slot.

The procedure for the formant enhancement can be quite time-consuming if a radius reduction of only 0.004 is used each time. An alternate procedure is to solve for the roots of the polynomial $A(z)$ in the cases where spectral enhancement is required since all resonances, independent of bandwidth or appearances of spectral merging, will then be obtained.

The amplitudes of the peaks are always reset to the amplitudes in the original spectrum. In addition, if the empty slot was S_3, and enhancement failed to yield a peak, then the peak in S_4 is moved down to S_3; i.e., it is assumed that F_3 was mistakenly called F_4.

7.5.3 Final Smoothing

There will still be the possibility that a formant slot has not been filled or that the formant value is grossly out of line for one or several frames. The final smoothing algorithm for resolving these situations is given below:

Step 1: If a single formant slot is empty, fill in its frequency and amplitude with the average of the values in the previous and following frames.

Step 2: If a formant is grossly out of line or missing in one, two, or three frames, but well aligned in the two previous and two following frames, correct the misaligned frames by interpolation as follows.

Let the frequency location of a formant in the nth frame be L_n.

Define $D_{a,b} = L_a - L_b$, a measure of the alignment of a particular formant. Also a threshold θ is defined as 240 Hz. If $D_{n,n-1} < \theta$, frame n is considered smooth. If $D_{n,n-1} > \theta$, an attempt is made to smooth frame n, but only if either a), b), or c) is true.

a) If $D_{n-1,n-2} < \theta, D_{n+1,n-1} < \theta$, and $D_{n+2,n+1} < \theta$, then replace L_n with $(L_{n+1} + L_{n-1})/2$, and move to frame $n+1$ (One frame out of line).

b) If $D_{n-1,n-2} < \theta, D_{n+2,n-1} < \theta$, and $D_{n+3,n+2} < \theta$, then replace L_n with $(L_{n+2} + L_{n-1})/2$, and move to frame $n+1$ (Two frames out of line).

c) If $D_{n-1,n-2} < \theta, D_{n+3,n-1} < \theta$, and $D_{n+4,n+3} < \theta$, then replace L_n with

$(L_{n+3}+L_{n-1})/2$, and move to frame $n+1$ (Three frames out of line).
The new L_n is used in evaluating frame $n+1$.

Step 3: Smooth each formant track twice using the following filter:

$$F_i'(n)=1/4\,F_i(n-1)+1/2\,F_i(n)+1/4\,F_i(n+1),$$

but only at those frames where $|F_i'(n)-F_i(n)|<100$ Hz.

7.5.4 Results and Discussion

This algorithm is quite successful in handling formant merging problems. McCandless reports that enhancement was applied in about 15 percent of the voiced frames and was 90 percent successful. Difficulties still occur in the analysis of nasalized vowels due to pole-zero interaction. Unfortunately, at present there does yet not seem to be any workable procedure for resolving these kinds of problems. Several examples of this algorithm are shown in Fig. 7.12. The photographs labeled A and B correspond to the data and formant estimates, respectively. In "they are one" it is seen that several formant mergers occur along with several extraneous peaks where F_2 and F_3 are widely separated. The algorithms appear to resolve these difficulties quite satisfactorily. Similar results are seen for "heard" and "I want".

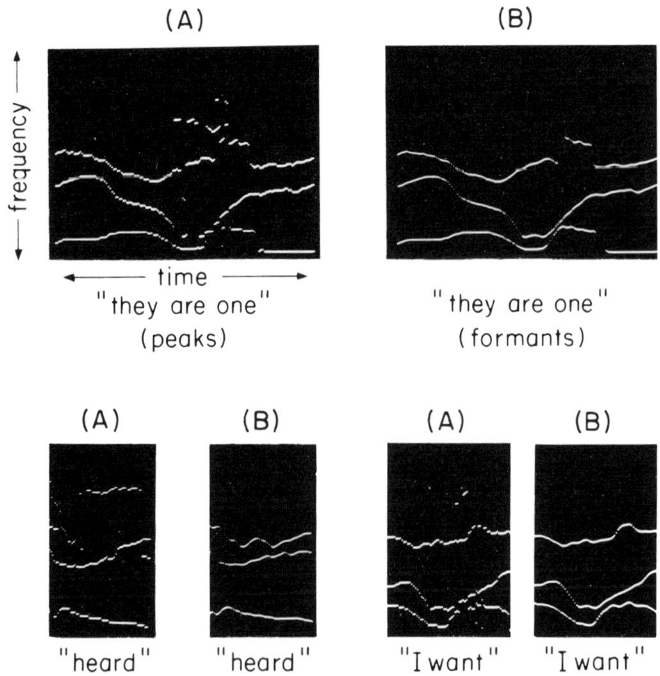

Fig. 7.12 Formant tracking examples. A) raw data. B) formant data.

The utterance "linear prediction" was analyzed with results as shown in Fig. 7.13. The upper part of the trace is the output of the segmentation system [Weinstein, et al., 1975a]. The highest and lowest levels are defined as vowel and silence segments respectively. The small markers in the middle of each vowel segment define the anchor points.

As compared to the algorithm of the previous section, this algorithm results in essentially error-free smoothed formant trajectories. The ending of "linear" is seen to be very smooth without gross errors. Furthermore, the third formant movement appears more like what would be expected for the phoneme combination /ri/ of p*re*diction.

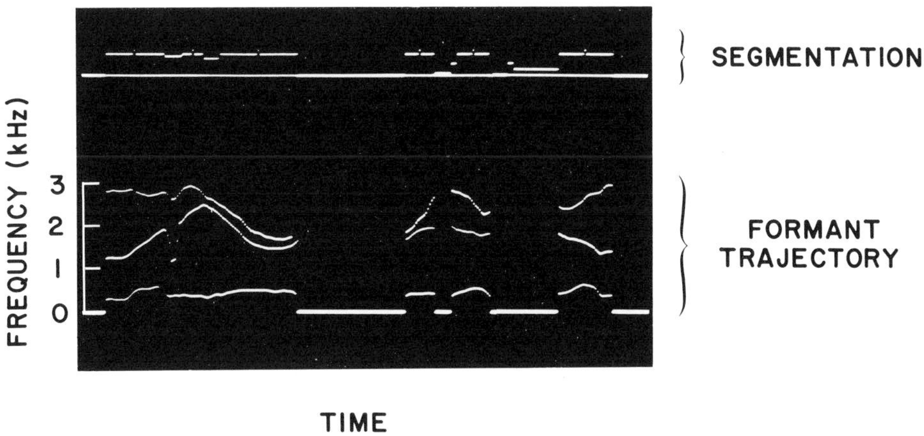

Fig. 7.13 Automatic formant trajectory estimation for the utterance "linear prediction".

7.6 Formant Estimation Accuracy

It is of considerable interest to know, in an absolute sense, how accurately formant parameters can be estimated. Based upon the theoretical development of Chapters 2 and 3, it would be expected that the covariance method would produce better results than the autocorrelation method if single pitch periods are analyzed and if the speech sample can be accurately modeled within the analysis interval as being of complex exponential behavior. Chandra and Lin [1974] have performed a detailed experimental comparison of the covariance and autocorrelation methods for both pitch-synchronous and asynchronous conditions with synthetic and real speech. Several examples from their work are now presented.

7.6.1 An Example of Synthetic Speech Analysis

A synthetic speech example from Chandra and Lin [1974] is shown in Fig. 7.14. A synthetic vowel /i/ as in h*ee*d was generated from a digital vowel synthesizer.

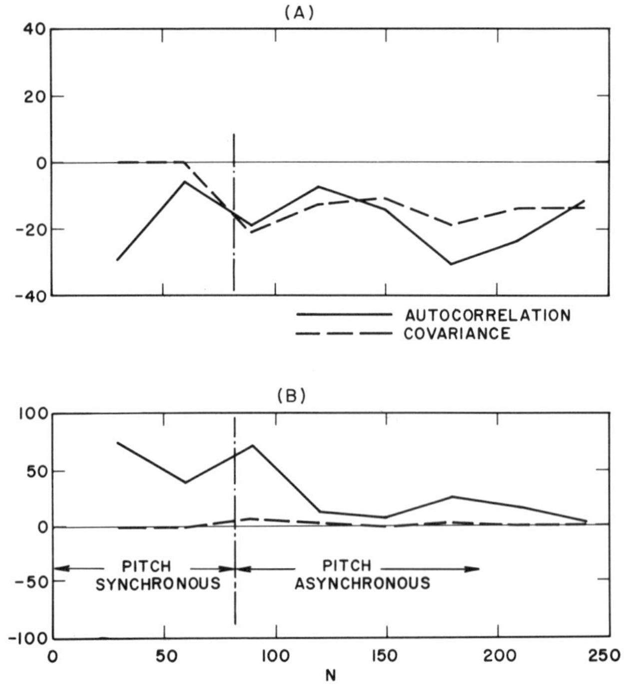

Fig. 7.14 Estimation errors in analysis of synthetic vowel. A) F_1 estimation versus N. B) B_1 estimation error versus N. [After Chandra and Lin, 1974]

The synthesizer used 5 complex poles for the formants and 2 real poles for the glottal and lip radiation source. The driving function to the synthesizer was a series of unit samples every 83 samples, and the sampling frequency was 10 kHz. Since 12 complex exponentials exist in the model (including the 2 real poles), the filter length M was chosen as 12. The starting position of the analysis coincided with an input unit sample pitch pulse. The number of samples N used in the analysis was increased from 30 to 240 in steps of 30.

The figure shows the estimation error for both the first formant frequency and bandwidth as a function of N. For $N < 83$ the analysis is within a single pitch period and corresponds to pitch-synchronous analysis. For $N > 83$ the analysis window exceeds one period and is therefore pitch-asynchronous analysis. For pitch-synchronous analysis the covariance method results in zero error for both the formant frequency and bandwidth. As remarkable as this may at first seem, it is a simple consequence of the basic property of Prony's method for complex exponential representation discussed in Chapter 2. The autocorrelation method results are relatively poor for small numbers of samples but then begin to stabilize when greater than 2 pitch periods (166 samples) are analyzed. The bandwidth estimates are very stable for all N using the covariance method with only small errors in the estimation using the autocorrelation method. It is important to note that as soon as the analysis interval exceeds one pitch period, the covariance and autocorrelation methods give rather similar errors in formant frequency estimation. These results very nicely illustrate the fact that if the speech has strictly complex exponential behavior within a single pitch period, then very accurate formant estimation is possible with the covariance method.

The advantage to studying synthetic speech is that all the parameter values are known. The disadvantage is that the results are meaningful only if the synthesis model output closely resembles real speech.

7.6.2 An Example of Real Speech Analysis

A representative example of real voiced speech analysis by Chandra and Lin [1974] is shown in Fig. 7.15. The pitch period was estimated as 68 samples and 12 coefficients were used for $N \leq 60$. The Hamming window was not used in the autocorrelation method. The graph shows the normalized squared error $0 \leq \eta = \alpha_M/\alpha_0 \leq 1$ as a function of N for both methods. In Chapter 3 it was shown that $\alpha_M = 0$ or equivalently, the inverse filter norm $\|A(z)\| = 0$, only if the complex exponential behavior was exactly represented. The non-zero behavior of η for both methods illustrates the fact that the real speech waveform is not precisely composed of 12 complex exponentials. Although knowledge of η is insufficient to determine absolute formant accuracy, it is certainly clear from the results that the covariance method is superior to the autocorrelation method out to $N \leq 180$. Past this point, both methods will produce rather similar results. The pitch period influence on the autocorrelation method estimates is rather dramatically shown in the graph. Relatively large errors are obtained when the analysis window is near

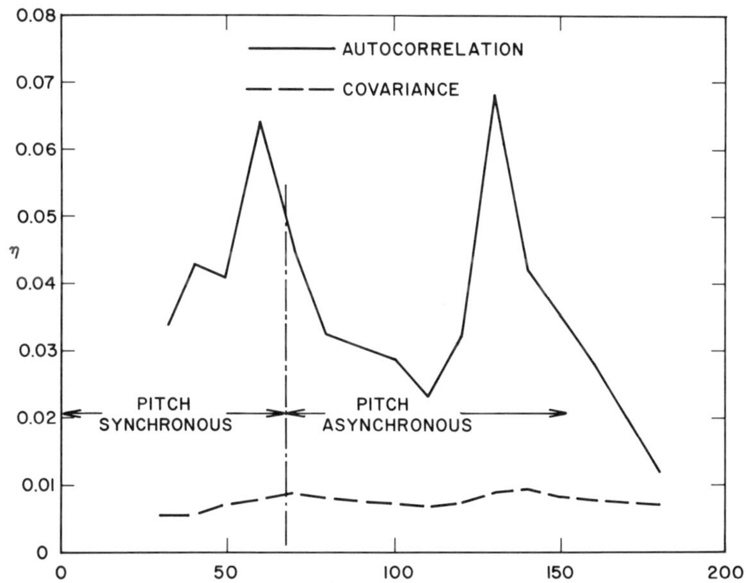

Fig. 7.15 Normalized squared error for real speech. [After Chandra and Lin, 1974]

one or two pitch periods in length. It is important to keep in mind, however, that even the maximum error is around 0.07 with respect to a normalized error of unity.

7.6.3 Influence of Voice Periodicity

Atal [1974b] has also studied the influence of voice periodicity on formant frequency and bandwidth estimation. Periodic signals composed of one formant (a complex pole pair) having a varying center frequency and a fixed 50 Hz bandwidth were synthesized and then analyzed using linear prediction. The total squared error was minimized over exactly one period. The estimated formant frequency F_1 and bandwidth were obtained by solving for the roots of $A(z)$. The results for the formant frequency estimation are shown in Fig. 7.16, where the fundamental frequency F_0 takes on three different values. The maximum errors were reported as

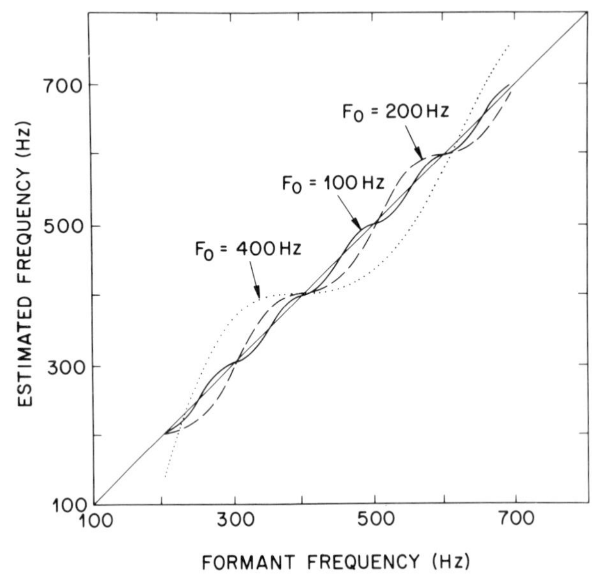

Fig. 7.16 Comparison of the estimated formant frequency with the actual frequency used in the synthesis for three different pitch frequencies. [From Atal, 1974b]

11, 30, and 67 Hz for $F_0 = 100$, 200, and 400 Hz, respectively. An increase in error with increasing F_0 would be expected since for a fixed bandwidth, the decaying exponential will have a much greater interaction with the next period. The error is observed to oscillate about the exact value as the formant frequency increases for a constant F_0. It can be seen that when $F_1 = KF_0$ for $K = 1, 2, \ldots$, the error is also near or equal to zero. This result is obtained because the decaying oscillation from one period will then precisely reinforce the oscillations in following periods.

The results obtained in estimating the formant bandwidth are shown in Fig. 7.17. The maximum errors reported were 25, 95, and 250 Hz for $F_0 = 100$, 200, 400 Hz, respectively. The maximum errors in the direction of overestimating bandwidth appear to be at $F_1 = KF_0 + F_0/2$. This result is explained by noting that, for this condition, the major oscillation within a period will have interference from the decaying oscillation of the previous period out of phase for nearly one-half of each F_1 cycle, on the average. Therefore, the apparent envelope decay would be increased and the bandwidth would be widened. In addition, interference would be greatest for the largest F_0's, and Atal's experiment shows this to be true. The maximum errors in the direction of underestimating the bandwidth occur for $F_1 = KF_0$, since reinforcement from previous periods will cause the period of data being analyzed to have a decreased envelope decay with respect to a single isolated formant oscillation.

In summary, for formant frequency estimation under pitch-synchronous conditions, Atal's experiment suggests that minimum error is expected when the formant frequency is a multiple of F_0. For the bandwidth estimation, however, this condition is expected to lead to a maximum error in underestimating the corresponding bandwidth. The worst case error in overestimating formant bandwidth would be expected when the formant frequency is at $F_0/2$ plus multiples of F_0.

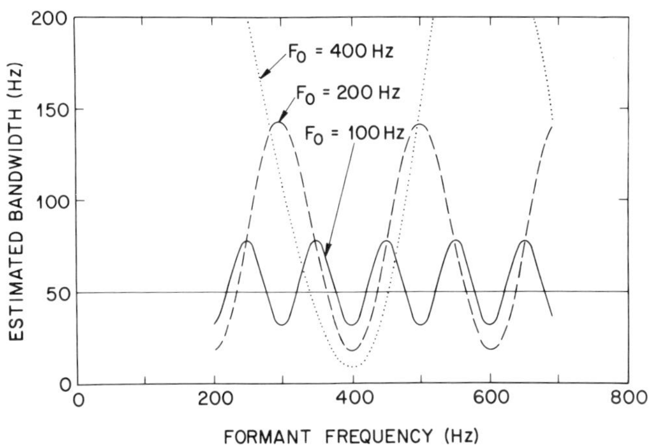

Fig. 7.17 Plots showing the dependence of the estimated formant bandwidth on the formant frequency and the pitch. [From Atal, 1974b]

8. Fundamental Frequency Estimation

8.1 Introduction

The *fundamental frequency* (F_0) is a basic parameter in acoustical studies of speech. It is also a necessary parameter for low bit rate speech coding systems. It is generally considered to be one of the acoustical correlates to the perceived intonation pattern of speech. If the fundamental frequency of a speaker is constant, the speech would be perceived as being machine-like or monotone. If the speaker is excited, the fundamental frequency generally increases. It is the acoustical correlate to the rate at which the vocal folds open and close (or vibrate). If the folds are vibrating rapidly, a high fundamental frequency will be measured. In the linear speech production model, the fundamental frequency is the rate at which glottal volume velocity pulses are applied to the vocal tract, i.e., the driving function to the model is periodic with a period of $1/F_0$.

The reciprocal of the fundamental frequency is the *pitch period P*. The pitch period measures the time interval between successive complete cycles of vocal fold opening and closing. Although a fundamental frequency value (averaged over some appropriate time interval) can be directly estimated from the speech spectrum, generally the average pitch period P is estimated over some appropriate length of data. The corresponding fundamental frequency value F_0 is then defined by $1/P$ for the specific analysis frame. Specific reference to frame or time intervals such as $F_0(k)$ and $F_0(k-1)$ will be used only where confusion would otherwise result.

A thorough discussion of published techniques for fundamental frequency or pitch period estimation would probably be as long as this book. In this chapter, just those techniques based upon linear prediction are considered.

First, the spectral flattening property of linear prediction is discussed within the context of fundamental frequency estimation. Correlation techniques for automatically estimating the pitch period either implicitly or explicitly from a spectrally flattened signal are then discussed.

Finally, an efficient linear prediction method of pitch estimation for moderate fundamental frequency ranges is presented in detail.

8.2 Preprocessing by Spectral Flattening

Several preprocessing techniques have been suggested for reducing or eliminating errors due to interaction between the first formant and the fundamental frequency information. Three possibilities are spectral flattening [Sondhi, 1968], center clipping [Sondhi, 1968], and signal cubing [Atal, 1968a]. The application of linear prediction for spectral flattening will be discussed in this section. Techniques which correlate the data and thus destroy phase information will be discussed in the following section.

The basic idea in spectral flattening is that if harmonics of the fundamental frequency can be readjusted to a constant value and forced to have zero phase, then the resulting waveform will have sharp pulses at the pitch period intervals without any formant structure between pitch periods. For unvoiced intervals there will be no harmonic structure and thus the waveform will have a random character due to the flattened spectrum. Sondhi suggested several procedures based upon processing the speech through a bank of bandpass filters. The resultant accuracy is dependent upon the number of bandpass filters and their characteristics. Linear prediction techniques can easily be used to accomplish the same goals but without any requirement for bandpass filter analysis or elimination of phase information.

8.2.1 Analysis of Voiced Speech with Spectral Regularity

Atal and Hanauer [1971b] suggested computing a prediction filter $F(z)$ and then applying the speech signal to an inverse filter $A(z)=1-F(z)$ as shown in Fig. 8.1. The output $\{e(n)\}$ was then filtered by a 1 kHz, low-pass filter and used in pitch period estimation. In Chapter 6 it was shown that *all of the approaches that result in inverse filters by solving autocorrelation equations are inherently maximizing the spectral flatness of the signal for the specified number of filter coefficients.*

Although spectral criteria for designing an inverse filter do not exist for the covariance method or for Kalman filtering techniques, the end result is quite similar for these techniques. The resultant filter will still behave as an inverse filter in the sense that if speech is passed through it, the output spectrum will be a flattened version of the input. Several examples are now presented to illustrate time

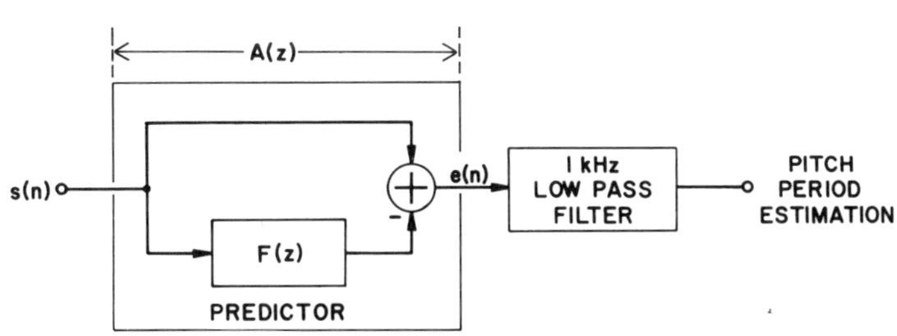

Fig. 8.1 Block diagram showing computation of error signal from linear prediction. [After Atal and Hanauer, 1971b]

and frequency domain behavior with unprocessed and spectrally flattened speech signals. The speech signal and error signal for a segment /ʃe/ of the utterance "*sha*de", spoken by a male with low average fundamental frequency, is shown in Figs. 8.2A and B, respectively. Note that the scale in Fig. 8.2B is two times that of

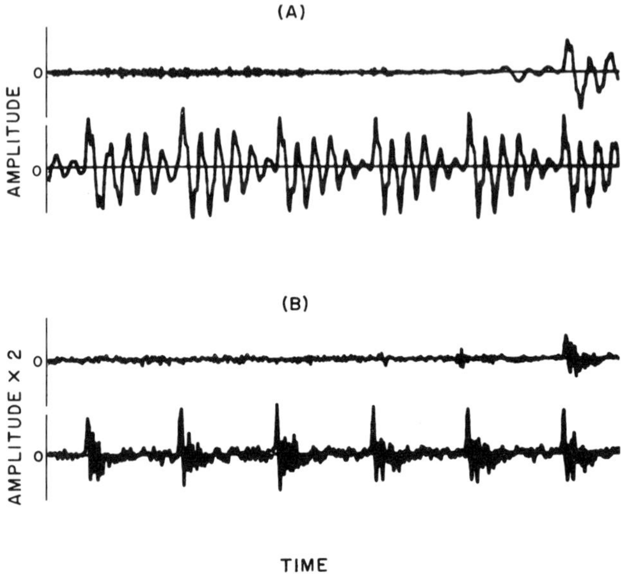

Fig. 8.2 Speech and error signals for segment /ʃe/ of "shade". A) speech signal. B) error signal.

Fig. 8.2A. The analysis parameters used in computing the error signal (the spectrally flattened signal) were $N=128$, $M=10$, and $f_s=6.5$ kHz. The analysis window was shifted every 64 samples to obtain a new set of filter coefficients which were then applied to the inverse filter equation

$$e(n+64l)=\sum_{i=0}^{M} a_i(l)\, s(n+64l-i)$$

for $n=1, 2, \ldots, 64$, where l denotes the frame number $l=0, 1, \ldots,$ and $a_0(l)=1$. The term $a_i(l)$ denotes the ith inverse filter coefficient for frame l. Each frame requires M speech samples from the previous frame. Although the speech is windowed before applying the autocorrelation method, for obvious reasons the error signal is based upon the unwindowed speech data. No pre-emphasis is used.

The speech waveform shows a very strong first formant influence over all of the voiced portions. The error signal, however, shows a dominant spike-like behavior at the initiation of each pitch period with the first formant influence almost completely removed. In place of the first formant structure, essentially uncorrelated signal components spanning the total frequency range from 0 to $f_s/2$ occur (except for the highly correlated fundamental frequency components).

Figure 8.3A shows the windowed speech spectrum (inverse filter input) and the error spectrum (inverse filter output) for a 128 sample interval of /e/ in sh*a*de. A very strong first formant (in the sense that the peak has relatively high energy) is seen in the input spectrum. The first three formants are very obvious. In Fig. 8.3B the inverse filter has removed essentially all formant structure, leaving a spectrally flattened version of the input signal. The log spectrum of the error signal $LM(E)$ shows extremely strong and consistent periodicity over the complete frequency range, with only moderate deviations from total flatness.

8.2.2 Analysis of Voiced Speech with Spectral Irregularities

As a second example of spectral flattening with linear prediction, a problem area in preprocessing for the reduction of formant influence is illustrated. The transition

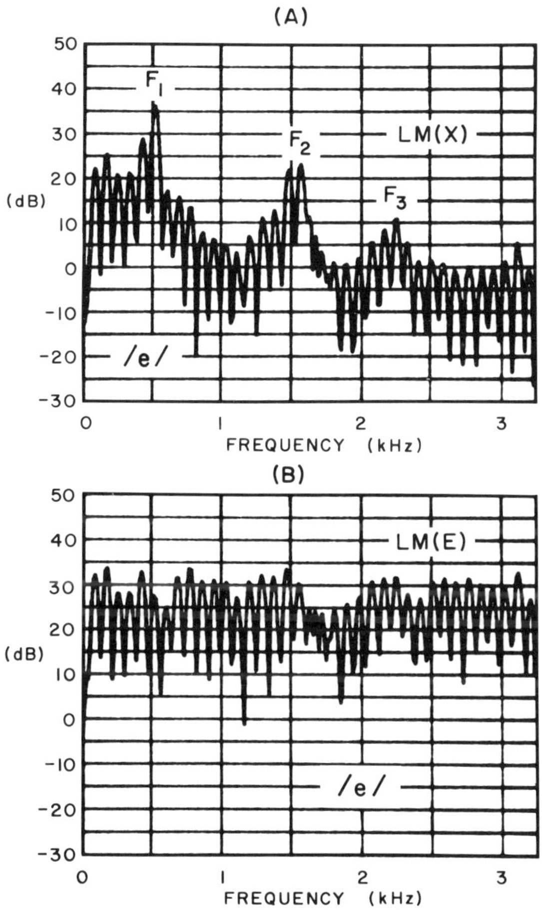

Fig. 8.3 Spectra from one analysis frame in /e/ of "shade". A) speech spectrum. B) error signal spectrum.

into and out of the nasal /n/ from li*n*ear is shown in Fig. 8.4. The waveform is the same one used to illustrate formant tracking in Fig. 7.10 of the preceding chapter. Figure 8.4A shows the speech waveform, while Fig. 8.4B shows the corresponding error signal obtained from the autocorrelation method. The analysis parameters were $N=160$, $M=12$, and $f_s=10\,\text{kHz}$. The 16 ms analysis window was shifted every 6.4 ms. The total duration of the utterance shown is 96 ms. There appear to be three distinctly different acoustical waveform characteristics corresponding to the nasal /n/ and the vowels /ɪ/ and /i/ preceding and following the nasal /n/. Due to coarticulation effects, both vowels include nasalization. The first formant structure, most easily seen in the preceding vowel, is essentially eliminated in the error signal, with the net effect that the spikes at the occurrence of each pitch period

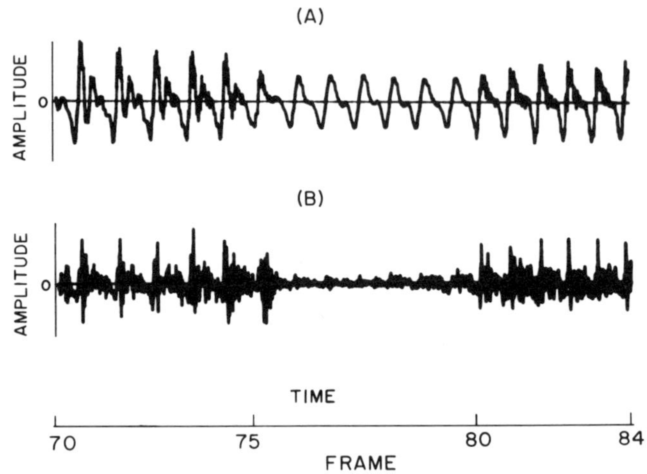

Fig. 8.4 Speech and error signals in segment /n/ of "linear". A) speech signal. B) error signal from autocorrelation method.

are sharpened. A similar trend is observed for the following vowel. The amplitude of the nasal portion has been greatly reduced with respect to the /ɪ/ and /i/ portions. The normalized squared error for the nasal portion is thus substantially smaller than that of the other portions. The speech signal log spectra $LM(S)=LM(X)$ and the error signal $LM(E)$ are shown for the vowel /ɪ/ and for the nasal /n/ in Figs. 8.5 and 8.6. In Figs. 8.5A and 8.6A, the formant frequency locations are marked. The abrupt F_2 discontinuity shown by the raw data in Fig. 7.10 is clearly seen in terms of the differences between these two spectra. Both error spectra have flat trend characteristics, although they show very complex structure. The error spectra show the introduction of a nasal zero very clearly. In Fig. 8.5B, the velum is beginning to open, causing a zero to appear near 1400 Hz. In Fig. 8.6B, the velum has completely opened, causing a shift in the nasal zero location to the vicinity of 900 Hz. The effect of nasalization is to cancel a significant amount of the spectral periodicity which is necessary to obtain sharp spikes in the error signal. It was stated in Chapter 6 that linear prediction (with a relatively few number of coeffi-

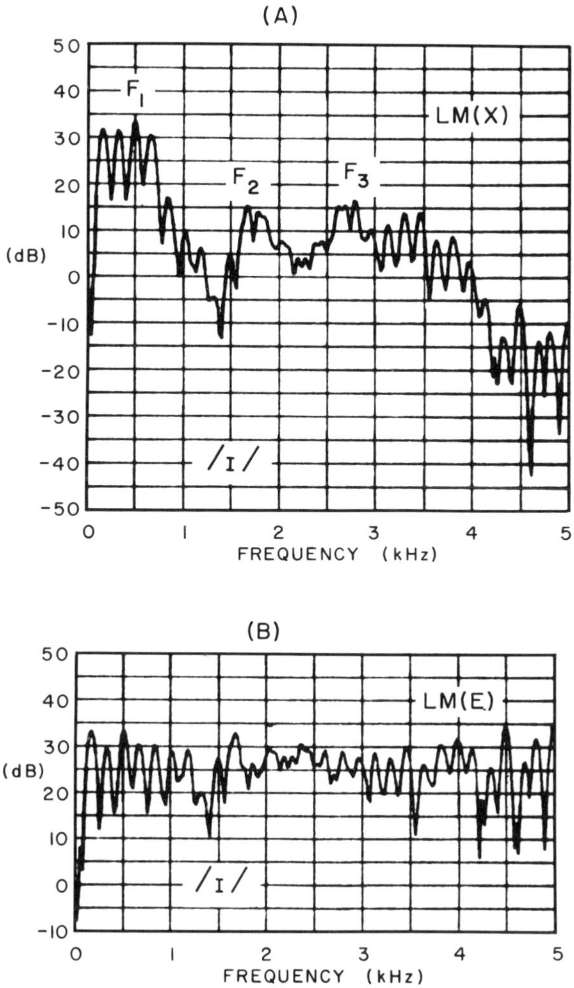

Fig. 8.5 Log spectra from the /ɪ/ segment of "linear". A) input speech. B) error signal.

cients as used here) tends to favor the peaks over the valleys (the formants over the zeros). Although, based upon speech perception, this is a desirable characteristic, it does cause difficulty for F_0 estimation in the presence of spectral zeros. A solution to this problem is obtained by filtering the error signal to the most reliable region of F_0 information where nasalization effects and spectral irregularities at higher frequencies are minimized. Applying a simple low-pass filter produces the filtered error signal shown in Fig. 8.7. The first formant influence has been reduced while at the same time only the very periodic, low-frequency region of the nasal sound contributes to the filtered error signal, allowing strong waveform periodicity to be seen. The filtering was performed by differencing the speech signal, computing the inverse filter, and then passing the non-pre-emphasized

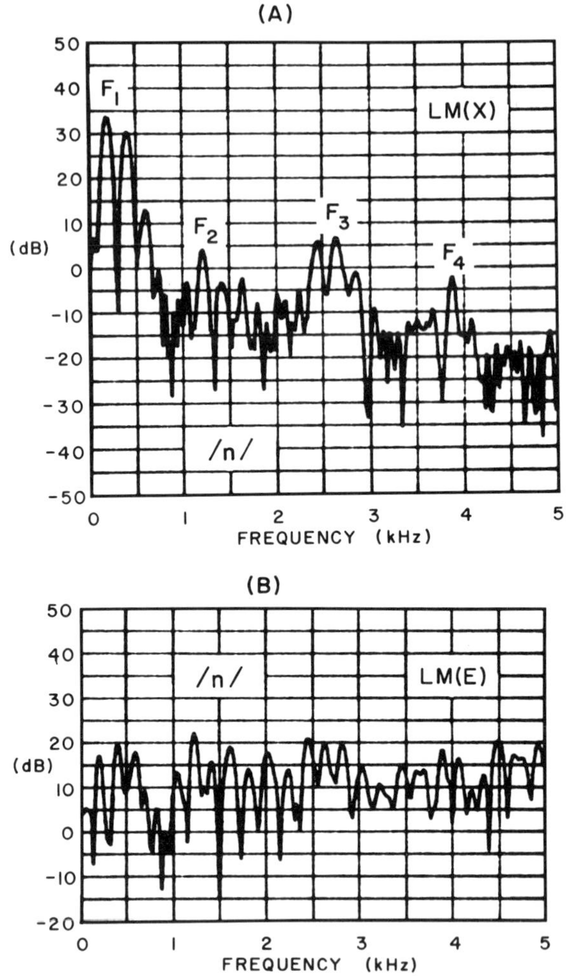

Fig. 8.6 Log spectra from nasal segment /n/ of "linear". A) input speech. B) error signal.

speech signal through the filter. (This procedure effects a $1/(1-z^{-1})$ integrator without low frequency bias buildup.)

It is important to realize that filtering of the error signal is a justifiable and necessary operation based upon the acoustical properties of speech, and is not

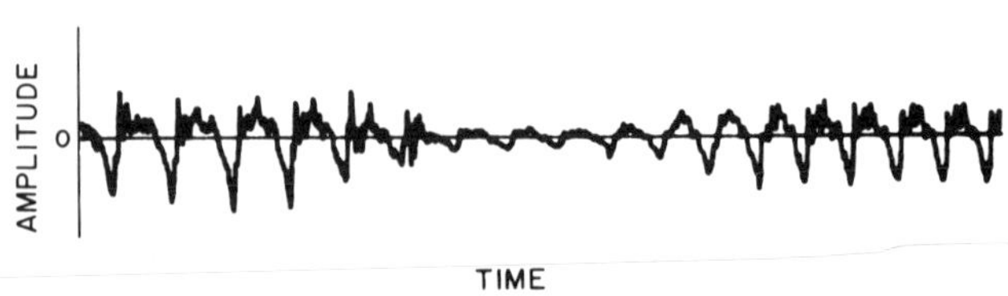

Fig. 8.7 Filtered error signal for /n/ segment from the utterance "linear".

just an ad hoc modification. Fujimura [1968] has noted that there are many voiced speech segments that appear to be either partially or completely unvoiced in the frequency range above 1 kHz, whereas voicing components above 1 kHz without accompanying voicing components below 1 kHz are seldom noted. When portions of the spectrum appear unvoiced, the effect is to lower the peak amplitudes of the error signal used for defining voicing. If spectral flattening is applied in this situation, incorrect voicing decisions may be made. By down-sampling the speech to a bandwidth of approximately 1 kHz, even partial voicing will cause the voicing threshold to be exceeded. The choice of filter is an open question. In most instances, a simple one-pole filter $1/(1-\mu z^{-1})$, with μ near unity, is adequate to reduce the higher frequency influence.

8.2.3 The STREAK Algorithm

Another approach for spectrally flattening the speech signal is called STREAK, for a *s*implified *t*echnique for *r*ecursively *e*stimating *a*utocorrelation *k*-parameters [Boll, 1974]. Here an error signal is generated in a manner similar to the PARCOR lattice discussed in Chapter 2, but the reflection coefficients (k-parameters) are updated for each index n. The forward and reverse error signals of Fig. 2.5 satisfy the recursion relations

$$x_m^+(n) = x_{m-1}^+(n) + k_m x_{m-1}^-(n) \tag{8.1a}$$

and

$$x_m^-(n+1) = k_m x_{m-1}^+(n) + x_{m-1}^-(n) \tag{8.1b}$$

for $m = 1, 2, \ldots, M$, where

$$x_0^+(n) = x(n) \quad \text{and} \quad x_0^-(n) = x(n-1). \tag{8.1c}$$

The output error signal is then

$$e(n) = x_M^+(n). \tag{8.1d}$$

The PARCOR approach minimizes the sum of the squares of the error signals, whereas the STREAK approach first defines the coefficient k_m as a function of time $t = nT$ by

$$k_m = k_m(n),$$

and then minimizes the instantaneous value of the sum of the squares of $x_m^+(n)$ and $x_m^-(n+1)$, where

$$[x_m^+(n)]^2 + [x_m^-(n+1)]^2 = 2k_m x_{m-1}^+(n) x_{m-1}^-(n) + (1+k_m^2)\{[x_{m-1}^+(n)]^2 + [x_{m-1}^-(n)]^2\}.$$

A direct minimization of the above expression gives the result

$$k_m = k_m(n) = \frac{-2x^+_{m-1}(n)\, x^-_{m-1}(n)}{[x^+_{m-1}(n)]^2 + [x^-_{m-1}(n)]^2}. \tag{8.2}$$

As in the PARCOR approach, the magnitudes of these coefficients will be bounded by unity. A direct calculation shows that

$$1 - k_m^2 = \left(\frac{[x^+_{m-1}(n)]^2 - [x^-_{m-1}(n)]^2}{[x^+_{m-1}(n)]^2 + [x^-_{m-1}(n)]^2}\right)^2 \tag{8.3}$$

where the right-hand side must be bounded by zero and one, giving $0 \le k_m^2 \le 1$ or $|k_m| < 1$.

The forward prediction errors have an instantaneous energy that is monotonically decreasing with increasing m, since (8.1), (8.2), and (8.3) can be combined to produce the result

$$[x^+_m(n)]^2 = [x^+_{m-1}(n)]^2 [1 - k_m^2].$$

This expression is of the same form as shown in Chapter 3 for the autocorrelation and covariance methods,

$$\alpha_m = \alpha_{m-1}(1 - k_m^2).$$

The equivalent relationship for the backward prediction errors includes a time (or index) lag,

$$[x^-_m(n+1)]^2 = [x^-_{m-1}(n)]^2 [1 - k_m^2].$$

A Fortran subroutine STREAK which implements (8.1) and (8.2) is shown in Fig.

```
C
      SUBROUTINE STREAK(XIN,XOUT,M,D)
      DIMENSION D(1)
      EP=XIN
      EM=D(1)
      D(1)=EP
      MP1=M+1
      DO 10 I=2,MP1
      DEN=EM*EM+EP*EP
      IF(DEN.EQ.0.) DEN=1.
      XK=-2.*EM*EP/DEN
      EMOLD=EM
      EM=D(I)
      D(I)=XK*EP+EMOLD
   10 EP=EP+XK*EMOLD
      XOUT=EP
      RETURN
      END
```

Fig. 8.8 Fortran subroutine STREAK for performing error signal preprocessing.

8.8. The parameters are $XIN = x(n)$, $XOUT = e(n)$, M, the number of filter sections, and D, an $M+1$ length array used as the delay or buffer elements. At the program initiation, $D(J) = 0$, $J = 1, 2, \ldots, M+1$. The results obtained by analyzing the same transition into and out of the /n/ from li*n*ear as in Figs. 8.4 are shown in Fig. 8.9.

The error signal spikes during the vowel /ɪ/ are remarkably impulse-like with essentially no residual between the periods. During the nasal /n/, the amplitude is substantially reduced as in Fig. 8.4, but the initiation of each period is still clear. Only during the final portion of /i/ does the periodicity of the STREAK error signal show moderate signs of irregularity.

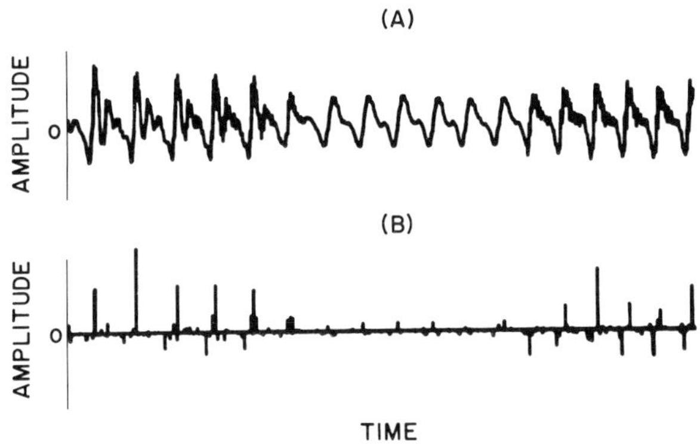

Fig. 8.9 Speech and error signals in segment /n/ of "linear". A) speech signal, B) error signal from STREAK algorithm.

In Chapter 10, it is shown that this method can be considered as a special case of a PARCOR implementation scheme suggested by Itakura and Saito [1971a]. Although the physical interpretation of this algorithm is not well understood, it does appear to have considerable potential for pre-processing if computation time is not a significant restriction.

One other spectral flattening method based upon a feedback loop with a one-bit quantizer and a linear predictor filter has been suggested [Maksym, 1973]. Although the necessary sampling rate is quite high (on the order of 40–60 kHz) it can be implemented quite easily in hardware. An analog implementation of this type of system has been developed by Morris [1975]. The techniques for F_0 measurement discussed so far are based on the detection of peaks in the time-domain representation of the error signal.

8.3 Correlation Techniques

Now several correlation methods for pitch estimation are considered. The autocorrelation function which can be applied directly to the speech signal or to the spectrally flattened error signal is first discussed. The *modified autocorrelation method* [Itakura and Saito, 1968] which performs spectral flattening and autocorrelation using linear prediction methods is then discussed. Finally, an autocorrelation analysis based upon the filtered linear prediction error signal is presented.

8.3.1 Autocorrelation Analysis

One of the oldest methods for estimating the fundamental frequency of voiced speech is autocorrelation analysis. If $\{x(n)\}$ is a finite length data sequence, its autocorrelation is defined by

$$r(j) = \sum_{n=-\infty}^{\infty} x(n)\,x(n+j). \tag{8.4}$$

When $x(n)$ is truncated so that $x(n)=0$ for $n<0$ and $n>N-1$, (8.4) reduces to

$$r(j)=0 \quad \text{for} \quad |j| \geq N \tag{8.5a}$$

and

$$r(j)=r(-j)=\sum_{n=0}^{N-1-j} x(n)\,x(n+j) \quad \text{for} \quad j=0,1,\ldots,\ N-1. \tag{8.5b}$$

A Fortran program for directly implementing the above equation is included in the upper portion of the subroutine AUTO in Chapter 3.

In the estimation of F_0, the sequence $\{r(j)\}$ is usually calculated for values of j out to about $N/2$. For this condition the FFT can be used to make the autocorrelation more efficient. First zeros are appended to the N data points, $x(n)=0$ for $n=N, N+1, \ldots, L-1$, where L is a power of two. Then an L-point FFT of the data sequence is taken giving the FFT sequence $\{X(k)\}$. The magnitude square of each sample is computed and the inverse FFT is applied (or since the sequence is real, one can apply the FFT and divide by L), resulting in the cyclic autocorrelation sequence $\{r_c(j)\}$. As $\{x(n)\}$ is real, $\{X(k)\}$ can be obtained through the use of an $L/2$-point FFT, and as the sequence $\{|X(k)|^2\}$ is both real and even, $\{r_c(n)\}$ can be obtained through the use of an $L/4$-point FFT [Cooley, et al., 1970].

The sequence obtained through the use of FFTs is a cyclic autocorrelation of the data sequence, and is related to the true autocorrelation by

$$r_c(j)=r(j)+r(L-j) \quad \text{for} \quad j=0,1,\ldots,L-1. \tag{8.6a}$$

As $r(j)$ is zero for $j>N-1$, the two sequences are equal only for values of j up to $L-N$ (the number of appended zeros), i.e.,

$$r_c(j)=r(j) \quad \text{for} \quad j=0,1,\ldots,\ L-N. \tag{8.6b}$$

Thus, the number of appended zeros must equal at least the number of points for which accurate values of $r(j)$ are desired.

The autocorrelation sequence has several important properties. The sequence can always be normalized to unity at the origin since $|r(j)| \leq r(0)$ and $|r_c(j)| \leq r_c(0) = r(0)$ for all values of j. In autocorrelation analysis, the initial pitch period estimate P is generally defined by the location of the maximum autocorrelation value within some specified range. The dynamic range of pitch period peaks in the autocorrelation sequence is generally less than 10 dB. In contrast, pitch extraction based directly upon the acoustic waveform, a common approach with analog devices, may require peak detection over more than a 30 dB range.

There are several variations of the autocorrelation calculation. One is the use of the *circular correlation* [Skinner, 1973] which in effect does not attempt to append sufficient zeros to obtain accurate values of $r(j)$, but rather evaluates $r_c(j)$. From (8.6a) it is seen that if $\{r(j)\}$ has a relatively high peak at j_p, then $\{r_c(j)\}$ will also have a peak at j_p. Another variant is called *contained correlation* where the products $x(n)x(n+j)$ are summed from $n=0$ through $n=L-1$ (where L is an integer less than N and $j=0, 1, \ldots, N-L$) giving

$$\sum_{n=0}^{L-1} x(n)x(N+j) = r(j) - \sum_{n=L}^{N-j-1} x(n)x(N+j). \tag{8.7}$$

The summation on the right-hand side is smallest for values of j near $N-L$ and largest for values of j near zero. The contained correlation uses the same N data points with fewer terms in the sum to be evaluated than for direct calculation of $r(j)$; however, it may yield smaller peak correlations near the pitch peak.

The normalized autocorrelation sequences obtained from the segment /ʃe/ of Fig. 8.2A are shown in Fig. 8.10. The correlation was computed every 64 samples (9.85 ms) using a window of 256 samples (39.4 ms) multiplied by a Hamming window. The unvoiced portion /ʃ/ corresponds to the first four frames with the remaining frames corresponding to the voiced portion /e/. As can be seen, the formant structure significantly influences the estimation of the pitch period. In several frames, the pitch period is completely masked. By observing the initial portion of /e/ in Figs. 8.2A and 8.3A, the difficulty is seen to be due to a combination of the first formant having narrow bandwidth (slow decay of the oscillations in each period) and the relatively long pitch period with respect to $1/F_1$ (resulting in five first formant cycles per pitch period). Several linear prediction methods that resolve the formant and fundamental frequency interaction problem are now discussed.

8.3.2 Modified Autocorrelation Analysis

Itakura and Saito [1968] suggested a modified autocorrelation technique for eliminating the formant frequency influence in the autocorrelation sequence. In effect, this calculation is the autocorrelation sequence of the linear predictor error sequence. The inverse filter output is given by

$$e(n) = \sum_{i=0}^{M} a_i x(n-i), \tag{8.8}$$

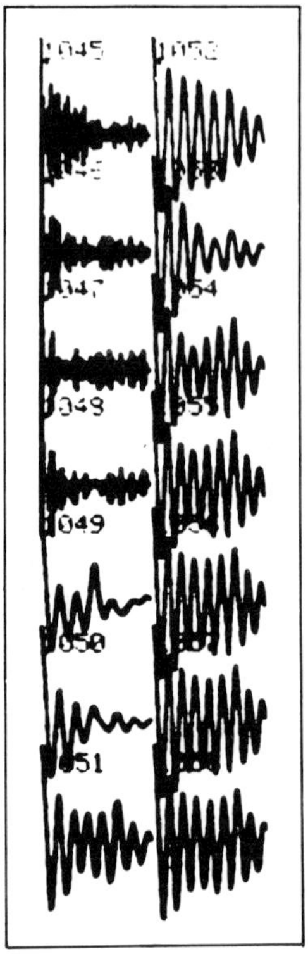

Fig. 8.10 Input speech autocorrelation sequences based upon segment /ʃe/ of "shade".

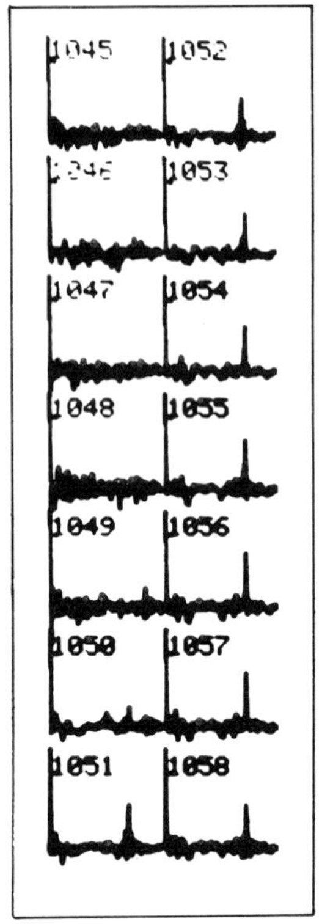

Fig. 8.11 Preprocessed (error signal) autocorrelation sequences based upon segment /ʃe/ of "shade".

or, since $a_i = 0$ for $i < 0$ and $i > M$, (8.8) can be expressed as

$$e(n) = \sum_{i=-\infty}^{\infty} a_i x(n-i).$$

A direct evaluation for the autocorrelation function gives

$$r_e(k) = \sum_{n=-\infty}^{\infty} e(n) e(n+k) = \sum_{n=-\infty}^{\infty} \sum_{i=-\infty}^{\infty} \sum_{j=-\infty}^{\infty} a_i a_j x(n-i) x(n+k-j).$$

The summation over n can be evaluated in terms of the autocorrelation sequence for the original data, giving

$$r_e(k) = \sum_{i=-\infty}^{\infty} \sum_{j=-\infty}^{\infty} a_i a_j r_x(k-j+i). \tag{8.9}$$

If the index of summation on the inner sum is changed from j to $l = j - i$, (8.9) is equal to

$$r_e(k) = \sum_{i=-\infty}^{\infty} \sum_{l=-\infty}^{\infty} a_i a_{i+l} r_x(k-l).$$

The modified autocorrelation method is then based upon the calculation of

$$r_e(k) = \sum_{l=-\infty}^{\infty} r_a(l)\, r_x(k-l), \tag{8.10a}$$

where

$$r_a(l) = \sum_{i=-\infty}^{\infty} a_i a_{i+l} \,. \tag{8.10b}$$

The autocorrelation sequence for the error signal can be computed either from (8.8) directly, or from the convolution (8.10). While the limits on the summations of (8.10) are shown as infinite, they are in fact finite summations, because $a_i = 0$ for $i < 0$ and $i > M$, $x(n) = 0$ for $n < 0$ and $n > N-1$, and thus $r_a(n)$ is zero for $|n| > M$ and $r(n)$ is zero for $|n| > N-1$. The error signal autocorrelation sequence can also be computed using (8.5) in the same manner as stated for the cepstrum with the only difference being that the logarithm is not used. The pitch and voicing detection procedure suggested by Itakura and Saito [1968] is as follows:

The normalized autocorrelation sequence $\{r_e(n)/r_e(0)\}$ is searched over the time interval corresponding to $3-15$ ms, for the peak location n_p. If $r_e(n_p)/r_e(0) \leq .18$, the frame is defined as unvoiced ($P=0$). If $r_e(n_p)/r_e(0) \geq .25$, the frame is defined as voiced with the period $P = n_p/f_s$. A transition curve (see Chapter 10) was used for estimating partial voicing.

The error signal autocorrelation sequences obtained from the speech signal of Fig. 8.2A are shown in Fig. 8.11. The same conditions as in Fig. 8.10 were used along with $M = 10$ coefficients for the inverse filter. By comparing Figs. 8.11 and 8.10, it is seen that the error signal autocorrelation function has in this example removed nearly all of the formant influence, allowing correct pitch and voicing without ad hoc tests and corrections.

The normalized autocorrelation sequences obtained from the speech signal of Fig. 8.4A are shown in Fig. 8.12. The autocorrelation peaks for frames 71–76 are now much less obvious, although the threshold setting of 0.18 is still sufficient to allow correct extraction of the pitch period peaks for this example. Note the wide variation of correlation peak values. The low correlation values in frames 71–76 are a consequence of having reliable harmonic structure only below about 800–1000 Hz. Frames 71 and 76 correspond to the inverse Fourier transforms of the spectra shown on log magnitude scales in Figs. 8.5B and 8.6B, respectively.

8.3.3 Filtered Error Signal Autocorrelation Analysis

By low-pass filtering the error signal to about 800 Hz, significantly higher correlation peaks can be obtained without the influence of the formant structure.

The autocorrelation sequences of the filtered error signal based upon the segment /n/ of l*ine*ar are shown in Fig. 8.13. By comparing Fig. 8.12 with Fig. 8.13, the effect is seen to be dramatic. The filtered error signal autocorrelation sequence appears to be similar to a direct autocorrelation of the speech signal with the important exception that the formant interaction is eliminated. This fact is demon-

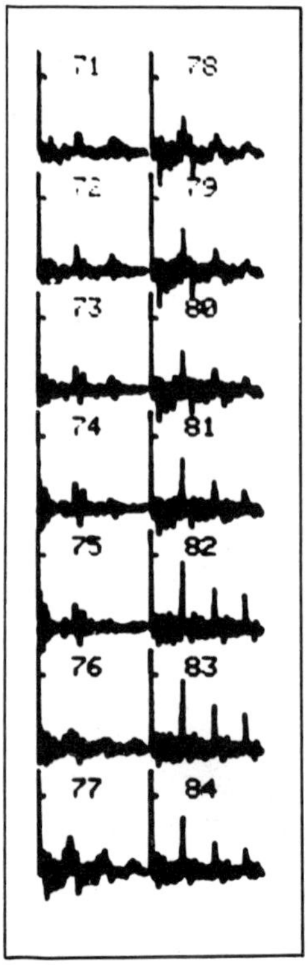

Fig. 8.12 Error signal autocorrelation sequences based upon segment /n/ of "linear".

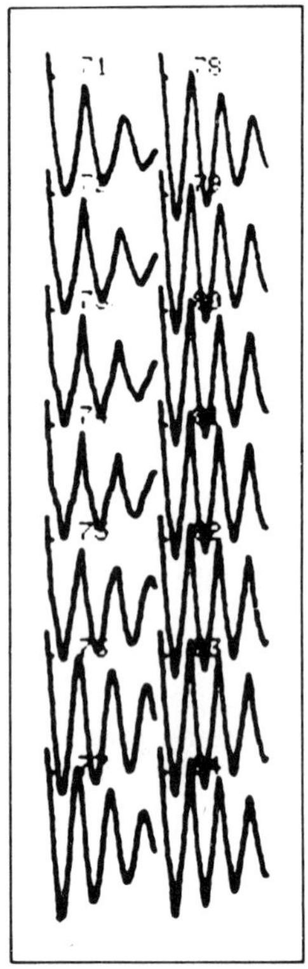

Fig. 8.13 Filtered error signal autocorrelation sequences based upon segment /n/ of "linear".

strated in Fig. 8.14 where the filtered error signal autocorrelation sequences from the segment /ʃe/ of *sha*de are presented. By comparing Figs. 8.10 and 8.14, it is seen that the filtered error signal registers the locations of each pitch period with essentially complete elimination of the formant interaction effects seen in Fig. 8.10.

A block diagram of the filtered error signal autocorrelation method is shown in Fig. 8.15. Linear prediction analysis using the autocorrelation method is performed with the same considerations discussed in Chapter 6 for the pre-emphasis factor, number of coefficients, etc. The data, before windowing or pre-emphasis, are then processed by an inverse filter $A(z)$ which is updated at the coefficient analysis frame rate, f_r. The filtered error sequence $\{f(n)\}$ is then computed. One sample $f(n)$ is obtained for each speech sample $s(n)$, to ensure that no discontinuities exist.

As soon as a sufficient number of samples have been obtained to fill a pitch buffer of length N', an N'-length Hamming window is applied to produce a new N'-length sequence. An autocorrelation is then performed on the windowed sequence to produce $\{r_w(n)\}$. This sequence is searched from $0 < n_{\min}$ to $n_{\max} < N'/2$ to obtain the peak location, n_p. If $r_w(n_p)/r_w(0) \geq t_h$, where t_h is the voicing threshold, then frame k is said to be voiced with a pitch period $P = n_p/f_s$. Based upon a sampling

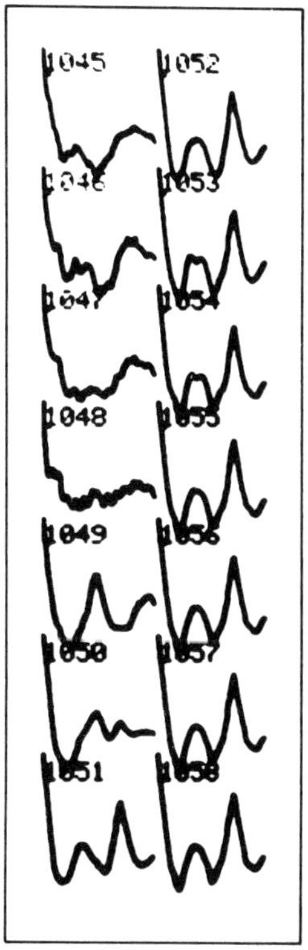

Fig. 8.14 Filtered error signal autocorrelation sequences based upon the segment /ʃe/ of "shade".

frequency of 10 kHz, the following conditions were found to result in accurate pitch period and voicing estimates: pre-emphasis of the input signal using the filter $1-z^{-1}$, an analysis window of length $N=160$ (16 ms analysis window), $M=12$, and a pitch window of length $N=380$ (38 ms). The voicing threshold was set at $t_h=0.4$.

One of the interesting features of this structure is that, as a byproduct, it defines a complete linear prediction analysis for both inverse filter coefficients or reflection coefficients along with the pitch period and voicing parameters.

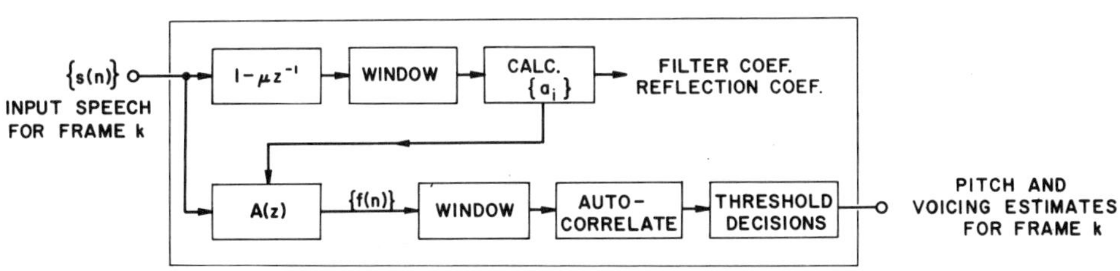

Fig. 8.15 A block diagram of the filtered error signal autocorrelation method.

Fortran programs have been presented in earlier chapters for performing the necessary operation. The inverse filter operation can be inefficiently implemented using DIRECT of Chapter 5 by setting $A(1)=1$, $A(2)=\ldots=A(M+1)=0$, $P(1)=1$, and $P(J)=a_{J-1}$ where $J=2, 3, \ldots, M$, or more efficiently, by implementing (8.8).

8.3.4 Practical Considerations

It should be noted that all of the correlation procedures discussed are computationally expensive due to the calculation of the autocorrelation function, the interval to be searched, and the sampling frequency. For example, if $f_s=10$ kHz, and pitch periods from $2-17$ ms are desired (implying a window of at least 340 samples), then nearly 25,000 multiply-add operations per analysis frame are necessary to implement the correlation operation. For F_0 ranges below 250 Hz, the total number of operations can be substantially reduced by using the SIFT algorithm, described in the next section. For F_0 ranges above 250 Hz, the operations can be substantially reduced by searching only an interval of, for example, 2–4 ms. Unfortunately, there do not appear to be any accurate F_0 estimation procedures that can handle the complete range of voices from around 50 to 500 Hz (2–20 ms) without the amount of computation described above.

The reader should not be misled into thinking that the procedures described lead to perfect pitch period or F_0 estimation by testing only the correlation peak against a specified threshold. The problem of F_0 estimation with linear prediction techniques is analogous to that of formant estimation, namely, simple threshold testing will produce correct results for a high percentage of frames analyzed. However, gross errors will be generated a small percentage of the time. Error correction is an art form that necessarily involves some degree of ad hoc decision-making, e.g., if three consecutive frames are analyzed and only the center frame causes the threshold to be just barely crossed, it should be reset to unvoiced, assuming that no other information is available.

As a general rule, the best and most complex error correction is attainable with the largest number of frames used as reference information, both to the right and left of the initial pitch period in question. In a non-real time system, the anchor-point concept of Chapter 7 could be used to advantage. For real-time systems where transmission delay time must be considered, l-frames of delay (with $l=2, 3$, or 4) are usually introduced so that at least a few of the least delayed pitch period estimates can be used for error correction of the new estimate before transmitting pitch. A simple error correlation scheme will be presented in the next section.

8.3.5 The SIFT Algorithm

An efficient and accurate pitch extraction method based upon linear prediction principles for the range $50-250$ Hz is the simplified inverse filter tracking (SIFT) algorithm [Markel, 1972c]. A down-sampling procedure is used so that the effective sampling frequency for F_0 analysis is about 2 kHz. Therefore, only the most reliable frequency range out to about 1 kHz is processed and, in addition, the necessary number of operations is substantially reduced. A block diagram of the SIFT algorithm, represented in two steps, is shown in Fig. 8.16. Efficient preprocessing to reduce formant and fundamental frequency interaction is performed in step 1. A sequence of NP samples corresponding to frame k is prefiltered with a cutoff close to $f_s/I = 2$ kHz, where I is the integer down-sampling factor.

Down-sampling is performed to reduce the effective sampling rate to f_s/I. The samples are differenced to accentuate the region of the second formant, and multiplied by a Hamming window. A fourth-order inverse filter $A(z)$ is then designed using the autocorrelation method. Due to the fact that at most two formants can reside in the range (0, 1 kHz), four coefficients have been demonstrated to be sufficient.

Although the inverse filter was designed on the basis of differenced windowed data the output is obtained by applying the unwindowed non-differenced data. The effect of this operation is to produce a low-pass filtered error signal without low-frequency bias. This signal is then multiplied by a second Hamming window.

In step 2, an autocorrelation sequence is obtained and then the peak within the minimum-to-maximum desired pitch range is obtained. Parabolic interpolation is applied to provide greater pitch period resolution. (Without interpolation, the maximum resolution would be I/f_s). A variable threshold has been found to be of significant utility with a filtered error signal. The threshold is defined by two linear segments intersecting at some quiescent threshold location. As the peak location becomes smaller, the threshold is raised, since proportionally more pitch periods will be obtained per analysis interval. As the peak location increases, the threshold is lowered. If a peak crosses the variable threshold, its location becomes

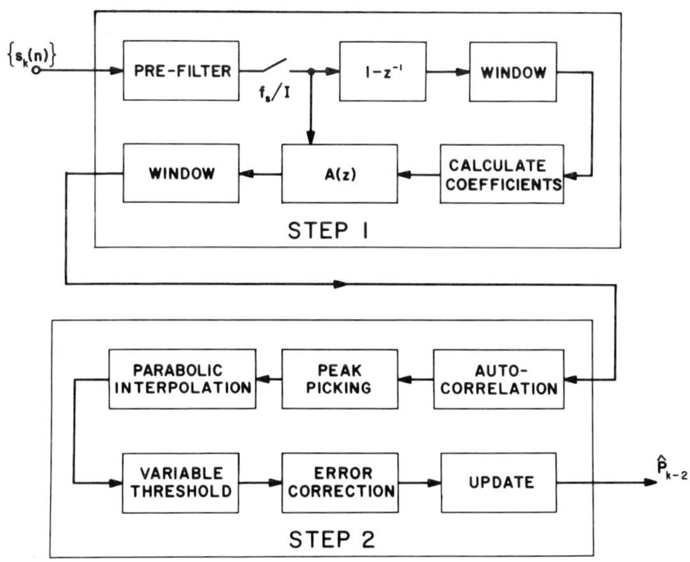

Fig. 8.16 Block diagram of the SIFT algorithm.

the pitch period candidate for that frame. Otherwise the frame is defined as unvoiced ($P=0$). An attempt at error detection and correction is made by storing several pitch period candidates. After this operation, the pitch period estimate with maximum delay is output.

To clarify the concepts of the SIFT algorithm and to also provide the reader with a useful tool for pitch period or fundamental frequency estimation, Fortran subroutines STEP1 and STEP2 are provided in Figs. 8.17 and 8.18.

Due to the number of variables involved, pitch extraction programs which attempt to be general are also usually quite lengthy. As a compromise between length and usability, the programs shown in Figs. 8.17 and 8.18 assume the following specific conditions.

A sampling frequency, $f_s=10$ kHz, is assumed for obtaining the original speech samples $\{s(n)\}$. The samples are prefiltered by a third-order elliptic filter having frequency response and coefficient values as shown in Fig. 8.19, and then they are down-sampled to 2 kHz ($I=5$). A program for designing elliptic filters is presented in Gray and Markel [1975b].

The input to STEP1 is a 400 sample (40 ms) sequence $\{s_k(n)\}$ for frame k. The analysis frame rate is defined by the user, but should be restricted to about $f_r=50$ Hz. For example, if $f_r=100$ Hz, there will be a 4 to 1 overlap of input data since 40 ms windows will be processed in 10 ms increments (implying the reading of data from a file structured device). Pitch period estimation occurs for the range $(6-1)I/f_s=2.5$ ms to $(32-1)I/f_s=15.5$ ms. The variable threshold is defined so that 9 ms is the quiescent location with a threshold value of 0.35. At the extreme left end, the peak must exceed 0.7 to be an acceptable pitch candidate, whereas at the extreme right end, the peak must only exceed 0.315. Two frames of delayed pitch and voicing information are retained for error detection and correction. Although the rules implemented here are very simple, they do illustrate the important kinds of ad hoc tests that are occasionally needed. For example, if an unvoiced frame is surrounded by two voiced frames that have reasonably close pitch period values, it is reasonable to assume that an isolated unvoiced frame (corresponding to a 10 ms interval, for example) really does not exist. Therefore the frame is reset as voiced with a period equal to the average of the two adjacent frames. After updating the frames, the pitch period estimate for frame $k-2$, in terms of the Fortran variable PITCH(·), will be

$$p_{k-2}=(\text{PITCH}(3)-1)I/f_s \quad \text{(ms)}$$

where $I/f_s=0.5$.

The prefiltering and inverse filtering both make use of the subroutine DIRECT from Chapter 5. The calculation of the filter coefficients is based upon the subroutine AUTO from either Chapter 3 or Chapter 9. The program was written in a form which most easily illustrates the down-sampling and inverse filtering operations. By expanding the program length, a number of modifications are possible which increase the computational efficiency and generality for other sampling frequencies. Most importantly, by updating a down-sampled pitch buffer, only f_s/f_r samples per frame need to be prefiltered. Also, DIRECT can be rewritten to implement only the numerator terms used in the inverse filtering operation. The

```
C
C       SIFT ALGORITHM PROCESSING  -  STEP1
C
C       INPUT PARAMETER:    SPCH(J)   (J=1,2,...,400)
C               THE SPEECH SIGNAL TO BE PROCESSED FOR PITCH
C
C       OUTPUT PARAMETER:  PBUF(J)   (J=1,2,...,80)
C               THE DOWN-SAMPLED, FILTERED ERROR SIGNAL
C
C       NOTE: PARAMETERS FIXED FOR FS=10 KHZ
C
        SUBROUTINE STEP1(SPCH,PBUF)
        DIMENSION SPCH(1),PBUF(1),AF(4),PF(4),DF(5),D(5)
        DIMENSION U(80),A(5),P(5),AL(5),RC(4),RA(5)
        DATA AF/1.,-2.34036589,2.01190019,-.614109218/
        DATA PF/.0357081667,-.0069956244,-.0069956244,.0357081667/
        DATA P/1.,4*0./
C
C       INITIALIZE MEMORY OF DIRECT TO ZERO
C
        DO 10 J=1,5
        DF(J)=0.
   10   D(J)=0.
C
C       PRE-FILTER, DOWN-SAMPLER, DIFFERENCER AND HAMMING WINDOWER.
C
        UPREV=0.
        DO 20 J=1,400
        CALL DIRECT(AF,PF,3,DF,SPCH(J),SOUT)
        IF (MOD(J,5).NE.0) GO TO 20
        K=J/5
        PBUF(K)=SOUT
        U(K)=(SOUT-UPREV)*(.54-.46*COS((K-1.)*6.28318/79.))
        UPREV=SOUT
   20   CONTINUE
C
C       COMPUTE INVERSE FILTER COEFFICIENTS
C
        CALL AUTO(80,U,4,A,AL,RC)
C
C       PERFORM INVERSE FILTERING AND HAMMING WINDOW
C
        DO 30 J=1,80
        CALL DIRECT(P,A,4,D,PBUF(J),FOUT)
        IF (J.LE.4) GOTO 30
        PBUF(J-4)=FOUT*(.54-.46*COS((J-5)*6.28318/75.))
   30   CONTINUE
        RETURN
        END
```

Fig. 8.17 Fortran subroutine STEP 1 for implementing the first step of the SIFT algorithm.

major computational items are the implementation of the autocorrelation calculation, DIRECT, and the cosine calculations, in that order. If these are implemented with assembly language or table lookup (for the cosine), considerable efficiency will be gained.

One example of the SIFT algorithm as implemented by STEP 1 and STEP 2 is shown in Fig. 8.20 where the utterance "linear prediction" was analyzed. The acoustic waveform and pitch contour obtained by straight line interpolation of the individual frame results were automatically analyzed without error in this instance. The minimum and maximum pitch periods are 4.5 and 14.1 ms. The fundamental

```
C
C       SIFT ALGORITHM PROCESSING  -  STEP2
C
C       INPUT PARAMETER:  PBUF(J)  (J=1,2,...,76)
C               THE DOWN-SAMPLED, FILTERED ERROR SIGNAL
C
        SUBROUTINE STEP2(PBUF,PITCH)
        DIMENSION PBUF(1),PITCH(1),ABUF(33)
C
C       PERFORM AUTOCORRELATION ON PITCH BUFFER
C
        DO 20 JJ=1,33
        J=JJ-1
        NMJ=76-J
        SUM=0.
        DO 10 I=1,NMJ
        IPJ=I+J
   10   SUM=SUM+PBUF(I)*PBUF(IPJ)
   20   ABUF(JJ)=SUM
C
C       OBTAIN PITCH VALUES FROM LAST 3 FRAMES
C
        P1=PITCH(1)
        P2=PITCH(2)
        P3=PITCH(3)
C
C       GET PEAK WITHIN RANGE [6,32]
C
        L=6
        AMAX=ABUF(L)
        DO 30 J=6,32
        IF (ABUF(J).LE.AMAX) GO TO 30
        AMAX=ABUF(J)
        L=J
   30   CONTINUE
C
C       TEST FOR MAX = ZERO
C
        IF (AMAX.EQ.0.) GO TO 60
C
C       TEST FOR LEFT HAND EDGE.
C       IF ABUF(L) IS NOT A PEAK, SET UNVOICED.
C
        IF (ABUF(L).LT.ABUF(L-1)) GO TO 60
C
C       PERFORM PARABOLIC INTERPOLATION
C       ABOUT LOCATION L
C
        AA=ABUF(L-1)-ABUF(L)
        AA=(AA+ABUF(L+1)-ABUF(L))/2.
        BB=(ABUF(L+1)-ABUF(L-1))/4.
        AP=ABUF(L)-BB*BB/AA
        AL=L-BB/AA
        V=AP/ABUF(1)
C
C       TEST WITH VARIABLE THRESHOLD
C
        IF (L.GE.19) GO TO 40
        D=-1.*(L-6.)/13.+2.
        GO TO 50
   40   D=-.1*(L-19.)/13.+1.
   50   V=V/D
C
C       DECISIONS
C
        IF (V  .GE. .35)  GO TO 70
        IF (P1 .EQ. .00)  GO TO 60
        IF (V  .GE. .30)  GO TO 70
   60   P0=0.
        GO TO 80
   70   P0=AL
C
   80   IF ( ABS(P1-P3) .LE. .375*P3 ) P2=(P1+P3)/2.
C
C       IF (P0 AND P1 ARE CLOSE) AND (P2 NOT 0)
C       BUT P3=0,THEN USE LINEAR EXTRAPOLATION FOR P2
C       (COMING OUT OF UNVOICED).
C
        IF (P3.NE.0.) GO TO 90
        IF (P2.EQ.0.) GO TO 90
        IF (ABS(P0-P1).GT.0.2*P1) GO TO 90
        P2=2.*P1-P0
C
C       TEST FOR ISOLATED "VOICED" AND
C       INCORRECT END OF "VOICED"
C
   90   IF (P1.NE.0.) GO TO 100
        IF (ABS(P2-P3).GT.(.375*P3)) P2=0.
C
C       UPDATE FRAMES
C
  100   PITCH(3)=P2
        PITCH(2)=P1
        PITCH(1)=P0
C
C       TRUE PITCH DELAYED BY TWO
C       FRAMES EQUALS (PITCH(3)-1)*5
C
        RETURN
        END
```

Fig. 8.18 Fortran subroutine STEP 2 for implementing the second step of the SIFT algorithm.

frequency contour for this utterance is the reciprocal of the pitch contour for $P \neq 0$. Not only is the voiced interval correctly obtained for the nasal /n/ region, but in addition, the rapidly changing final transition in "prediction" is correctly tracked without error. Furthermore, the fundamental frequency of the low amplitude voiced consonant /d/ is correctly extracted. It should be noted that the STREAK algorithm of Section 8.2 can also be applied with the same down-sampling procedure as used in the SIFT algorithm. The subroutine STEP 2 can be used without any modifications.

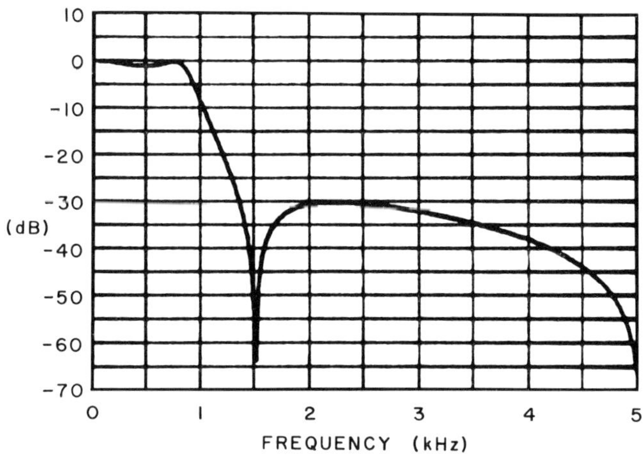

J	A(J)	P(J)
1	1.000000	0.0357082
2	-2.340366	0.0069956
3	2.011900	0.0069956
4	-0.614109	0.0357082

COEFFICIENT VALUES

Fig. 8.19 Frequency response of a third-order elliptic down-sampling filter with coefficient values.

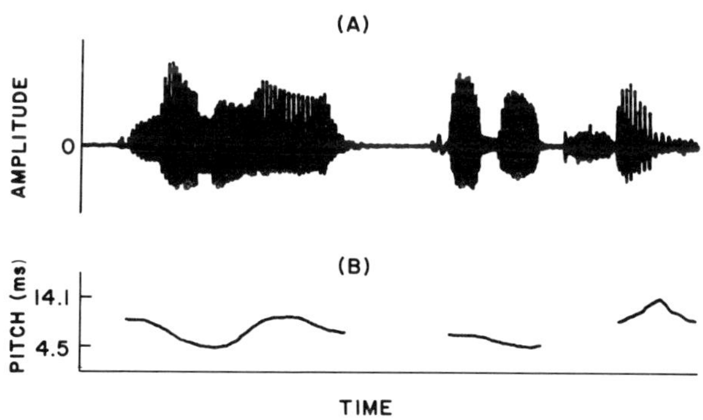

Fig. 8.20 Example output of SIFT algorithm in analysis of utterance "linear prediction".

9. Computational Considerations in Analysis

9.1 Introduction

The implementation of linear prediction techniques for all but trivial examples requires digital computation. In nearly all of the material presented thus far, an implicit assumption was made that the computation occurred without error. For example, the norm squares α_m and β_m of Chapter 3 are theoretically non-negative, yet when they are evaluated using the computer programs presented they may yield negative results due to computer errors. Numerical results are a function of the algorithms used, the form of computer arithmetic (e.g., floating or fixed point), and the computer word length.

When computers are applied to signal processing problems, it is desirable to strive for rapid and efficient implementation while retaining sufficient numerical accuracy and ease of programming. In the application of linear prediction to speech processing problems, there are four principal areas that encompass most of the computational effort and nearly all of the numerical problems. These areas are the computation of autocorrelation or covariance coefficients (matrix loading), solution of the autocorrelation or covariance equations (matrix solution), speech synthesis filter implementation, and discrete spectral computation.

The topic of synthesis filter structures and their implementation was covered in Chapter 5. Several computational considerations within the context of linear prediction vocoding are presented in Chapter 10. A general theory covering the numerical accuracy of the various synthesis structures has been developed [Gray and Markel, 1973, 1975a; Markel and Gray, 1975a, b], but will not be treated here due to the complexity of the error analysis.

The topic of discrete spectral computation using the FFT has been given extensive coverage in the literature [e.g., Gentleman and Sande, 1966; Welch, 1969; Brigham, 1974], and therefore will not be covered here.

The purpose of this chapter is to focus on computational considerations of the remaining two areas, matrix loading and solution. The topics to be considered are ill-conditioning, efficiency of programming, speed of calculation, and fixed-point arithmetic implementation.

9.2 Ill-Conditioning

In solving the autocorrelation or covariance equations for the coefficients of an inverse filter $A(z)$, the matrix equation

$$\mathbf{C}\mathbf{a} = -\mathbf{c} \tag{9.1}$$

from (3.6) is implicitly being solved. The matrix **C** and the transpose of the column vectors **a** and **c** are given by (3.7). The coefficients of the inverse filter make up the column vector a, obtained from (9.1) as

$$\mathbf{a} = -\mathbf{C}^{-1}\mathbf{c}, \tag{9.2}$$

with the assumption that $\mathbf{C}^{-1}$, the inverse of **C**, exists.

In carrying out the solution algorithms, the norm square β_m of the orthogonal polynomial set $\{B_m(z)\}$, given by

$$\beta_m = \langle B_m(z), B_m(z) \rangle, \tag{9.3}$$

is of computational importance. In the autocorrelation method β_m is identical to the norm square of the $A_m(z)$ polynomial, α_m. From (3.63) it is seen that the norm squares are related to the determinant of **C** by the product

$$\boxed{|\mathbf{C}| = \prod_{m=0}^{M-1} \beta_m} \quad . \tag{9.4}$$

This result shows that if any norm square equals zero, **C** is singular and does not have an inverse. Theoretically this condition cannot occur in the autocorrelation method. It can occur in the covariance method if and only if the input data sequence $\{x(n)\}$ exactly satisfies a homogeneous difference equation whose characteristic polynomial is $B_m(z)$.

As the coefficient β_m is used in the denominator of the reflection coefficient calculation (3.55), a small value of β_m can amplify numerical errors made in the numerator calculation and result in large errors in the reflection coefficient. As the coefficient β_m is computed from (3.52) as the sum of $m+1$ products, a small value indicates that β_m is relatively inaccurate since cancellation in the summation will have removed most of the numerical significance.

In the solution procedure, β_m should be inspected at each recursion step. If β_m is negative an error has been made and the process should terminate. If β_m is zero the process should also terminate, for theoretically, $B_m(z)$ gives the characteristic polynomial for the difference equation which the data sequence $\{x(n)\}$ satisfies, i.e., the $\{x(n)\}$ is predicted with zero error over the interval $[M, N-1]$. As a practical matter, if computation is performed with v significant figures, the computation should cease if β_m/β_0 is less than about 10^{1-v}.

This discussion does not depend upon the particular algorithm used, other than the fact that in the autocorrelation method, β_m equals α_m. Even if general linear simultaneous equation techniques such as the Gauss-Siedel method are used to solve (9.1), the same characteristics appear, for the set of coefficients $\{\beta_m\}$ then become the pivot elements at the separate steps. After analyzing numerous frames of speech data we have found that the accuracy is not a strong function of the particular algorithm. The norm square β_m will now be related to a meaningful measure of ill-conditioning and to the spectral flatness measure.

9.2.1 A Measure of Ill-Conditioning

There are a number of measures of ill-conditioning of matrices [Faddeev and Faddeeva, 1963; Turing, 1948; Todd, 1949], the most common of which is the ratio of the maximum and minimum eigenvalues of the matrix (for the case of strictly positive definite matrices). This ratio has sometimes been called the condition number for a matrix. To calculate such ratios involves more computational effort than appears justified in speech analysis. In the autocorrelation method, however, certain qualitative results can be obtained.

In using the autocorrelation method, all of the eigenvalues lie within the dynamic range of the spectrum [Grenander and Szego, 1958], and from the correlation matching viewpoint of linear prediction, the spectrum can be either the original speech spectrum or the model spectrum. Ekstrom [1973] applied a theorem of Szego's to show that for large values of M, the condition number approaches the dynamic range of the spectrum, the ratio of maximum and minimum values. This provides a motivation for using the dynamic range of the spectrum as an ill-conditioning measure [Makhoul and Wolf, 1972; Makhoul, 1975a], for it is an upper bound on the condition number and approaches the condition number as M increases.

An alternate measure which can be applied to both the covariance and the autocorrelation methods without spectral computation uses the ratio of the geometric and arithmetic means of the eigenvalues,

$$\varrho_M = \frac{\left[\prod_{i=1}^{M} \lambda_i\right]^{1/M}}{\frac{1}{M}\sum_{i=1}^{M} \lambda_i}. \tag{9.5}$$

Here ϱ_M defines the ill-conditioning measure for an Mth order analysis and the eigenvalues of $\mathbf{C}$ are λ_i. This ratio can be readily calculated, for the product of the eigenvalues is simply the determinant of the matrix, given by (9.4), and the sum of the eigenvalues is the trace, i.e.,

$$\prod_{i=1}^{M} \lambda_i = |\mathbf{C}| = \prod_{i=0}^{M-1} \beta_i, \tag{9.6a}$$

and

$$\sum_{i=1}^{M} \lambda_i = \sum_{i=1}^{M} c_{ii}. \tag{9.6b}$$

Therefore,

$$\boxed{\varrho_M = \frac{\left[\prod_{i=0}^{M-1} \beta_i\right]^{1/M}}{\frac{1}{M}\sum_{i=1}^{M} c_{ii}}} \quad . \tag{9.6c}$$

As the geometric mean must be less than or equal to the arithmetic mean, the ill-conditioning measure ϱ_M always lies between zero and one. It can equal zero if and only if $|\mathbf{C}|=0$, i.e., if and only if $\mathbf{C}$ is singular. A value of ϱ_M equal to one implies that all of the eigenvalues are equal. This result can be used with the fact that $\mathbf{C}$ is symmetric to prove that $\varrho_M=1$ if and only if $\mathbf{C}$ is proportional to an identity matrix which is trivial to invert (and thus perfectly conditioned).

In the autocorrelation method, ϱ_M can be related to the spectral flatness measure. In this case $\beta_i=\alpha_i$, and all of the diagonal elements of $\mathbf{C}$ are equal,

$$c_{ii}=r(0)=\beta_0=\alpha_0$$

leading to the result

$$\varrho_M = \left[\prod_{i=0}^{M-1} (\alpha_i/\alpha_0)\right]^{1/M}. \tag{9.7}$$

It has been pointed out [Gray and Markel, 1974b] that ϱ_M decreases with increasing M, approaching the limit

$$\boxed{\lim_{M\to\infty} \varrho_M = \varrho_\infty = \alpha_\infty/\alpha_0 = \Xi(X)} \tag{9.8}$$

where $\Xi(X)$ is the spectral flatness measure of the input data sequence, as discussed in Chapter 6. *Thus the spectral flatness measure is both a lower bound and a limiting value of ϱ_M, and as such can itself be considered a measure of ill-conditioning.* As such it leads to a more quantitative explanation for several qualitative statements:

1) It takes more accuracy to analyze voiced sounds than unvoiced sound (for voiced speech $\Xi(X)$ is generally smaller).

2) Increasing the sampling rate increases the amount of computational accuracy needed (since the higher frequency energy of speech generally decreases as frequency increases, $\Xi(X)$ will decrease).

3) Proper pre-emphasis can decrease the amount of computational accuracy needed (pre-emphasis will tend to whiten voiced spectra, thus increasing $\Xi(X)$).

9.2.2 Pre-emphasis of Speech Data

Differencing or approximate differencing is important for accenting higher formants [Markel, 1972b] and obtaining reasonable vocal tract area functions [Wakita, 1973b]. A more general first-order pre-emphasis of speech data can be effected by the filter of the form $1-\mu z^{-1}$. It has been shown that for the purpose of reducing ill-conditioning, the pre-emphasis filter should be a first-order inverse filter [Gray and Markel, 1974b; Makhoul, 1975a]. Therefore, an optimum choice for μ in the sense of maximizing the spectral flatness at the output of the pre-emphasis filter is

$$\mu = r(1)/r(0). \tag{9.9}$$

For most voiced sounds μ lies near one, giving an approximate differencer, whereas for most unvoiced sounds, μ is relatively small and the pre-emphasis filter has little effect.

The pre-emphasis filter coefficient can be severely quantized, since any value of μ between zero and twice the value $r(1)/r(0)$ will enhance the spectral flatness [Gray and Markel, 1974b].

9.2.3 Prefiltering before Sampling

It is necessary to filter an analog speech signal before sampling to ensure against aliasing. In many signal processing methods it is customary to use an analog filter with a cutoff substantially below $f_s/2$. If the filter cutoff is appreciably lower than $f_s/2$, it will both increase the spectral dynamic range and decrease the spectral flatness, thus increasing both ill-conditioning measures. It has been suggested that the analog filter have a cutoff frequency at or only slightly below $f_s/2$ to avoid ill-conditioning [Markel and Gray, 1974b]. However, both the spectral dynamic range and spectral flatness measures are measures of ill-conditioning for large filter orders. The spectral flatness measure is the lower bound and limiting value of ϱ_M as indicated by (9.8). For low-order filters it can be extremely conservative. If a signal was severely filtered so that the dynamic range was effectively infinite (because of energy removal in a filter stop band) and the spectral flatness measure was effectively zero, the solution (9.2) for finite values of M is not meaningless as the ill-conditioning measures based on the spectrum alone indicate, for finite values of M. It appears that a narrow portion of the spectrum can be filtered out with little change in actual ill-conditioning, provided that the order of the filter remains small. This does not seem unreasonable based upon the fact that the spectral model from linear prediction weights the spectral peaks more heavily than the valleys or null regions.

9.3 Implementing Linear Prediction Analysis

In this section, efficient computer techniques are presented for implementing the autocorrelation and covariance methods. Most of the discussion will focus on evaluation of the autocorrelation and covariance coefficients, also referred to as matrix loading. A direct count of operations, either from the equations or from the computer programs of Chapter 3, shows that most of the computational effort is spent in the matrix loading operation, i.e., the evaluation of the coefficients $r(i)$ in the autocorrelation method for $i=0, 1, \ldots, M$, or the evaluation of the coefficients c_{ij} in the covariance method for $j=0, 1, \ldots, i$, and $i=0, 1, \ldots, M$.

9.3.1 Autocorrelation Method

While the FFT is often used to evaluate autocorrelation values, the relatively small number of values needed, M, compared to the number of data samples, N, does not make it attractive for this application. A minimum of M zeros would have to be appended to the data, followed by an FFT, squaring the magnitudes of each of the FFT samples, and finally, an inverse FFT.

Several techniques for reducing the required number of multiplications in a direct calculation of the autocorrelation coefficients based upon factorization procedures have been introduced [Pfeifer, 1973; Blankinship, 1974; Kendall, 1974]. What is desired is an evaluation of the autocorrelation coefficients

$$r(i)=\sum_{n=0}^{N-1-i} x(n)x(n+i) \quad \text{for} \quad i=0, 1, \ldots, M.$$

The direct computation requires a sum of $N-1$ products in order to evaluate a single term $r(i)$. By factoring out common terms, the number of multiplications can be reduced for values of i in the range $0<i<N/2$. To illustrate this concept, $r(3)$ for $N=19$ can be expressed in the form

$$\begin{aligned}r(3)=&\{x(3)[x(0)+x(6)]+x(4)[x(1)+x(7)]+x(5)[x(2)+x(8)]\}\\&+\{x(9)[x(6)+x(12)]+x(10)[x(7)+x(13)]+x(11)[x(8)+x(14)]\}\\&+x(15)[x(12)+x(18)]+x(16)x(13).\end{aligned}$$

A reduction of one multiplication for each set of square brackets $[\cdot]$ can be seen, resulting in seven multiplications plus an additional term, instead of the seventeen required for a direct computation. The terms of the summation are shown in two groups of three in the brackets $\{\cdot\}$, to indicate the logical scheme behind this form of factoring, based on the approach by Pfeifer.

Figure 9.1 shows the multiplication count for a factored and unfactored form of $r(i)$ where $N=256$. As a result of factoring, the number of multiplications is reduced by 45 to 46 percent over the range $15<M<63$. While factoring reduces the number of multiplications by a significant percentage, it does not provide an equivalent percentage reduction in computation time. A factored autocorrelation

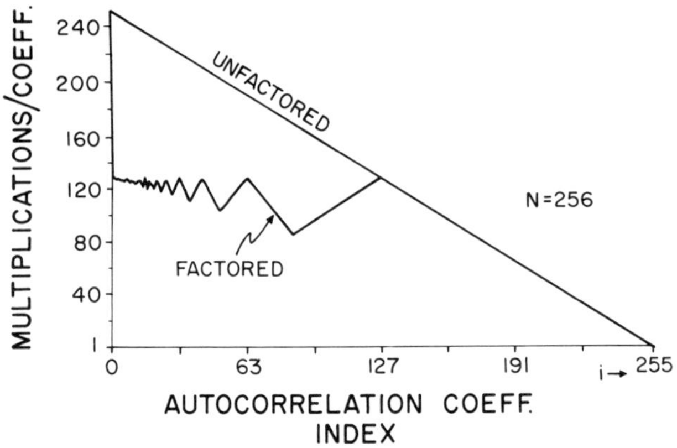

Fig. 9.1 Multiplication count necessary for factored and unfactored form as a function of the index. [From Pfeifer, 1973]

has a more complicated indexing scheme and therefore more overhead than the direct computation. Although the factored autocorrelation appears rather complicated when written in equation form, it lends itself to a simple computer program FACTAC as shown in Fig. 9.2. This program allows for computation of M over the range $1 \leq M < N/3$. Most of the complicated indexing is accounted for in

```
C
C       FACTORED AUTOCORRELATION
C       (VALID FOR M < N/3)
C
        SUBROUTINE FACTAC (M,N,R,X)
        DIMENSION R(1), X(1)
        RR=0.
        MP=M+1
        DO 10 I=1,N
   10   RR=RR+X(I)*X(I)
        R(1)=RR
        DO 60 I=2,MP
        IM1=I-1
        MOD=IM1+IM1
        NTERM=N-IM1
        RR=0.
        DO 20 J=I,MOD
        KR=J-IM1
        DO 20 K=J,NTERM,MOD
        KL=KR
        KR=K+IM1
   20   RR=RR+X(K)*(X(KL)+X(KR))
        IF(KR.EQ.N) GO TO 60
        IF(N-KR.LT.I) GO TO 30
        KSTRT=N-MOD+1
        KSTOP=KR
        GO TO 40
   30   KSTRT=KL+I
        KSTOP=NTERM
   40   DO 50 K=KSTRT,KSTOP
        KK=K+IM1
   50   RR=RR+X(K)*X(KK)
   60   R(I)=RR
        RETURN
        END
```

Fig. 9.2 Fortran subroutine FACTAC for implementing the factored autocorrelation calculation. [From Pfeifer, 1973]

Fortran DO loops. A comparison using a computer with a multiply-add time ratio of 1.1 resulted in a 26 percent faster computation than the unfactored form over the range $15 < M < 60$ with $N = 256$. For larger multiply-add time ratios or assembler programming (to reduce overhead operations) the computational efficiency can increase to 40 or 45 percent.

Figures 9.3 shows a slightly more efficient version of AUTO using the unfactored autocorrelation evaluation. The dimensioned variable $B(\cdot)$ of Chapter 3 has been removed since it contains the same information as the dimensioned variable $A(\cdot)$ but in reversed order. Other slight variations have been introduced with the net result being a reduction in multiplies from approximately $M^2 + 2M$ to $M^2 - (3/2)M$ and a reduction in adds from $M^2 + M + 1$ to $M^2 - (3/2)M - 1$. The subroutine FACTAC can be inserted in the subroutine AUTO in place of *DO* loop 10 to obtain greater efficiency at the expense of a slightly larger program.

9.3.2 Covariance Method

In the subroutine COVAR from Chapter 3, a dominant amount of computational effort is spent in evaluation of the covariance coefficients

$$c_{ij} = \sum_{n=M}^{N-1} x(n-i)x(n-j).$$

```
C
      SUBROUTINE AUTO(N,X,M,A,ALPHA,RC)
      DIMENSION X(1),A(1),RC(1)
      DIMENSION R(21)
      MP=M+1
      DO 10 K=1,MP
      R(K)=0.
      NK=N-K+1
      DO 10 NP=1,NK
   10 R(K)=R(K)+X(NP)*X(NP+K-1)
      RC(1)=-R(2)/R(1)
      A(1)=1.
      A(2)=RC(1)
      ALPHA=R(1)+R(2)*RC(1)
      DO 40 MINC=2,M
      S=0.
      DO 20 IP=1,MINC
   20 S=S+R(MINC-IP+2)*A(IP)
      RC(MINC)=-S/ALPHA
      MH=MINC/2 + 1
      DO 30 IP=2,MH
      IB=MINC-IP+2
      AT=A(IP)+RC(MINC)*A(IB)
      A(IB)=A(IB)+RC(MINC)*A(IP)
   30 A(IP)=AT
      A(MINC+1)=RC(MINC)
      ALPHA=ALPHA+RC(MINC)*S
      IF(ALPHA) 50,50,40
   40 CONTINUE
   50 RETURN
      END
```

Fig. 9.3 An efficient Fortran subroutine AUTO for implementing the autocorrelation method of linear prediction.

The calculations can be performed more efficiently by noting [McDonough, 1963] that

$$c_{ij}=c_{i-1,j-1}+x(M-i)x(M-j)-x(N-i)x(N-j). \tag{9.10}$$

As a result, only the coefficients c_{0i} require a full summation of $N-M$ products, since the c_{ij} terms which do not have zero subscripts can be evaluated in terms of $c_{i-1,j-1}$ and two additional multiplies and adds. The number of calculations can be further reduced [Blankinship, 1973] by factoring the c_{0i} terms in a manner similar to that described for evaluating the $r(i)$ terms in the autocorrelation method.

Storage requirements in the COVAR subroutine of Chapter 3 can be reduced as follows. First, it can be noted that only one column (or row) of the covariance matrix is needed at each step in the recursion. Thus, that column (or row) can be stored as a one-dimensional array, and after it has been used to update the solution it can be used as indicated by (9.10) to generate the next column (or row) which can be stored in the same array. Second, the two-dimensional array which stores the coefficients of $B_i(z)$, b_{ij} is wasteful, since $b_{ij}=0$ for $j>i+1$. The two-dimensional array can be converted to a one-dimensional array by storing the coefficients in the order

$$b_{01}, b_{11}, b_{12}, b_{21}, b_{22}, b_{23}, b_{31}, b_{32}, b_{33}, b_{34}, b_{41}, \ldots .$$

The storage can be further reduced by using the fact that $b_{i,i+1}=1$.

A Fortran subroutine which implements these modifications (except for a factored evaluation of the c_{0i} terms) is shown in Fig. 9.4. The covariance coefficients are stored in the array $CC(\cdot)$ and the coefficients of the polynomials $B_i(z)$ are stored in the array $B(\cdot)$. The arguments are identical to those described for COVAR of Chapter 3. The matrix solution procedure is effectively unchanged from that presented in Chapter 3 with a reduction of only two multiplications and one addition.

9.3.3 Computational Comparison

The approximate number of operations necessary to implement the autocorrelation method (AUTO) and the covariance method (COVAR) as functions of N and M, are shown in Table 9.1, for the programs in Chapter 3 and those just presented in Section 9.3. Since the number of adds is nearly equal to the number of multiplies, an operation is defined as one multiply and one add. The number of operations for a typical case of $N=128$ and $M=10$ is shown in brackets along with the results as a function of N and M. It is seen that the computational effort required to actually solve the equations is substantially less than the effort in computing the coefficients for the equations. For the efficient programs in this chapter, the equation solution takes only 6 percent of the computational effort for AUTO and 27 percent of the effort for COVAR.

With efficient computation of the covariance coefficients, COVAR requires

```
C
      SUBROUTINE COVAR(N,X,M,A,ALPHA,GRC)
      DIMENSION X(1),A(1),GRC(1)
      DIMENSION B(210),BETA(20),CC(21)
      MP=M+1
      ALPHA=0.
      CC(1)=0.
      CC(2)=0.
      DO 10 NP=MP,N
      NP1=NP-1
      ALPHA=ALPHA+X(NP)*X(NP)
      CC(1)=CC(1)+X(NP)*X(NP1)
   10 CC(2)=CC(2)+X(NP1)*X(NP1)
      B(1)=1.
      BETA(1)=CC(2)
      GRC(1)=-CC(1)/CC(2)
      A(1)=1.
      A(2)=GRC(1)
      ALPHA=ALPHA+GRC(1)*CC(1)
      MF=M
      DO 130 MINC=2,MF
      DO 20 J=1,MINC
      JP=MINC+2-J
      N1=MP+1-JP
      N2=N+1-MINC
      N3=N+2-JP
   20 CC(JP)=CC(JP-1)+X(MP-MINC)*X(N1)-X(N2)*X(N3)
      CC(1)=0.
      DO 30 NP=MP,N
   30 CC(1)=CC(1)+X(NP-MINC)*X(NP)
      MSUB=(MINC*MINC-MINC)/2
      MM1=MINC-1
      B(MSUB+MINC)=1.
      DO 70 IP=1,MM1
      ISUB=(IP*IP-IP)/2
      IF (BETA(IP)) 140,70,40
   40 GAM=0.
      DO 50 J=1,IP
   50 GAM=GAM+CC(J+1)*B(ISUB+J)
      GAM=GAM/BETA(IP)
      DO 60 JP=1,IP
   60 B(MSUB+JP)=B(MSUB+JP)-GAM*B(ISUB+JP)
   70 CONTINUE
      BETA(MINC)=0.
      DO 80 J=1,MINC
   80 BETA(MINC)=BETA(MINC)+CC(J+1)*B(MSUB+J)
      IF (BETA(MINC)) 140,120,90
   90 S=0.
      DO 100 IP=1,MINC
  100 S=S+CC(IP)*A(IP)
      GRC(MINC)=-S/BETA(MINC)
      DO 110 IP=2,MINC
      M2=MSUB+IP-1
  110 A(IP)=A(IP)+GRC(MINC)*B(M2)
      A(MINC+1)=GRC(MINC)
  120 CONTINUE
      S=GRC(MINC)*GRC(MINC)*BETA(MINC)
      ALPHA=ALPHA-S
      IF (ALPHA) 140,140,130
  130 CONTINUE
  140 RETURN
      END
```

Fig. 9.4 An efficient Fortran subroutine COVAR for implementing the covariance method of linear prediction.

Table 9.1. A comparison of computational effort for the autocorrelation (AUTO) and the covariance (COVAR) methods.

		Matrix load operations	Matrix solve operations	Total operations $\left(\begin{matrix} N=128 \\ M=10 \end{matrix}\right)$	Ratio load to total
AUTO	CH3	$(M+1)(2N-M-2)$ [1342]	M^2+2M [120]	[1462]	.92
	CH9	$(M+1)(2N-M-2)$ [1342]	$M^2-\frac{3}{2}M$ [85]	[1427]	.94
COVAR	CH3	$(N-M)(M+1)(M+2)$ [7788]	$\frac{1}{3}M^3+2M^2+\frac{8}{3}M-2$ [558]	[8346]	.93
	CH9	$2(M+2)(N-1)$ [1524]	$\frac{1}{3}M^3+2M^2+\frac{8}{3}M-4$ [556]	[2080]	.73

only 1.4 times longer computation for $N=128$ and $M=10$, even though the solution requires $O(M^3)$ operations as opposed to $O(M^2)$ operations for the autocorrelation method. *Therefore, except under stringent computational conditions, the choice of the autocorrelation method or covariance method for linear prediction analysis can be based upon the inherent properties of the methods, e.g., accuracy of complex exponential representation versus stability, instead of computer speed restrictions.* By direct application of standard linear simultaneous equation solution subroutines and direct coefficient evaluation, the covariance method would require approximately six times greater computational effort.

9.4 Finite Word Length Considerations

Results from a study of fixed-point implementation problems in the autocorrelation method are now presented [Markel and Gray, 1974b]. These results can be extrapolated to the covariance method in the sense that the results for the autocorrelation method will generally present a lower bound for the necessary requirements in the covariance method. The total squared error (for voiced sounds) will generally be lower for the covariance method which implies a larger numerical dynamic range. In addition, the Gram-Schmidt orthogonalization procedure for obtaining the $B_m(z)$ polynomials is more numerically involved for the covariance method.

For these experiments, two representative sampling frequencies ($f_s=6.7$ kHz and $f_s=10.0$ kHz) and pre-emphasis factors ($\mu=0$ and $\mu=1$) were used. Two figures of merit were also used for quantitatively judging the results for various word lengths. Since the reflection coefficients $\{k_m\}$ are theoretically bounded by unity, the number of frames N_I or relative number of frames, in which an instability occurred ($|k_m| > 1$), was recorded as a possible figure of merit. A more physically meaningful figure of merit also used was the mean square error (MSE) between $ln|\sigma/A[\exp(j\theta)]|^2$ obtained with full precision floating-point arithmetic and that obtained using β-bit fixed-point arithmetic. The MSE term was evaluated by applying a 256-point DFT to the filter coefficients in the fixed-point and floating-point cases in the manner discussed in Chapter 6. If any $|k_m| \geq 1$ was measured, the filter coefficients in the previous recursion step were used along with $M = m-1$ since $1/A_{m-1}(z)$ is then stable and $1/A_m(z)$ is unstable.

Finally, the experiments were performed using two's complement fixed-point arithmetic with truncation since this is the arithmetic form generally available in computers. The results using truncation arithmetic place a lower bound on necessary word lengths when the more desirable form of rounding arithmetic is used. For these experiments, simulation programs were written to allow variable word length implementation.

9.4.1 Finite Word Length Coefficient Computation

Results obtained by computing the autocorrelation coefficients exactly with double precision integer accumulation and then truncating to the most significant β bits are shown in Fig. 9.5. Figures 9.5A and B show the number of instabilities N_I measured as a function of β for the conditions $f_s = 6.7$ kHz and $f_s = 10.0$ kHz, respectively, with pre-emphasis as a parameter. The all voiced utterance was divided into 125 equally spaced analysis frames for both sampling frequency conditions.

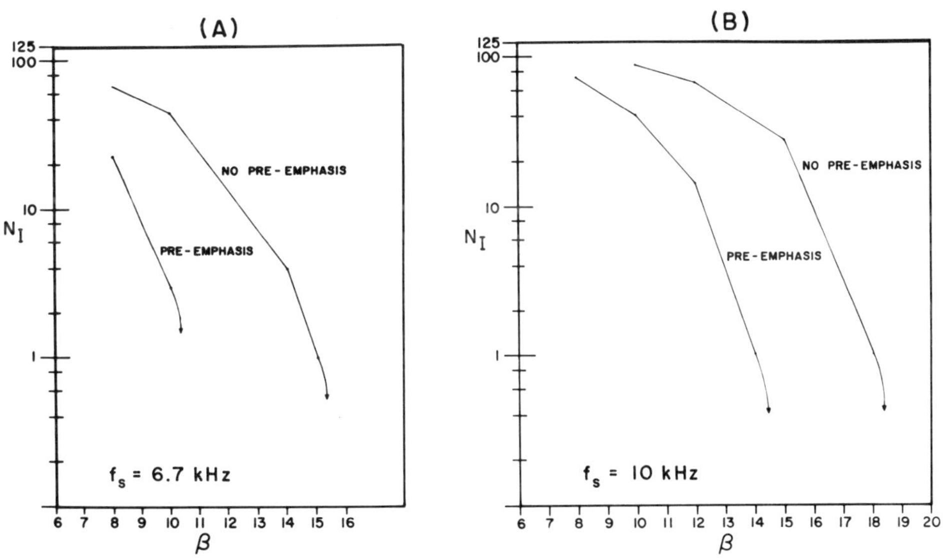

Fig. 9.5 Effects of coefficient truncation only. A) $f_s = 6.7$ kHz. B) $f_s = 10$ kHz.

For $f_s = 10$ kHz and no pre-emphasis, completely unsatisfactory results are obtained by truncating the coefficients to 16 bits (a common word length for small computers). On the other hand, if f_s is lowered to 6.7 kHz with pre-emphasis, zero instabilities are observed for 11 bits. For each choice of sampling frequency, pre-emphasis leads to an advantage of approximately 4 bits over a large range of β (that is, the curves are displaced with respect to one another by about 4 bits). The lower sampling frequency results in a 3-bit advantage for either pre-emphasis or no pre-emphasis. The $f_s = 6.7$ kHz curve with pre-emphasis lies almost precisely 7 bits to the left of the $f_s = 10$ kHz curve with no pre-emphasis over the range of N_I from 2 to 20.

The effects of solving the autocorrelation equations using β-bit arithmetic with exact autocorrelation coefficients, and with autocorrelation coefficients obtained using β-bit arithmetic, are now considered.

9.4.2 Finite Word Length Solution of Equations

The scaling necessary for a fixed-point solution of the autocorrelation equations is trivial. First, the reflection coefficient is theoretically bounded by unity. Since the denominator term α_m satisfies $\alpha_m \leq \alpha_{m-1} \leq r(0)$, the numerator term must be bounded by unity (see Chapter 3, Eq. (3.55) with $\beta_m = \alpha_m$). By normalizing the autocorrelation sequence so that, in effect, $1/2 \leq r(0) < 1$, the magnitudes of all parameters except the filter coefficients are theoretically bounded by unity. Although the filter coefficients are theoretically bounded only by the binomial coefficients, experimentally they are conservatively bounded by 4 for $f_s = 6.7$ kHz and 8 for $f_s = 10.0$ kHz. By initially scaling a_0 to 1/4 or 1/8, and inverse scaling in (3.55), a fractional, fixed-point algorithm is obtained.

The robustness of the fixed-point algorithm is demonstrated by the experimental results shown in Fig. 9.6. The same conditions were used to obtain Fig. 9.5 except that now the autocorrelation equations are also solved with β-bit arithmetic.

The surprising result of this experiment was that for either pre-emphasis or no pre-emphasis, $f_s = 6.7$ or 10 kHz, and for the full range of β, there was at most a 1.5-bit degradation due to finite word length effects in the equation solution stage.

If overflow occurred, the procedure was terminated, the frame was defined as unstable, and the last polynomial $A_m(z)$ obtained without overflow was used. This procedure was found to be necessary for analyzing with small word lengths where over a third of the frames were unstable. This range of β would obviously not be used under normal conditions.

It is important to realize that the results just presented are equivalent to performing the complete autocorrelation method where the autocorrelation coefficients are computed using double precision accumulation with the final results truncated to the most significant β-bits, and the autocorrelation equations are computed using β-bit single precision fixed-point arithmetic. This result is made obvious by noting that each of the autocorrelation coefficients is stored in, at most, a β-bit register for the solution.

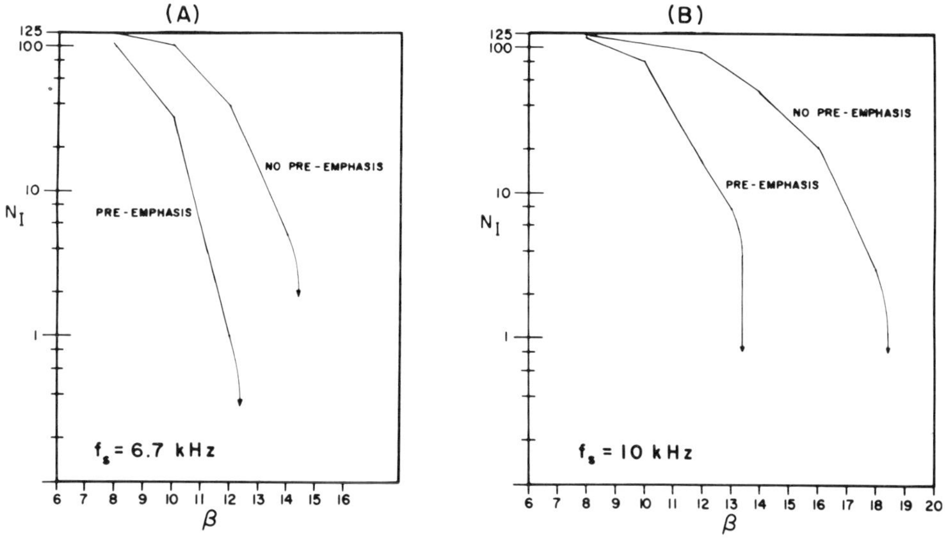

Fig. 9.6 Effects of truncation in solution of auto-correlation equations. A) $f_s = 6.7$ kHz. B) $f_s = 10$ kHz.

9.4.3 Overall Finite Word Length Implementation

The number of instabilities as a function of β for a complete fixed-point implementation of both the autocorrelation coefficients and the solution of the autocorrelation equations is shown in Fig. 9.7. For $f_s = 6.7$ kHz and pre-emphasis, 15 bits

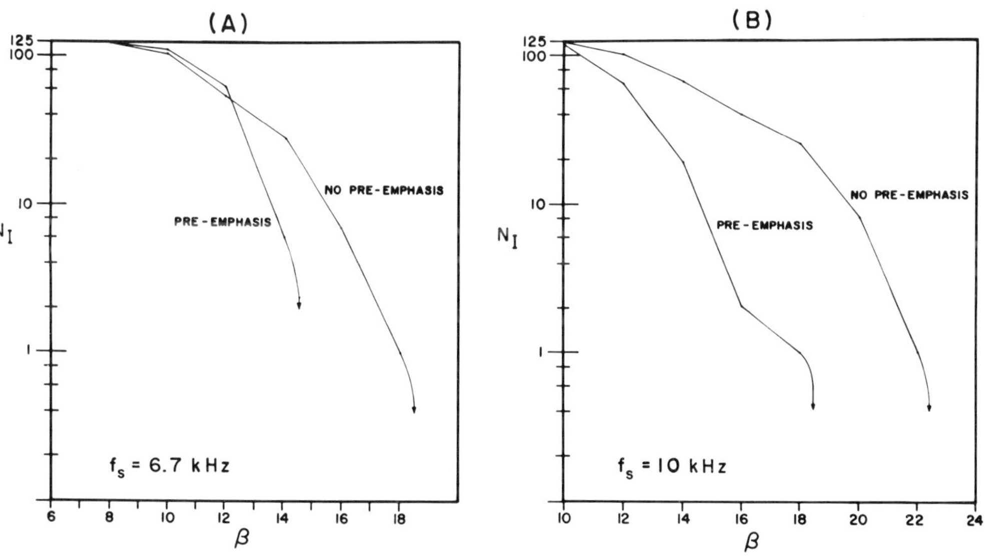

Fig. 9.7 Effects of truncation in overall autocorrelation method implementation. A) $f_s = 6.7$ kHz. B) $f_s = 10$ kHz.

result in a completely stable analysis. However, if $f_s = 10$ kHz, a 23-bit word length is necessary to ensure stability in all frames.

By comparing Figs. 9.6 and 9.7, it is seen that, depending upon the particular condition, the fixed-point computation of the autocorrelation coefficients alone degrades the system by 2 to 3 bits.

In general, pre-emphasis is very beneficial. For $\beta = 10$ and 12 in Fig. 9.7A, however, there are a larger number of instabilities for the pre-emphasis case. This effect was caused by several unusual overflow conditions in the autocorrelation coefficient calculations. These results are of little consequence due to the fact that an intolerable number of instabilities is obtained for practical application. Also for Fig. 9.7B, the apparently strange behavior from 16 to 18 bits for the pre-emphasis case is not statistically significant since the difference between a smoothly descending curve at $\beta = 16$ without an inflection point and the actual result is one unstable frame.

In Fig. 9.8, the resulting MSE is shown for comparison with Fig. 9.7. There is a surprisingly close correlation between the relative frequency of occurrence of instabilities N_I/N_F and the MSE where N_F is the total number of frames. In particular, MSE (dB^2) $= 75\ N_I/N_F$ is a very good order of magnitude approximation between all corresponding curves for $B = 12$. In view of the non-linear nature of the problem, it is not clear that any simple explanation of this relationship exists.

General conclusions from these experiments for computer implementation of linear prediction with fixed-point arithmetic are: 1) pre-emphasis should be applied, 2) the sampling frequency should be as low as possible, and 3) the calculation of the autocorrelation (or covariance coefficients) should be performed with full precision and only the final results should be truncated (or rounded) to β-bits, for maximal accuracy.

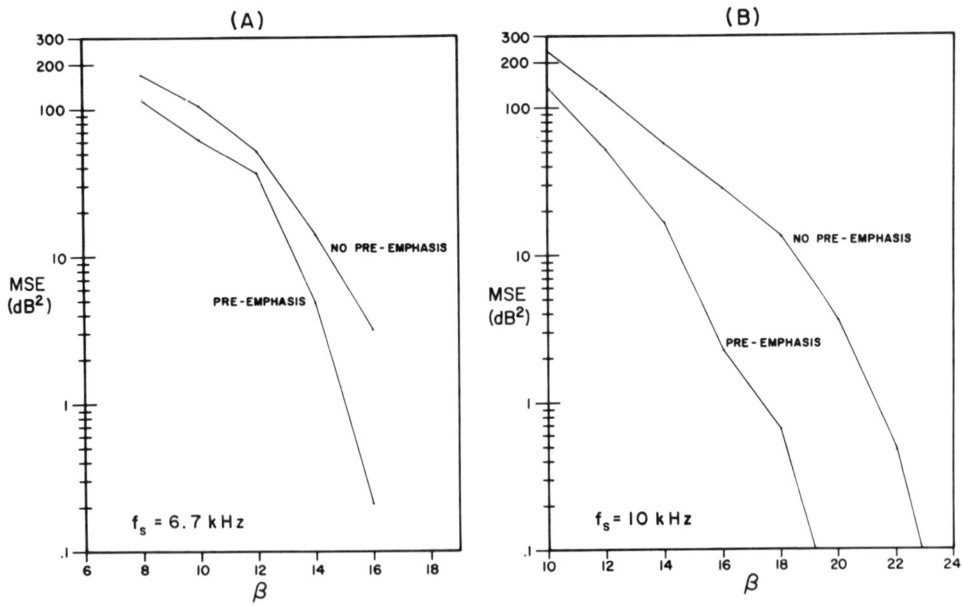

Fig. 9.8 Mean square error due to truncation. A) $f_s = 6.7$ kHz. B) $f_s = 10$ kHz.

10. Vocoders

10.1 Introduction

The term *vocoder* is a contraction of *voice-coder* as defined by Dudley [1939]. Since the inception of the vocoder great effort and time have been expended on improving the synthesis quality of devices that code speech for efficient transmission. An excellent summary of that effort up to 1966 is presented by Schroeder [1966]. One of the major problems was the difficulty in obtaining accurate separation of the driving function behavior (fundamental frequency) from the vocal tract behavior (formant structure). The first published reference to linear prediction of speech (using maximum likelihood formulation) was in the same year [Saito and Itakura, 1966]. Linear prediction techniques have made a major impact on the development of vocoder systems, since source-tract interaction problems are efficiently resolved, resulting in improved synthetic speech quality, and since the technology is available for the necessary implementation in real-time hardware.

Many new techniques have been developed which allow reduced transmission bit rates while retaining acceptable synthesis quality. An initial reaction to linear prediction vocoding was that it might be a good laboratory tool for studying speech behavior but since linear simultaneous equations had to be solved, too much numerical accuracy and complexity would be necessary for a real-time vocoder. Based upon finite word length considerations such as those discussed in the previous chapter, and the rapid development of hardware technology, however, several sophisticated real-time linear prediction vocoder systems have been developed.

There are two main purposes to this chapter. First, general techniques, and properties of various transmission parameters, are presented outside the scope of specific vocoder implementations. Specific linear prediction vocoder simulations or systems are then discussed. Although there have been a number of waveform coding techniques developed that make use of linear prediction, only those without feedback loop quantizers will be discussed here. An example of systems using feedback loop quantizers would be the adaptive predictive coding method [Atal and Schroeder, 1968 b, c, 1970 c]. The characteristics of these systems are predominantly determined by the quantizer and sampling frequency and, to a much lesser degree, the characteristic of the linear predictor.

The basic form of a pitch-excited linear prediction vocoder is illustrated in Fig. 10.1. The speech signal $s(t)$ is filtered to no more than one-half the system sampling frequency and then analog-to-digital conversion is performed. The speech is processed on a frame-by-frame basis where the analysis frame length can be variable, but the analysis frame rate is generally fixed. For each frame a pitch period estimation is made along with a voicing decision. A linear prediction coefficient analysis is performed to obtain an inverse model of the speech spectrum $A(z)$. In addition, a gain parameter σ, representing some function of the speech energy, is computed. An encoding procedure is then applied for transforming the analyzed parameters into an efficient set of transmission parameters with the goal of minimizing degradation in the synthesized speech for a specified number of bits. Knowing the transmission frame rate and the number of bits used for each transmission parameter, one can compute a noise-free channel transmission bit rate. Coding techniques for noisy channels will not be considered here except to note that a specified noise-free bit rate may have to be substantially increased to protect against parameter errors in a noisy channel.

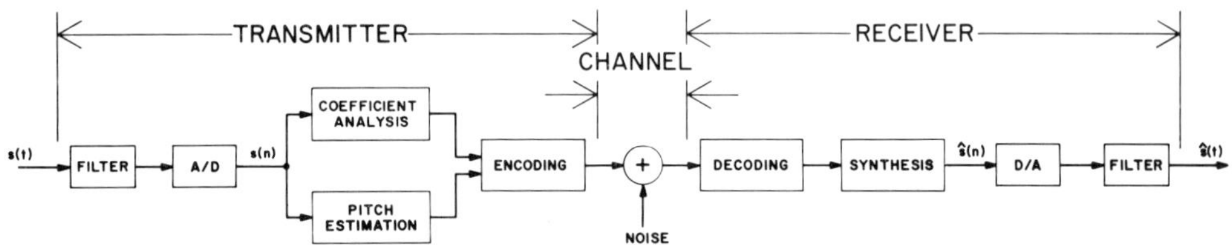

Fig. 10.1 Block diagram of a pitch-excited linear prediction vocoder.

At the receiver, the transmitted parameters are decoded into quantized versions of the coefficient analysis and pitch estimation parameters. An excitation signal for synthesis is then constructed from the transmitted pitch and voicing parameters. The excitation signal then drives a synthesis filter $1/A(z)$ corresponding to the analysis model $A(z)$. The digital samples $\hat{s}(n)$ are then passed through a digital-to-analog converter and low-pass filtered to generate the synthetic speech $\hat{s}(t)$. Either before or after synthesis, the gain term is used to match the synthetic speech energy to the actual speech energy. The digital samples are then converted to an analog signal and passed through a filter similar to that used at the input of the system.

An elementary visualization of the analysis-synthesis model was shown in Chapter 1. Methods of performing coefficient analysis were presented in Chapters 2 and 3. Considerations in the choice of analysis conditions and spectral characteristics of the resultant analysis model $A(z)$ were covered in Chapter 6. Pitch estimation with linear prediction techniques has been discussed in Chapter 8. Synthesis structures for implementing $1/A(z)$ were presented in Chapter 5.

In the next section, techniques which are specifically of importance to linear prediction vocoding and which have not been covered in previous chapters will be discussed.

10.2 Techniques

10.2.1 Coefficient Transformations

An obvious transmission parameter set based upon the linear predictor formulations is the M coefficients of the filter $A(z)$ from each analysis frame. These coefficients were in fact used in Itakura and Saito's maximum likelihood vocoder system [1968] with each parameter coded to 9 bits. Atal and Hanauer [1971b] discussed the use of filter coefficients for transmission and noted that not only is relatively high accuracy needed (8 – 10 bits per coefficient), but in addition, linear interpolation of the parameters at the receiver cannot guarantee stability. By solving for and transmitting the roots of $A(z)$, it was possible to accurately represent the spectral information contained in $A(z)$ with an average of 5 bits per parameter. In addition, as long as the transmitted roots correspond to stable synthesis filters, linear interpolation will guarantee stability.

Numerous other transformations have been used. A very important transformation is the set of PARCOR coefficients or reflection coefficients $\{k_m\}$ [Itakura and Saito, 1969, 1972b; Atal and Hanauer, 1971b]. They are directly obtained in the autocorrelation method as described in Chapters 2 and 3, and can be recursively obtained through the step-down relations of Chapter 5 if the covariance method is used. As a necessary and sufficient condition for stability, these parameters must be bounded in magnitude by unity and, furthermore, linear interpolation between stable filters by means of these parameters will result in stable filters.

The parameters $\{k_m\}$ have been related to the areas of a non-uniform acoustic tube $\{\mathscr{A}_m\}$ [Atal and Hanauer, 1971b; Wakita, 1972]. This transformation was discussed in Chapter 4, with the direct transformation from the parameter k_m to the area function ratio. $\mathscr{A}_m/\mathscr{A}_{m-1}$ as given by (4.17). Both the log area function ratios $\{ln(\mathscr{A}_m/\mathscr{A}_{m-1})\}$ and the area functions themselves $\{\mathscr{A}_m\}$ have been used as transmission parameters. Linear interpolation between two sets of these parameters leads to stable filters.

Another transformation which assures synthesis filter stability after interpolation is based upon the autocorrelation coefficients [Atal and Hanauer, 1971b]. In the autocorrelation method, the analysis can be separated into the autocorrelation coefficient computation at the transmitter and the solution of the autocorrelation equations at the receiver. In the covariance method the step-down and step-up procedures of Chapter 5 can be followed including the application of (2.64) during the step-up procedure to obtain the autocorrelation coefficients $\{r(m)\}$. At the receiver, the autocorrelation equations are solved as though the procedure were identical to the autocorrelation method. The stability of the synthesis filter based upon interpolation of the autocorrelation coefficients arises from the fact that linear interpolation on the elements of two positive definite Toeplitz matrices yields another positive definite Toeplitz matrix. This guarantee of stability assumes perfect computations, since errors in the calculations can destroy the positive definite property of the resulting matrices.

The parameter sets which guarantee stability with interpolation between stable frames are the roots of $A(z)$, the parameters $\{k_m\}$, the area functions, the area ratios, the log area ratios, and the autocorrelation coefficients. The parameter sets in which instabilities introduced by numerical errors can be most easily detected are the polynomial roots (magnitudes must be less than one for stability), the parameters $\{k_m\}$ (magnitudes must be less than one for stability), the area functions and area ratios (values must be positive for stability). Log area ratio errors cannot result in instability since exponentiation will always produce a positive area ratio.

Another transformation which may be of value in variable frame rate transmission (see Section 10.2.3) is based upon the cepstrum of the synthesis filter [Gold and Rader, 1969, p. 246; Atal, 1974a]. The cepstrum of a stable synthesis filter can be found recursively by using the fact that $A(z)$ has all of its roots inside of the unit circle. As a result, $ln[A(1/z)]$ is analytic inside the unit circle, and can be represented in a Taylor series, leading to the Laurent expansion

$$\begin{aligned} ln[\sqrt{\alpha}/A(z)] &= ln[\sqrt{\alpha}] - ln[A(z)] \\ &= ln[\sqrt{\alpha}] + \sum_{i=1}^{\infty} c(k) z^{-k}. \end{aligned} \tag{10.1}$$

Differentiating both sides of this equation with respect to z^{-1}, and multiplying by z^{-1}, gives

$$-\sum_{k=1}^{M} k a_k z^{-k} = \sum_{i=0}^{M} a_i z^{-i} \sum_{k=1}^{M} k c(k) z^{-k}. \tag{10.2}$$

If the coefficients of the powers of z^{-1} are equated, then

$$\boxed{-nc(n) - na_n = \sum_{k=1}^{n-1} (n-k) c(n-k) a_k \quad \text{for} \quad n>0}\;, \tag{10.3a}$$

where

$$a_0 = 1 \quad \text{and} \quad a_k = 0 \quad \text{for} \quad k > M. \tag{10.3b}$$

This result can easily be solved for either a_n or $c(n)$ in terms of the lower ordered coefficients, thus producing recursive relationships. In terms of the log magnitude spectrum, the Taylor series can be written as

$$ln[\alpha/|A(e^{j\theta})|^2] = ln[\sqrt{\alpha}/A(e^{j\theta})] + ln[\sqrt{\alpha}/A(e^{-j\theta})]$$

so that

$$\boxed{ln[\alpha/|A(e^{j\theta})|^2] = \sum_{k=-\infty}^{\infty} c(k) e^{-jk\theta}}\;, \tag{10.4a}$$

where

$$c(0) = ln(\alpha) \quad \text{and} \quad c(-n) = c(n). \tag{10.4b}$$

The Fortran subroutine LPTRAN of Fig. 10.2 recursively implements each of the important linear prediction transformations. The steps involved have been described at various appropriate places in the book. Only the use of the subroutine is described here.

The inputs to the subroutine are I, an index representing which parameter set is to be considered as an input set, M, the order of the filter, and the input parameter set. All other parameter sets then become the program output. The separate choices are listed in Table 10.1. It is important to note that $a_0 = A(1) = 1$ for both input and output. In addition, $\mathscr{A}_m = AREA(M+1) = 1$ at the output. The input

Table 10.1. Definitions of text and Fortran variables for subroutine LPTRAN

I	Name	Text	*FORTRAN*	Subscript Range
1	filter coefficients	a_j	$A(j+1)$	$j=0, 1, \ldots, M$
	gain term	α	$ALPHA$	
2	cepstral coefficients	$c(j)$	$C(j+1)$	$j=0, 1, \ldots, M$
3	autocorrelation	$r(j)$	$R(j+1)$	$j=0, 1, \ldots, M$
4	reflection coefficients	k_j	$RC(j)$	$j=0, 1, \ldots, M$
	input energy	$r(0)$	$R(1)$	
5	log area ratios	$ln(\mathscr{A}_j/\mathscr{A}_{j-1})$	$ALAR(j)$	$j=0, 1, \ldots, M$
	input energy	$r(0)$	$R(1)$	
6	area functions	$\mathscr{A}_j$	$AREA(j+1)$	$j=0, 1, \ldots, M$
	input energy	$r(0)$	$R(1)$	

energy $r(0)$ is necessary for $I=4$, 5, or 6 if transformations to parameters $I=1$, 2, or 3 are desired with the correct α, $c(0)$, or $r(0)$, respectively. Otherwise, $r(0)$ can be chosen as unity. If only certain transformations are required, then the subroutine can be reduced in size by eliminating unnecessary portions. There is only one additional dimensioned variable used by the subroutines, a shifted set of filter coefficients,

$$a_j = A(j+1) = SA(j) \quad \text{for} \quad j=1, 2, \ldots, M. \tag{10.5}$$

This variable is used in some of the intermediate calculations so that the input remains unchanged.

No specific stability test is built into the program. If the filter is unstable, then at least one reflection coefficient has a magnitude of one or greater, at least one area function is negative, and, what is more important, there will be an attempt to take the logarithm of a negative number in the evaluation of $ALAR(j)$ for some argument. A stability test is easily added if it is desired.

The program can be used to test itself by changing the index I after each use. An example test program is shown in Fig. 10.3A. Here the test program generates the first parameter set to be used as an input, for $I=1$, using the filter coefficients $a_0=1$, $a_1=-.45$, $a_2=.81$, $a_3=0$ with $M=3$, and $\alpha=10$. By incrementing I from 1 through 6, the roles of the inputs are changed according to Table 10.1. Each printout will be the same, except for possible computer roundoff errors and the index I. A sample printout is shown in Fig. 10.3B for I=1.

```
C
      SUBROUTINE LPTRAN(I,M,A,C,R,RC,
     1                  ALAR,AREA,ALPHA)
      DIMENSION A(1),C(1),R(1),RC(1)
      DIMENSION ALAR(1),AREA(1)
      DIMENSION SA(50)
      MP=M+1
      IF(I-2) 40,10,40
C
C     GENERATES A(.),ALPHA, FROM C(.)
   10 ALPHA=EXP(C(1))
      A(1)=1.
      DO 30 K=1,M
      KP=K+1
      SUM=0.
      DO 20 J=1,K
      JB=K-J+2
   20 SUM=SUM+A(J)*C(JB)*(JB-1)
   30 A(KP)=-SUM/K
C
   40 GO TO(100,100,150,50,190,210),I
C
C     ..GENERATES SA(.), R(.), & ALPHA
C     .. FROM RC(.) & R(1)
C
   50 DO 60 J=1,M
   60 SA(J)=RC(J)
      R(2)=-RC(1)*R(1)
      ALPHA=R(1)*(1.-RC(1)*RC(1))
      DO 90 J=2,M
      MH=J/2
      Q=RC(J)
      ALPHA=ALPHA*(1.-Q*Q)
      DO 70 K=1,MH
      KB=J-K
      AT=SA(K)+Q*SA(KB)
      SA(KB)=SA(KB)+Q*SA(K)
   70 SA(K)=AT
      SUM=0.
      DO 80 L=1,J
      LB=J+1-L
   80 SUM=SUM+SA(L)*R(LB)
   90 R(J+1)=-SUM
C
      IF(I-4) 230,230,250
C     ......
  100 DO 110 J=1,M
  110 SA(J)=A(J+1)
C
C     ..GENERATES RC(.),R(1),
C     ..FROM SA(.) & ALPHA
C
      DO 120 J=1,M
  120 RC(J)=SA(J)
      ALT=ALPHA
      DO 140 J=2,M
      JB=M+1-J
      MH=(JB+1)/2
      RCT=RC(JB+1)
      D=1.-RCT*RCT
      DO 130 K=1,MH
      KB=JB-K+1
      Q=(RC(K)-RCT*RC(KB))/D
      RC(KB)=(RC(KB)-RCT*RC(K))/D
  130 RC(K)=Q
  140 ALT=ALT/D
      R(1)=ALT/(1.-RC(1)*RC(1))
      GO TO 50
C
C     ..GENERATES RC(.),SA(.)
C     ..AND ALPHA FROM R(.)
C
  150 RC(1)=-R(2)/R(1)
      SA(1)=RC(1)
      ALPHA=R(1)*(1.-RC(1)*RC(1))
      DO 180 J=2,M
      MH=J/2
      JM=J-1
      Q=R(J+1)
      DO 160 L=1,JM
      LB=J+1-L
  160 Q=Q+SA(L)*R(LB)
      Q=-Q/ALPHA
      RC(J)=Q
      DO 170 K=1,MH
      KB=J-K
      AT=SA(K)+Q*SA(KB)
      SA(KB)=SA(KB)+Q*SA(K)
  170 SA(K)=AT
      SA(J)=Q
  180 ALPHA=ALPHA*(1.-Q*Q)
      GO TO 50
C
C     ..GENERATES RC(.) AND
C     ..AREA(.) FROM ALAR(.)
C
  190 AREA(MP)=1.
      DO 200 J=1,M
      JB=M+1-J
      AR=EXP(ALAR(JB))
      RC(JB)=(1.-AR)/(1.+AR)
  200 AREA(JB)=AREA(JB+1)/AR
      GO TO 50
C
C     ..GENERATES ALAR(.) AND
C     ..RC(.) FROM AREA(.)
C
  210 DO 220 J=1,M
      AR=AREA(J+1)/AREA(J)
      ALAR(J)=ALOG(AR)
  220 RC(J)=(1.-AR)/(1.+AR)
      GO TO 50
C
C     ..GENERATES AREA(.) AND
C     ..ALAR(.) FROM RC(.)
C
  230 AREA(MP)=1.
      DO 240 J=1,M
      JB=M+1-J
      AR=(1.-RC(JB))/(1.+RC(JB))
      ALAR(JB)=ALOG(AR)
  240 AREA(JB)=AREA(JB+1)/AR
      IF(I-2) 270,300,250
C
  250 DO 260 J=2,MP
  260 A(J)=SA(J-1)
      A(1)=1.
      IF(I-2) 270,300,270
C
C     ..GENERATE C(.) FROM A(.)
C     ..AND ALPHA
C
  270 C(1)=ALOG(ALPHA)
      C(2)=-A(2)
      DO 290 L=2,M
      LP=L+1
      SUM=L*A(LP)
      DO 280 J=2,L
      JB=L-J+2
  280 SUM=SUM+A(J)*C(JB)*(JB-1)
  290 C(LP)=-SUM/L
C
  300 RETURN
      END
```

Fig. 10.2 Fortran subroutine LPTRAN for performing various linear prediction coefficient transformations.

(A)

```
C
C        LINEAR PREDICTOR TRANSFORMATION TEST
         DIMENSION A(21),C(21),R(21),RC(21),ALAR(21)
         DIMENSION AREA(21)
         DATA A/1.,-.45,.81,18*0./
         M=3
         MP=M+1
         ALPHA=10.
         DO 20 I=1,6
         CALL LPTRAN(I,M,A,C,R,RC,ALAR,AREA,ALPHA)
   10    WRITE (5,11) I,ALPHA
   11    FORMAT(/,I10,F11.6,/)
         DO 20 J=1,MP
   20    WRITE (5,21) A(J),C(J),R(J),RC(J),ALAR(J),AREA(J)
   21    FORMAT(1X,6F11.6)
         END
C
C        I=1      INPUT=A(.) & ALPHA
C        I=2      INPUT=C(.)
C        I=3      INPUT=R(.)
C        I=4      INPUT=RC(.) & R(1)
C        I=5      INPUT=ALAR(.) & R(1)
C        I=6      INPUT=AREA(.) & R(1)
C
```

(B)

```
I   ALPHA
1   10.000000
```

A	C	R	RC	ALAR	AREA
1.000000	2.302585	30.993998	-0.248619	0.507880	5.732651
-0.450000	0.450000	7.705689	0.810000	-2.254058	9.526317
0.810000	-0.708750	-21.637579	0.000000	0.000000	1.000000
0.000000	-0.334125	-15.978518	0.000000	0.000000	1.000000

Fig. 10.3 A) Test program for LPTRAN. B) Numerical results.

10.2.2 Encoding and Decoding

The transmission parameters for pitch-excited vocoders are generally transformations of pitch P, gain σ, and the filter coefficients $\{a_i\}$. It is standard vocoding practice to code pitch and gain logarithmically. Typical values are 5-bit or 6-bit log coding of pitch and 5-bit log coding of gain. The focus here will be upon the reflection coefficients $\{k_i\}$ as a transformation of $\{a_i\}$ and various non-linear transformations, since their properties are not as well known.

The reflection coefficients (and transformations such as log areas) are widely used in linear prediction vocoding. They are easily obtained as a direct part of the analysis in the autocorrelation method, and from the step-down procedure in the covariance method. Necessary and sufficient conditions for stability of the synthesis filter are that they have a magnitude less than one. Therefore, linear interpolation between reflection coefficients of stable filters guarantees stable interpolated filters.

The reflection coefficients have a non-uniform spectral sensitivity, with the highest sensitivity near unity magnitude. This property was theoretically shown by Gray and Markel [1973]. They demonstrated that in the step-up procedure, the change at step m in the log spectra of the filter $1/A_m(z)$ due to a change in k_m of the amount Δk_m will oscillate (as frequency varies from zero to $f_s/2$) between the values

$$ln\ [1+\Delta k_m/(1+k_m)] \quad \text{and} \quad ln\ [1-\Delta k_m/(1-k_m)]. \tag{10.6}$$

Thus, values of k_m which have magnitudes nearest unity are most sensitive to slight variations. The non-uniform spectral sensitivity has also been studied in detail by Viswanathan and Makhoul [1975].

It is known that the first few reflection coefficients have skewed distributions for many voiced sounds (k_1 is near -1 and k_2 near $+1$) and tend toward more Gaussian-like distributions, centered about zero, for the higher ordered coefficients. This has been noted empirically by a number of researchers. It was analytically shown (using approximations) that this skewness is expected for k_1 and k_2 [Markel and Gray, 1974a], when no pre-emphasis is used to remove correlation. In addition, for reasonably low sampling frequencies, 10 kHz and less, the reflection coefficients $k_3, k_4, \ldots$, have been observed to have magnitudes less than 0.7, with high probability [Markel and Gray, 1974a].

Linear quantization of the reflection coefficients over the range $[-1, 1]$ is wasteful, for values near unity magnitude are usually found only for k_1 and k_2. In addition, due to the non-uniform spectral sensivity, non-linear quantization should be considered. A number of transformation and coding schemes have been utilized. Haskew, et al. [1973] experimented with numerous transformations and concluded that the most efficient encoding arrangement was logarithmic coding of the area ratios, i.e., $ln[(1-k_m)/(1+k_m)]$. Viswanathan and Makhoul [1975] drew a similar conclusion, based upon an experimental evaluation of the reflection coefficients' spectral sensitivity. Welch [1974] utilized a modified log area ratio, $ln[(F-k_m)/(F+k_m)]$ with values of F larger than one, due to the fact that for reflection coefficients near unity magnitude, the quantization for log area ratios can become so precise that the inherent accuracy of the data is exceeded.

Inverse sine coding of the reflection coefficients, $\theta_m = \sin^{-1}(k_m)$, has been suggested by Markel and Gray [1975c] for ease of synthesis using the normalized filter structure, because it allows for greater accuracy in quantization for reflection coefficients near unity, and is the only transformation that produces equally spaced angles for direct trigonometric table lookup of the filter parameters at the receiver (trigonometric tables are standard hardware items in high-speed array processor systems). While such coding does not fit the averaged sensitivity curves of Viswanathan and Makhoul [1975] as well as the log area ratio coding, it is still a reasonable fit and is superior to linear quantization of the reflection coefficients. Inverse sine coding has the same difficulty with excess precision as log area coding near unity magnitude.

Some of the approaches used in reducing the number of bits needed for transmission are relatively simple, while others are much more sophisticated. Markel and Gray [1974a] removed the bias from k_1 and k_2 (by adding and subtracting 0.3, respectively) and then quantized the unbiased results for all of the reflection coefficients uniformly from -0.7 through $+0.7$, using fewer bits for the higher order reflection coefficients. Itakura and Saito [1972b] used dynamic programming to allocate bits for the reflection coefficients. It was found that pre-emphasis of the speech signal greatly reduces the difference between dynamic programming bit allocation and uniform bit allocation. Makhoul, et al. [1974b] utilized Huffman coding on the log area function ratios to further increase the efficiency of representation. This procedure has the desirable feature that fewer bits are used with absolutely no degradation in performance.

McCandless [1974b] utilized an equal area coding method based upon statistically obtained histograms. This approach involves different (but efficient) coding for each separate reflection coefficient. The specific form of such coding depends upon statistical averages from many frames of data, and is a strong function of system conditions such as sampling rate, pre-emphasis, and types of recording apparatus.

It is doubtful that any unique optimum encoding/decoding scheme can be defined in the sense that the maximum subjective synthetic speech quality is obtained at the lowest encoding bit rate. In addition to the large number of factors involved, when the criterion of goodness is based upon perception, a very broad (and not necessarily unique) maximum will exist among listeners.

For simulation purposes, the unquantized parameters (floating-point or full precision integer values) are usually transformed into an integer set $\{0, 1, \ldots, 2^{\beta}-1\}$ where β is the number of bits used in the representation. This transformation is many-to-one and can be efficiently performed by table searching, using a binary-search, for example. These transmission parameters have a one-to-one correspondence with the decoded parameters and can therefore be used in a direct table-lookup at the receiver.

Specific examples of attainable bit rates with various parameter transformations will be presented in the discussion of vocoder systems and simulations.

10.2.3 Variable Frame Rate Transmission

A large portion of conversational speech contains pauses. In addition, the information necessary to accurately represent the incoming speech varies considerably. For example, the transitions between voiced and unvoiced sounds must be represented by closely spaced analysis frames (e.g., $f_r = 100$ Hz) or else a word such as "pea" may sound like "fee" when synthesized. However, for long interval sounds such as "ahh" the steady-state behavior of the speech can be represented by a much slower analysis frame update rate. By taking advantage of pauses and the time-varying information in packet switched systems such as the Arpanet [Roberts and Wessler, 1972], it should be possible to have average voice transmission bit rates substantially below those available in single-user dedicated circuit-switched systems (e.g., telephone systems) while retaining acceptable quality.

To take advantage of the time-varying characteristics of speech for low bit rate transmission, one must have some measure of change. Such a measure would compare the spectra or system parameters of each new data frame with the previously transmitted data frame. If the measure exceeds a given threshold, then frames are judged to have a large enough change to require transmission of a new set of parameters. As the largest percentage of the transmission bits is allocated to the parameters defining the spectrum (such as the reflection coefficients or log area ratios), major emphasis has been focussed on this area.

There are a large number of possible measures, each based upon some set of parameters that describes the analysis frame. For example, measures might be based upon averages or sums of absolute differences or squares of parameter sets, such as reflection coefficients, autocorrelation coefficients (possibly normalized to remove the effects of gain), inverse filter coefficients, or cepstral coefficients. If $\mathbf{p}_n$ is a column vector which contains a set of parameters for frame n, then one possible measure of change in the parameters between frames n and l would be the Euclidian distance $(\mathbf{p}_n - \mathbf{p}_l)^t\, (\mathbf{p}_n - \mathbf{p}_l)$. Such a measure has the advantage of being easy to calculate by summing the squares of the separate parameters; however, it has the disadvantage of weighting all parameters independently and equally.

A more general measure of distance might be $(\mathbf{p}_n - \mathbf{p}_l)^t\, \mathbf{W}^{-1} (\mathbf{p} - \mathbf{p}_l)$ where $\mathbf{W}^{-1}$ is the inverse of a positive definite weighting matrix $\mathbf{W}$. The weighting matrix $\mathbf{W}$ can allow heavier weighting of the more significant differences and can compensate for correlation between separate parameters (the elements of the parameter column vectors). The matrix $\mathbf{W}$ must be determined from statistical measurements, and to be most effective it should be chosen differently for different sounds and different speakers.

More absolute measurements of deviation based upon information from only two frames of data are available. Magill [1973] and Itakura [1975] proposed measurements based upon error signal energies which used predictor coefficients from one frame, applied to signals from another. Using superscripts to denote frame numbers, so that $A^{(n)}(z)$ and $c_{ij}{}^{(n)}$, respectively, represent the inverse filter and covariance (or autocorrelation) coefficients from frame n, one defines a norm square by

$$\boxed{\alpha(l,n) = \sum_{i=0}^{M} \sum_{j=0}^{M} a_i{}^{(n)}\, c_{ij}^{(l)}\, a_j^{(n)}} \quad . \tag{10.7}$$

From the minimizing property of linear prediction (i.e., $\alpha_M^{(l)} = \alpha(l,l)$ is the minimum attainable value for frame l with M coefficients),

$$\alpha(l,n) \geq \alpha(l,l) = \alpha_M^{(l)} = \alpha^{(l)}. \tag{10.8}$$

Thus a ratio of $\alpha(l,n)$ to $\alpha(l,l)$ can be used as a measure of difference between the frames. The ratio will always be larger than or equal to unity. This measure depends upon the order of the items taken, for in general

$$\frac{\alpha(l,n)}{\alpha(l,l)} \neq \frac{\alpha(n,l)}{\alpha(n,n)} \,. \tag{10.9}$$

In detecting changes between frames, it is necessary only to save the predictor coefficients (or an equivalent parameter set) from the previous frame. Thus, if frame n represents a past frame of data, and frame l represents the present frame so that $l > n$, then following Magill [1973],

$$\boxed{d(l,n) = \alpha(l,n)/\alpha(l,l) \geq 1} \tag{10.10}$$

is defined as a measure of the distance between two frames. The denominator $\alpha(l,l)$ is simply the minimum predictor error for the present frame l as indicated by (10.8). The numerator $\alpha(l,n)$ can be evaluated from (10.7) as a double summation. In the autocorrelation method, the coefficients $c_{ij}^{(l)}$ are functions only of the subscript difference, and can be evaluated somewhat more efficiently. If

$$c_{ij}^{(l)} = r^{(l)}(i-j) = r^{(l)}(j-i), \tag{10.11}$$

then (10.7) takes on the form

$$\begin{aligned}\alpha(l,n) &= \sum_{i=0}^{M} \sum_{j=0}^{M} a_i^{(n)} r^{(l)}(i-j) a_j^{(n)} \\ &= \sum_{k=0}^{M} r^{(l)}(k)\; r_a^{(n)}(k),\end{aligned} \tag{10.12a}$$

where

$$r_a^{(n)}(k) = \sum_{i=0}^{M-k} a_i^{(n)} a_{i+k}^{(n)}. \tag{10.12b}$$

Thus the coefficients $r_a^{(n)}(k)$ for $k = 0, 1, \ldots, M$, can be stored and used to evaluate $\alpha(l,n)$ in each frame as l increases. When $d(l,n)$ exceeds a fixed threshold, the new coefficients are transmitted. This procedure has been referred to as the DELCO algorithm by Magill [1973], with a suggested threshold value of 1.4. Itakura [1975] derived the same distance measure (except that the logarithm of $d(l,n)$ is used), based upon a maximum likelihood approach as discussed in Chapter 2. The theoretical results depend upon an assumption of Gaussian statistics and $N \gg M$. Makhoul, et al. [1974b] have applied this log likelihood distance measure to fourteen sentences from ten speakers, with results as shown in Fig. 10.4, using a likelihood threshold of 1.4. The figure shows the relative frequency of occurrence of the different transmission interval sizes from 10 to 80 ms based upon a fixed analysis frame rate of 100 frames/s. The average frame rate using variable frame rate transmission was 37 frames/s.

A more physically meaningful measure using linear prediction might be based upon a direct Euclidian distance between the corresponding log spectra. The magnitude squared spectrum for frame n can be denoted as

$$V^{(n)}(e^{j\theta}) = |\sigma^{(n)}/A^{(n)}(e^{j\theta})|^2 \tag{10.13}$$

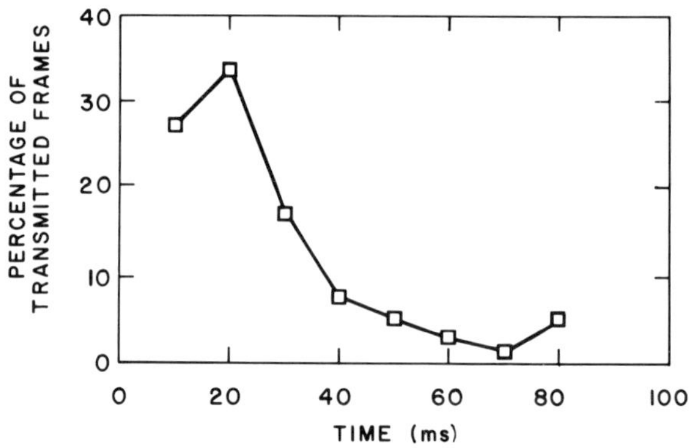

Fig. 10.4 Percentage of transmitted frames versus transmission interval for variable frame rate. [From Makhoul, et al., 1974b]

where $\sigma^{(n)}$ represents a gain term. If the absolute gain is to be included in the measurements, then $\sigma^{(n)}$ will equal $[\alpha_M^{(n)}]^{1/2}$ for the autocorrelation method. If the effects of absolute gain are to be removed, then there are two choices for $\sigma^{(n)}$. If the spectra are to be normalized so that their mean squares are identical, then $\sigma^{(n)}$ must be proportional to the square root of $\alpha_M^{(n)}/\alpha_0^{(n)}$. If the spectra are to be normalized so that the log spectra all have average values of zero, then $\sigma^{(n)}$ is chosen as unity. Using (10.13), different measures can be defined as

$$\boxed{d_p(n, l) = \int_{-\pi}^{\pi} \left| ln[V^{(n)}(e^{j\theta})] - ln[V^{(l)}(e^{j\theta})] \right|^p \frac{d\theta}{2\pi}} \quad . \tag{10.14}$$

For $p=1$, $d_1(n, l)$ defines the mean of the absolute difference between the log spectra at frame l and n. For $p=2$, $d_2(n, l)$ defines the mean square of the difference between the log spectra. If $d_p(n, l)$ is raised to the $1/p$ power, the limit as p approaches infinity will represent the maximum difference between the log spectra. In practice, $d_p(n, l)$ would be estimated by a finite summation. Two choices are available, one in the frequency domain and one in the time domain (applicable only when $p=2$). In the frequency domain, $V^{(n)}[\exp(j\theta)]$ can be evaluated with an N-point DFT at the frequencies $\theta=0$, $2\pi/N, \ldots. 2\pi(N-1)/N$, by appending $N-M-1$ zeros to the coefficients of $A^{(n)}(z)$. The integral of (10.14) can then be approximated by the summation

$$d_p(n, l) \simeq \frac{1}{N} \sum_{k=0}^{N-1} \left| ln[V^{(n)}(e^{j2\pi k/N})/V^{(l)}(e^{j2\pi k/N})] \right|^p \quad . \tag{10.15}$$

When $p=2, d_p(n, l)$ can be evaluated directly from the cepstral coefficients obtained in the previous section. Combining the series expansion for $1/A^{(n)}(z)$ with (10.13), the result

$$ln[V^{(n)}(e^{j\theta})] = \sum_{k=-\infty}^{\infty} c^{(n)}(k) e^{-jk\theta} \tag{10.16a}$$

is obtained, where

$$c^{(n)}(0)=2ln[\sigma^{(n)}], \tag{10.16b}$$

$$c^{(n)}(-k)=c^{(n)}(k), \tag{10.16c}$$

and $c^{(n)}(k)$ for $k=1,2,\ldots,$ is computed recursively from (10.3) using the filter coefficients for frame n. From (10.15) and (10.16a), the distance measure $d_2(n,l)$ can then be expressed as the summation

$$d_2(n,l)=\sum_{k=-\infty}^{\infty}[c^{(n)}(k)-c^{(l)}(k)]^2 \tag{10.17}$$

$$\boxed{d_2(n,l)=[c^{(n)}(0)-c^{(l)}(0)]^2+2\sum_{k=1}^{\infty}[c^{(n)}(k)-c^{(l)}(k)]^2}\ .$$

The major portion of the log spectrum energy lies in the lower ordered cepstral coefficients for speech signals. It was shown in Section 10.2.1 that all of the spectral shaping information lies in the coefficients $c(1),\ldots,c(M)$, since they uniquely describe the filter coefficients $\{a_i\}$. The zero*th* cepstral coefficient contains the gain information (10.16b). Thus it appears that a reasonable measure of the difference between the log spectra may be given by (10.17) with truncated summations. No discrete Fourier transform operations and no logarithm operations are necessary (excluding the gain term). Further study is needed, though it has been shown [Atal, 1974a] that use of the first few cepstral coefficients has led to very good results in a particular speaker recognition and verification experiment (see Chapter 11).

10.2.4 Excitation and Synthesis Gain Matching

Atal and Hanauer [1971b] proposed a method for matching the synthesized speech energy within a single pitch period to the corresponding speech signal energy based upon transmission of the input signal energy over a single pitch period. Although their discussion was limited to the covariance method without pre-emphasis, the technique applies to both covariance and autocorrelation methods and can easily be modified to allow for pre-emphasis as long as the direct form synthesis filter is used.

The basic idea is that each synthesized speech sample has two major components: 1) the decaying complex exponential values $\{q(n)\}$ from the previously synthesized pitch period and 2) the synthesizer output $\{u(n)\}$ in response to an excitation sequence $\{e(n)\}$, without any effects from the previous frame.

Here the excitation source is either a series of periodic unit samples followed by zeros for voiced synthesis or a series of output samples from a pseudo-random number generator for unvoiced sounds (the pitch period for an unvoiced sound is defined as some constant number). Introducing a gain factor g, the total synthesizer response $\{\hat{s}(n)\}$ for the new frame is then given by:

$$\boxed{\hat{s}(n)=q(n)+g\,u(n)} \quad . \tag{10.18}$$

Using an overbar "———" to denote the sum over N samples, e.g.,

$$\overline{u(n)}=\sum_{n=0}^{N-1} u(n), \tag{10.19}$$

equal speech and synthesis energy requires

$$\begin{aligned}\overline{s^2(n)}=\overline{\hat{s}^2(n)}&=\overline{[q(n)+g\,u(n)]^2}\\ &=g^2\,\overline{u^2(n)}+2g\,\overline{q(n)u(n)}+\overline{q^2(n)}.\end{aligned} \tag{10.20}$$

This quadratic is then solved for g. An algorithm for computing the direct form synthesis filter response is given below:

1) Compute the filter output $q(n)$ for $n=0, 1, \ldots, N-1$, without any excitation (only the memory from the previous period), or new coefficient update.

2) Compute the filter output $u(n)$ for $n=0, 1, \ldots, N-1$, with excitation $\{e(n)\}$, and the filter memory reset to zero.

3) Calculate the indicated terms and solve the quadratic

$$g^2\,\overline{u^2(n)}+2g\,\overline{q(n)u(n)}+\overline{q^2(n)}-\overline{s^2(n)}=0 \tag{10.21}$$

for g.

4) Assuming that g is real and non-negative, define the synthesizer output as

$$\hat{s}(n)=q(n)+g\,u(n) \qquad n=0, 1, \ldots, N-1. \tag{10.22}$$

5) Reset the direct form filter memory values to

$$\hat{s}(N-1), \hat{s}(N-2), \ldots, \hat{s}(N-M).$$

With reference to step 3), if $\overline{s^2(n)}$ is larger than $\overline{q^2(n)}$ the roots of the quadratic will have opposite signs, and there will always be a real positive root. This condition occurs when the true energy is greater than that due to the decaying transient. If this condition is not satisfied, as might arise with a signal whose amplitude is decreasing, then it is necessary that $\overline{q(n)\,u(n)}/\overline{u^2(n)}$ be negative, and in addition,

$$\frac{\overline{s^2(n)}}{\overline{u^2(n)}}+\left[\frac{\overline{q(n)u(n)}}{\overline{u^2(n)}}\right]^2>\frac{\overline{q^2(n)}}{\overline{u^2(n)}}\,. \tag{10.23}$$

If there is no real positive root, the model has no physical meaning. If a solution does not exist, Atal and Hanauer suggest setting $g=0$. This procedure requires performing a synthesis filter operation twice, in addition to three sums over N

samples and solution of a quadratic. The transmission gain σ is defined from $\sigma^2=\overline{s^2(n)}$. The algorithm described above is then implemented completely at the receiver.

This approach can be extended to allow for other synthesis filter structures. The only additional computation is in transforming the delayed synthetic speech values into memory elements of the specified filter. The necessary modification for pre-emphasis is to replace the coefficients of $A(z)$ by the coefficients of $A(z)(1-\mu z^{-1})$ where $(1-\mu z^{-1})$ defines the pre-emphasis filter.

A somewhat simpler (and less accurate) method is to drive the synthesis filter with the input sequence $\{e(n)\}$ to compute $u(n)$, where $u(n)$ now contains both the response due to the previous frame and the present excitation. Applying the same energy matching criterion

$$\overline{s^2(n)}=\overline{\hat{s}^2(n)}=g^2\,\overline{u^2(n)},$$

the term g is then computed as

$$\boxed{g=\left[\frac{\overline{s^2(n)}}{\overline{u^2(n)}}\right]^{1/2}} \quad . \tag{10.24}$$

Since the above approaches directly match the energy on an input-output basis, an accurate envelope match is expected between the original and the synthetic speech. One must, however, be careful to ensure that step discontinuities are minimized since the gain factors at the end of one period and the beginning of the next period are not constrained except indirectly by the energy criterion.

Based upon the analysis-synthesis model in Chapter 1, the inverse filter output (with energy $\sigma^2=\alpha$) drives the reciprocal inverse filter or synthesis filter to generate the speech output. A computationally efficient scheme for computing the driving function with the autocorrelation method based upon the transmitted gain $\sigma=\alpha^{1/2}$ was proposed by Markel and Gray [1974a]. Assuming N data samples per analysis frame, a total squared error of $\sigma^2=\alpha$ is obtained. If the pseudo-random number generator samples $\{g(n)\}$ used for unvoiced excitation have a variance σ_g^2, then to match energies over N samples, the driving function $\{e(n)\}$ is computed by

$$N e^2(n)=g^2(n)\sigma^2/\sigma_g^2$$

or

$$\boxed{e(n)=g(n)\sigma/(\sigma_g N^{1/2}) \;\; \text{(unvoiced)}} \quad . \tag{10.25}$$

If the samples $\{g(n)\}$ are uniformly distributed over $[-b,b]$ then $\sigma_g=b/\sqrt{3}$. To compute $e(n)$ for a voiced frame, the inverse filter output sequence is modeled as a train of unit samples separated by an integer number of samples I corresponding to the pitch period. Therefore, $e(n)$ on the average is computed by

$$e^2(n)\delta_{n,lI}N/I=\sigma^2$$

where $l=0, 1, 2, \ldots$, or

$$e(n)=\begin{cases}\sigma(I/N)^{1/2} & n=0, \mathrm{I}, 2\mathrm{I}, \ldots \\ 0 & \text{elsewhere}\end{cases} \quad \text{(voiced)} \qquad (10.26)$$

If pre-emphasis has been applied in the analysis, post-emphasis must be applied at the output of the synthesis filter. Synthesis with non-zero-mean excitation sequences will cause a bias buildup in the post-emphasis filter which becomes particularly severe as the fundamental frequency increases. The perceived effect of this bias is a low frequency thumping sound. If the zero valued samples are replaced by samples of amplitude C, a zero mean sequence is obtained for $\sigma(I/N)^{1/2}(N/I)+C(N-N/I)$ or $C=-\sigma(I/N)^{1/2}/(I-1)$. A zero mean excitation signal $e(n)$ can then be defined as

$$e(n)=\begin{cases}\sigma(I/N)^{1/2} & n=0, \mathrm{I}, 2\mathrm{I}, \ldots \\ -\sigma(I/N)^{1/2}/(I-1) & n\neq 0, \mathrm{I}, 2\mathrm{I}, \ldots\end{cases} \quad \text{(voiced)} \qquad (10.27)$$

This sequence has slightly more energy than σ^2; however, it does allow acceptable synthesis with minimal computation. If more accuracy is desired, a quadratic can be solved from $C_1^2(N/I)+C_2^2(N-N/I)=\sigma^2$ and $C_1(N/I)+C_2(N-N/I)=0$ to satisfy both the zero mean requirement and the total squared error requirement.

A final method for input-output energy matching has been suggested by Klayman, et al. [1973]. The input signal energy $\overline{s^2(n)}$ is computed over an interval I equal to one pitch period. The transmitted gain σ is then computed as

$$\sigma=[\overline{s^2(n)}/I]^{1/2} \qquad (10.28)$$

to obtain an energy per sample measure. For unvoiced speech, a fixed interval I is chosen. At the receiver, synthesis is performed pitch-synchronously based upon a fixed amplitude or standard deviation excitation conditions. The energy over each synthesized interval I is computed from the samples that are stored in a buffer. Each sample $s(n)$ is then replaced by $\hat{s}(n)C$ where

$$C=\sigma/\overline{[\hat{s}^2(n)/I]}^{1/2} \qquad (10.29)$$

10.2.5 A Linear Prediction Synthesizer Program

A Fortran program for implementing a linear prediction synthesizer is shown in Fig. 10.5. Features of this particular program are:

1)The synthesis filter uses the two-multiplier lattice form (TWOMUL) from Chapter 5.

```
C
C       ==============================
C       A LINEAR PREDICTION SYNTHESIZER
C       ==============================
C
        DIMENSION RCL(20),RCR(20),RC(20)
        DIMENSION RCBUF(20),Y(64),TAP(20)
        DATA RC,RCR,RCBUF,TAP/80*0./
        DATA N,M,IFLGTH,PE/128,8,64,0.9/
        DATA IFC,IVR,NN,GAIN/1,0,1,0./
        DATA IPTCHR,IPITCH,IPC/64,64,1/
        DATA YPREV,IRN,NB,LSTBLK/0.,33,1,10/
        DATA IPTCHL,IVL,DRVN,GAINL/0,1,2*0./
C
   10   IF (IPC.LE.IPITCH) GO TO 120
C
C       STARTNG A NEW PITCH PERIOD.
C
   20   IPC=1
   30   IF (IFC.LE.IFLGTH) GO TO 80
C
C       CROSSING ANALYSIS FRAME BOUNDRY.
C
        IFC=IFC-IFLGTH
        IF (NB.GT.LSTBLK) GO TO 140
        DO 40 J=1,M
   40   RCL(J)=RCR(J)
        IPTCHL=IPTCHR
        GAINL=GAINR
        IVLST=IVL
        IVL=IVR
        IF ((IVLST.EQ.0).OR.(IVL.EQ.1)) GO TO 60
C
C       ZERO BUFFER DURING TRANSITION FROM V TO UV
C
        DO 50 J=1,M
   50   RCBUF(J)=0.
   60   CONTINUE
C
C       ==============================
C       READ NEW PARAMETERS
C       RCR(1),....,RCR(M),SIGMA,IPTCHR
C       ==============================
C
        NB=NB+1
        IF (IPTCHR.GT.0) GO TO 70
        GAINR=SIGMA*SQRT(3./N)*10.
        IVR=0
        GO TO 30
   70   GAINR=SIGMA/SQRT(FLOAT(N))
        IVR=1
C
C       END OF THE CROSSING FRAME BOUNDRY WORK.
C
C       PERFORM INTERPOLATION
C
   80   IF ((IVR.EQ.1) .AND. (IVL.EQ.1)) GO TO 90
        IPITCH=IPTCHL
        IF (IVL.EQ.0) IPITCH=IFLGTH-IFC+1
        WLC=0.
        GO TO 100
   90   WLC=FLOAT(IFC-1)/FLOAT(IFLGTH-1)
        IPITCH=(IPTCHR-IPTCHL)*WLC+IPTCHL
  100   DRV=1.
        DRVN=-1./(IPITCH-1)
        DO 110 J=1,M
  110   RC(J)=(RCR(J)-RCL(J))*WLC+RCL(J)
        GAIN=(GAINR-GAINL)*WLC+GAINL
        IF (IVL.EQ.1) GAIN=GAIN*SQRT(FLOAT(IPITCH))
C
C       END OF SETUP FOR A NEW PITCH PERIOD.
C
  120   IF (IVL.EQ.1) GO TO 130
        DRV=(NOISE(IRN)/32768.)*.1
  130   TAP(1)=GAIN
        CALL TWOMUL(RC,TAP,M,RCBUF,DRV,YOUT)
        Y(NN)=YOUT+PE*YPREV
        DRV=DRVN
        YPREV=Y(NN)
        IFC=IFC+1
        NN=NN+1
        IPC=IPC+1
        IF (NN.LE.64) GO TO 10
        NN=1
C
C       ==============================
C       WRITE OUT 64 SYNTHESIZED SAMPLES
C       ==============================
C
        GO TO 10
  140   END
```

Fig. 10.5 Fortran program for implementing a linear prediction synthesizer.

2) Pitch-synchronous interpolation of gain and pitch and the reflection coefficients is applied.

3) The gain computation is based upon error signal energy transmission. Interpolation requires knowledge of two frames of transmitted parameters. These frames are referred to as left-hand and right-hand frames. The most important variables are shown in Table 10.2. For simplicity, the synthesizer starts with both left- and right-hand frames defined as unvoiced with zero gain. As long as the pitch counter *IPC* is less than *IPITCH*, then a new speech sample *Y*(*NN*) is

Table 10.2. Definition of variables used in linear prediction synthesizer program.

Variable	Definition (sample value)
RCL	Left-hand reflection coefficient array
RCR	Right-hand reflection coefficient array
RCBUF	Reflection coefficient buffer for lattice filter
Y	Lattice filter output array
TAP	Lattice filter tap parameters
N	Number of samples used in analysis frames ($N=128$)
M	Number of coefficients used in analysis ($M=8$)
IFLGTH	Integer analysis frame length in samples f_s/f_r ($IFLGTH=64$)
PE	Post-emphasis factor, equal to pre-emphasis factor in analysis ($PE=0.9$)
IFC	Integer frame counter
IPC	Integer pitch period counter
NN	Integer speech sample index
IVL	Integer voicing for left-hand frame
IVR	Integer voicing for right-hand frame
IPTCHL	Integer pitch for left-hand frame
IPTCHR	Integer pitch for right-hand frame
IPITCH	Interpolated pitch period in integer number of samples
SIGMA	Transmitted gain factor (σ)
GAINL	Synthesizer gain for left-hand frame
GAINR	Synthesizer gain for right-hand frame
GAIN	Interpolated synthesizer gain
IRN	Pseudo-random numbers uniformly distributed over $\lvert IRN\rvert < 32768$.

synthesized. Counters for *IFC*, *IPC*, and *NN* are updated. If *NN* exceeds 64, then the synthesized samples *Y*(1), *Y*(2), ..., *Y*(64) are output and the counter *NN* is reset.

If *IPC* exceeds *IPITCH*, *IPC* is reset to 1 and a new pitch period is obtained. If *IFC* does not exceed the frame length *IFLGTH*, then the left and right frames are tested to see if they are both voiced (statement 80). If so, linear interpolation is applied to define a new pitch period *IPITCH* and gain parameter *GAIN*. If one or both frames are unvoiced, then no interpolation is applied and *IPITCH* is set equal to the left-hand frame pitch value. After the interpolation, a new series of samples is synthesized until either $IPC > IPITCH$, or $NN > 64$.

If *IFC* exceeds *IFLGTH* it is time to update the left- and right-hand frames. The right-hand frame values are transferred to the left-hand side and then new right-hand parameters are read in. These parameters are the *M* reflection coefficients *RCR*(1), ..., *RCR*(*M*), the transmission gain $SIGMA=\sigma$, and the transmitted pitch $IPITCH = Pf_s$. If the frame is unvoiced, $IVR = 0$. If the frame is voiced, $IVR = 1$. The gain computations are made in accordance with (10.25) and (10.27). The reading of transmitter data is not shown in the program since this operation will be dependent upon the particular computer used. It is necessary only to insert the system-calling program at the point where the variables are read in. The counter *NB* counts the blocks or frames of input data. When the last block *LSTBLK* has been read, the program terminates.

A synthesized speech example is shown in Fig. 10.6A, based upon the parameters of the corresponding analysis frames shown in Fig. 10.6B. Eight reflection coefficients were obtained from an autocorrelation analysis. The gain *SIGMA* increases while the integer pitch *IPTCHR* decreases.

(A)

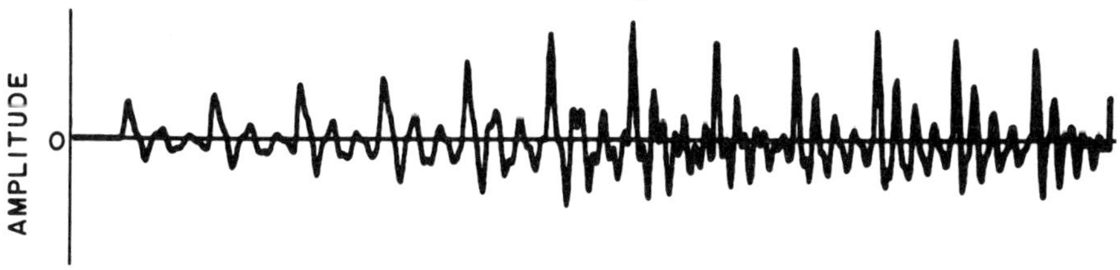

(B)

FRAME	RCR(1)	RCR(2)	RCR(3)	RCR(4)	RCR(5)	RCR(6)	RCR(7)	RCR(8)	SIGMA	IPTCHR
1	-0.79437	0.74130	-0.39279	0.01718	0.25439	-0.09662	0.21109	0.51923	26.	51
2	-0.75500	0.72168	-0.53149	0.33221	0.10782	-0.04651	0.35052	0.52917	32.	50
3	-0.69617	0.72559	-0.55615	0.39459	-0.02625	0.02689	0.46832	0.48801	47.	49
4	-0.64291	0.76331	-0.56036	0.31577	-0.04233	0.15195	0.40384	0.60056	60.	48
5	-0.60031	0.81363	-0.50970	0.09958	0.13388	0.20294	0.30560	0.49182	81.	48
6	-0.60739	0.76727	-0.34833	0.05301	0.21133	0.33029	0.29486	0.33658	110.	48
7	-0.58646	0.71616	-0.20697	0.11606	0.10236	0.39749	0.43729	0.22589	140.	47
8	-0.56000	0.73511	-0.13773	0.05637	0.16449	0.47031	0.39548	0.01706	162.	47
9	-0.53122	0.75671	-0.11603	0.12259	0.02786	0.67007	0.18893	0.07288	155.	46
10	-0.51172	0.77390	-0.08575	0.14377	-0.01614	0.69446	0.21368	0.01685	161.	46

Fig. 10.6 Example outputs from linear prediction synthesis program. A) synthetic waveform, B) input parameters for obtaining synthetic waveform.

10.3 Low Bit Rate Pitch Excited Vocoders

One of the most important potential applications of linear prediction is for low bit rate (2,400 – 3,600 bits/s) secure voice transmission over telephone-type bandwidths. Due to the fact that single-channel high-quality speech transmission requires from 40,000 to 200,000 bits/s, it should be obvious that certain characteristics of speech must be incorporated into the system model along with other trade-offs in order to reduce the bit rate by more than an order of magnitude. It is important to understand these considerations so that realistic judgments of the inherent capabilities and limitations of linear prediction vocoder systems can be obtained. Several of these realities are presented below.

An explicit pitch extraction value must be estimated for each frame (no voicing implies $P=0$). It is this feature alone that accounts for by far the greatest reduction in bit rate. If it is done well, a surprisingly small loss in naturalness occurs in the synthesized speech. Because of the explicit extraction, however, background noise must be kept to a minimum — music, dogs barking, or multiple speakers will cause considerable degradation. In addition, only a limited range of fundamental frequency is allowable, depending on how complex the pitch extraction rules are allowed to be.

Generally in the pitch extraction process, each frame is classified as all voiced ($V=1$) or all unvoiced ($V=0$). There are obviously speech sounds that would be more accurately classified as something in between, such as $V \simeq 0.7$ for the /v/ as in thie*v*es. The binary decision is usually made because of practical considerations. It is difficult to automatically compute the proper ratio of periodic component to turbulent (noise) component, and synthesis quality is often not greatly improved.

Unless variable bit rate schemes are applied which take advantage of silence intervals and other speech properties, the average bit rate cannot be lowered beyond a fairly sharp cutoff, in the neighborhood of 1,200 to 1,400 bits/s. With the fixed frame rate systems described in this section, the highest possible synthesis quality (no coding, quantization, or finite word length calculation) can be maintained down to around 3,300 bits/s with essentially negligible degradation in quality. From around 1,400 to 3,300 bits/s, the degradation may be negligible to significant depending upon the particular speech sounds and speaker characteristics. Below 1,400 bits/s, almost complete degradation of synthesis quality is obtained.

In this section, examples processed by vocoder systems are presented. An attempt is made to present an unbiased point of view that realistically shows both the capabilities and limitations of pitch-excited linear prediction vocoder systems.

10.3.1 Maximum Likelihood and PARCOR Vocoders

The first pitch-excited linear prediction vocoder system was developed in 1968 [Itakura and Saito, 1968]. The coefficient analysis portion was based upon the maximum likelihood formulation (see Chapter 2). The pitch extraction was performed by using a modified autocorrelation method (see Chapter 8). The transmission parameters are M filter coefficients $\{a_i\}$ (using the autocorrelation method), the gain $\sigma=\sqrt{\alpha_M}=\sqrt{r_e(0)}$, and a voicing energy variable V, and pitch P. The excitation signal is produced by mixing the outputs of a pulse generator and a white noise generator according to the scheme shown in Fig. 10.7A.

The normalized autocorrelation $r_e(n)/r_e(0)$ is searched for a maximum over the range 3–15 ms. If $r_e(n)/r_e(0) \leq 0.18$, the frame is defined as unvoiced and $V=0$. If $r_e(n)/r_e(0) \geq .25$, the frame is defined as voiced with $V=1$. A monotonically increasing curve is used to define the transition region $0.18 < r_e(n)/r_e(0) < .25$. By choosing the unvoiced energy factor U as $U=1-V$, a total energy $U+V=1$ is obtained.

Based upon a fixed frame rate analysis with N samples per frame, the amplitude of the pitch period pulses and noise generator (assuming a zero mean unity variance pseudo-random number generator) is shown in Fig. 10.7B with the mixing functions.

Parameters chosen for this system were; sampling frequency $f_s = 8$ kHz, frame rate $f_r = 50$ frames/s, $M = 10$ filter coefficients, each coded to 9 bits, pitch, gain, and V, each coded to 6 bits, with a resultant bit rate of 5.4 kb/s. A 30-ms analysis frame, multiplied by a Hamming window, was used for both filter coefficient analysis and pitch period estimation. This value is a compromise based upon the desire for a relatively long window for pitch extraction and a relatively short window for filter coefficient extraction.

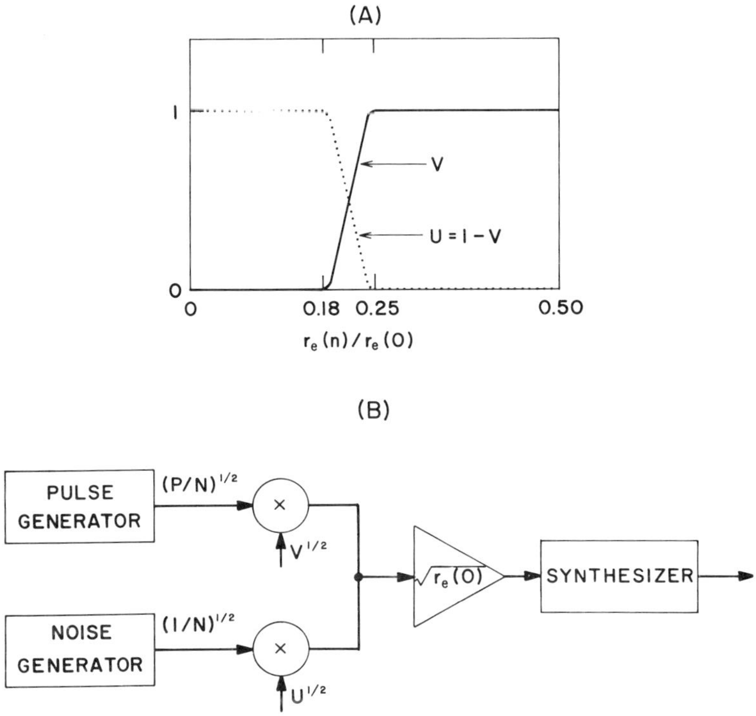

Fig. 10.7 Maximum likelihood vocoder. A) voiced-unvoiced mixing curve. B) synthesizer block diagram.

Although the idea of generalizing the voicing decision for mixing is good, it does not substantially improve synthetic speech compared with a binary voicing decision. To accurately represent voiced fricatives, for example, requires a somewhat more complex model whose parameters cannot easily be obtained from the speech waveform.

Itakura and Saito developed a second vocoder system based upon the PARCOR formulation [1969, 1972b, 1972c]. This formulation was presented in

Section 2.6. The structure of the analyzer showing the lattice form was presented as Fig. 2.5. The PARCOR coefficients as defined by Itakura and Saito are precisely equal to the reflection coefficients $\{k_m\}$ except for a sign change. The parameters $\{k_m\}$ are theoretically obtained by cross-correlating the residual error signals $x_m^+(n)$ and $x_m^-(n)$ at each stage of the analyzer from

$$k_m = \frac{-\sum_{l=-\infty}^{\infty} x_{m-1}^+(l)\, x_{m-1}^-(l)}{\left\{\sum_{l=-\infty}^{\infty} [x_{m-1}^+(l)]^2 \sum_{l=-\infty}^{\infty} [x_{m-1}^-(l)]^2\right\}^{1/2}} \tag{10.30}$$

where the input is $x_0^+(n) = x(n)$, and the output is $x_M^+(n) = e(n)$, as described in Chapter 2, Section 2.6. If $x(n) = 0$ for $n < 0$ and $n \geq N$, then the above equation results in precisely the same results as recursively solving the autocorrelation coefficients. The conceptual advantage of this system, however, is that the PARCOR coefficients can be computed on a sample-by-sample basis. Itakura and Saito [1971a] suggested an exponential filtering for weighting new samples as most important, with exponential decay into the past as shown by the correlator CORR in Fig. 10.8. First, the geometrical mean in the denominator is approx-

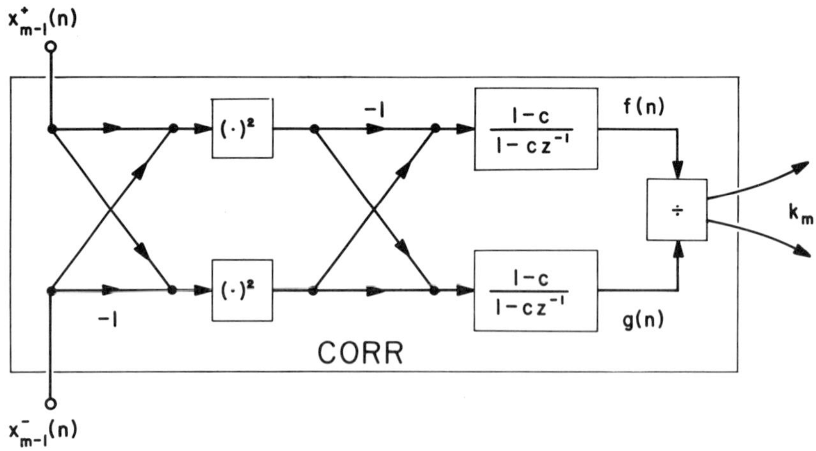

Fig. 10.8 One stage of PARCOR coefficient generation using first-order filters. [After Itakura and Saito, 1971a]

imated by an arithmetic mean. The inputs to the first-order filters on the top and bottom are given by $-4x_{m-1}^+(n)x_{m-1}^-(n)$ and $2\{[x_{m-1}^+(n)]^2 + [x_{m-1}^-(n)]^2\}$, respectively. Therefore, the filter outputs on a sample-by-sample basis are given by

$$f(n) = cf(n-1) - 4(1-c)x_{m-1}^+(n)\, x_{m-1}^-(n) \tag{10.31a}$$

and

$$g(n) = cg(n-1) + 2(1-c)\{[x_{m-1}^+(n)]^2 + [x_{m-1}^-(n)]^2\}, \tag{10.31b}$$

where $0 \leq c \leq 1$. The estimate for k_m at sample n is then

$$k_m(n) = f(n)/g(n). \tag{10.32}$$

As the input $\{x(n)\}$ passes through the analyzer, the correlation between adjacent samples is gradually removed. With a sufficiently large number of sections the spectral features of the input are removed and combined into the PARCOR coefficients. The pitch extraction is performed by autocorrelation of the final residual $\{e(n) = x_M^+(n)\}$ over some appropriate number of samples. The voicing decisions used are the same as stated in the maximum likelihood vocoder. The transmission parameters are $\{k_m\}$ for $m = 1, 2, \ldots, M$, pitch P, voicing V, and gain $\sigma = \sqrt{r_e(0)}$. The synthesis is performed in the same manner as previously described for the maximum likelihood vocoder except that the two-multiplier lattice form is used. This form, as described in Chapter 5, allows direct implementation in terms of the parameters $\{k_m\}$.

Analysis conditions used for this system were specified as analog pre-emphasis using a +6 dB/octave filter having a cutoff at 500 Hz, sampling frequency $f_s = 8$ kHz, and $M = 8$ [Itakura and Saito, 1971a].

Boll [1974] interpreted the problem of sample-by-sample estimation of k_m in a manner that shows the result (10.31) for $c = 0$ in a direct manner without having to approximate a geometrical mean by an arithmetic mean as shown in Chapter 8. The introduction of the filtering ($c \neq 0$) can then be interpreted as simply a way of averaging parameters over time since transmission of each $k_m(n)$ would result in an enormous bit rate. The insight gained in this manner is that one is minimizing the sum of the squared residuals $x_m^+(n)$ and $x_m^-(n)$ at every stage and every sample.

A hardware system for implementing the PARCOR analysis has been built and tested [Dunn, et al., 1973]. Suggested advantages of the system (with respect to the recursive implementation of the autocorrelation equations) solution are 1) greater numerical accuracy in the computations, 2) less complex sequence of arithmetic operations, and 3) simple stability checks. In a computer simulation with floating-point arithmetic, the PARCOR system will require a substantially larger number of arithmetic operations and will not be any more accurate. The above stated advantages, however, may be important for hardware implementation.

10.3.2 Autocorrelation Method Vocoders

Markel and Gray [1974a] presented the details of a linear prediction vocoder simulation using the autocorrelation method. The general form follows that specified by Fig. 10.1. The coefficient analysis procedure uses the approach described in Chapter 6 for spectral analysis, namely, sharp analog prefiltering at $f_s/2$, sampling of the signal, pre-emphasis, windowing, and then application of the autocorrelation method. The pitch extraction is performed using the SIFT algorithm described in Chapter 8. The transmission parameters are the pitch P, gain $\sigma = C\sqrt{\alpha_M}$, and reflection coefficients k_m, $m = 1, 2, \ldots, M$. The gain σ is adjusted for the effect of the

Hamming window by the factor $C=1.58$. This constant is chosen by forcing the rms value of the window to unity. The approximate 4 dB gain increase is necessary to retain absolute gain reference since windowing is not applied in the synthesis procedure. The reflection coefficients are linearly shifted into the interval $[-.7, .7]$ for purposes of efficient encoding as discussed in Section 10.2. To obtain accurate representation for a given number of bits with linear quantization, symmetrical quantization with rounding is used. Linear quantization of the shifted reflection coefficients results in greater than a 30 percent increase in coding efficiency. Pitch and gain are both logarithmically coded. Synthesis is performed using a two-multiplier lattice structure with pitch-synchronous synthesis. The details of the synthesis operation used have been presented in Section 10.2.5. The excitation was computed using (10.25) and (10.27).

The accuracy with which the simulation represents both spectral (frequency) and temporal (time) structure, based upon the simplified excitation function, is now considered. The acoustic waveform for the *italicized* portion of the utterance "*Thieves who rob frien*ds deserve jail" is presented on the same time scale with a spectrogram in Figs. 10.9 A and B, respectively. Analysis was performed with the

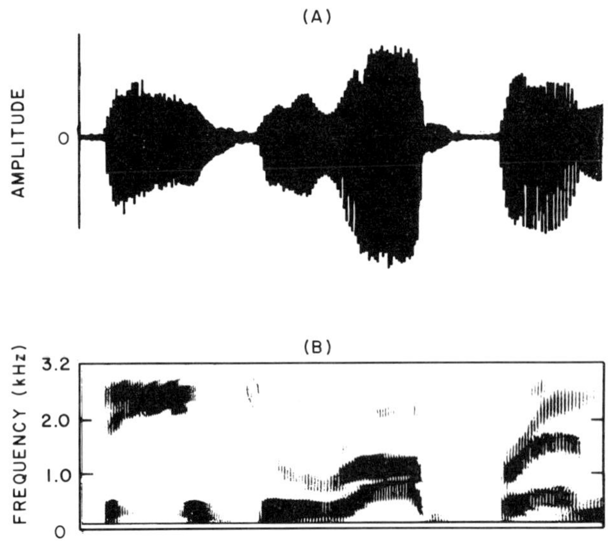

Fig. 10.9 Temporal and spectral structure of original utterance "Thieves who rob frien(ds deserve jail)".

following parameters: $f_s=6.5$ kHz, $N=128$, $M=10$, $f_r=51$ Hz. In addition, pre-emphasis based upon a first-order inverse filter, quantized to one bit, was used. The transmission parameters (10 reflection coefficients, pitch, and gain, respectively) were quantized and coded as follows: $\{7, 7, 5, \ldots, 5; 5, 4\}$ bits. The bit rate, therefore, is $B=(54+5+4+1)\,51=3{,}264$ bits/s. The synthetic speech corresponding to Fig. 10.9 is shown in Fig. 10.10.

Comparison of the spectrograms shows that the spectral structure is represented very accurately over the voiced regions. The only notable difference is in the representation of /v/ in thie*v*es where a forced error must be made, since only binary voiced-unvoiced decisions are made in the analyzer. It is very important in the system to be able to represent abrupt changes such as the transition from the /ɑ/ to /b/ in r*ob* and the /ɛ/ to /n/ in fri*en*ds. It is seen that these are fairly well represented, with the most error in the synthesis of "friends". This example illustrates a necessary tradeoff; namely, interpolation is necessary to effect good synthesis quality with a 50 Hz frame rate, but it will also cause a slight amount of smearing in situations such as these. The effect of choosing the simple synthesizer

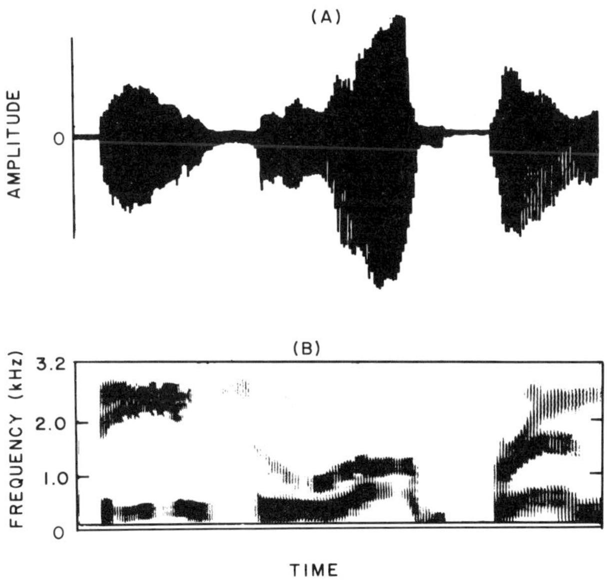

Fig. 10.10 Temporal and spectral structure of synthetic utterance "Thieves who rob frien(ds deserve jail)".

gain calculation for voiced and unvoiced frames is shown more clearly in the acoustic waveforms of Figs. 10.9A and 10.10A. The temporal representation is not as accurate as the spectral representation. The desirable feature of the system is that the accuracy of the temporal and spectral structure correlates with perception. Namely, during voiced sounds, it is far more important to have an accurate spectral envelope representation. If the envelope of Fig. 10.10A were forced to be a best match to Fig. 10.9A, a slight improvement would be perceived. If the same percentage changes were introduced in the location of the spectral peaks, the perceived difference would be substantial. Overall, the perceived quality and naturalness of this example is informally judged to be good.

There are, however, conditions for which this gain calculation produces undesirable results. Boll [1974, p. 23] has noted that an amplitude modulation of the synthesized waveform can be obtained, which is particularly noticeable in the synthesis of long, steady-state vowel utterances. The problem is due to the asynchronous relation between the analysis window location and the number of pitch periods (and thus energy) contained. The more complex calculation, (10.21), eliminates this difficulty.

As a general statement, pitch excited linear prediction vocoder systems will not produce synthetic speech indistinguishable from the original. This fact is illustrated here by studying the approximation process via the error signal $\{e_n\}$ obtained by passing the inverse filter input through the filter. In Figs. 10.11 A – D, various waveforms corresponding to the portion, "oak", spoken by a low-pitched male in the context "oak is strong ...", are presented in detail. Figure 10.11 A shows the original acoustic waveform. Figure 10.11 B shows the actual error signal which

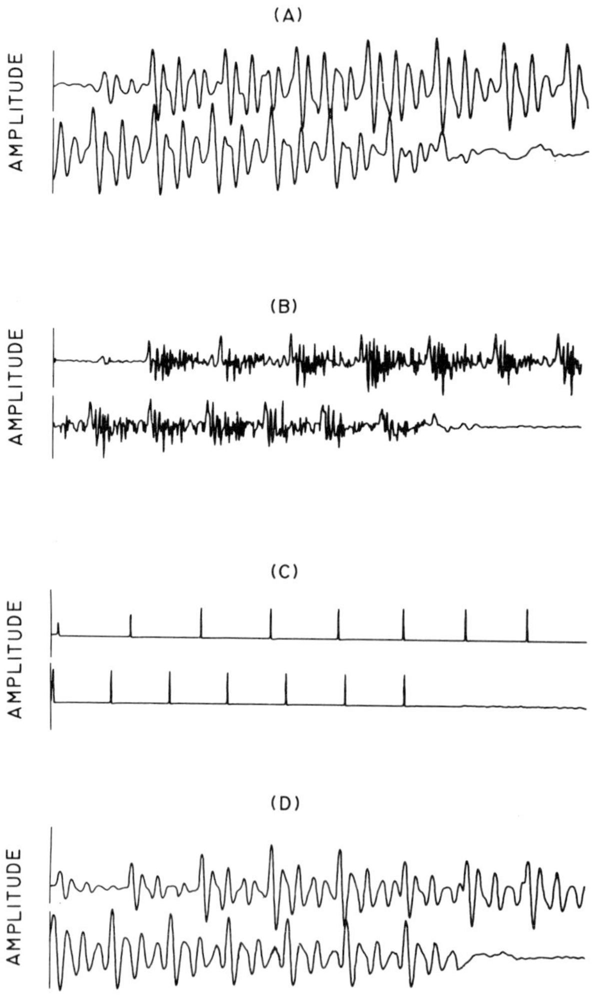

Fig. 10.11 Various waveforms from segment "oak". A) original. B) error signal. C) synthesizer driving function. D) synthesis.

has approximately flat spectral character with nearly all resonance structure removed. Figure 10.11 C shows the actual synthesizer driving signal (on the same scale as Fig. 10.11 B) reconstructed from interpolation of the transmitted gain parameters obtained during the fixed frame analysis. Finally, Fig. 10.11 D shows the synthesized speech based upon the driving sequence, Fig. 10.11 C, and five interpolated reflection coefficient sets. Figure 10.11 D was thus constructed from a total of 205 bits of information ($f_r = 34$ Hz). Even with this very small amount of information, a realistic approximation to both the temporal and spectral structure of the original signal can be obtained. This result is obtained in spite of the rather dramatic difference in temporal character of the actual error signal and the synthesizer driving signal. The quality is good, but there is a substantial difference in the timbre when compared directly to the original. If the actual error signal is used to drive the synthesizer, the resulting synthetic speech is essentially indistinguishable from the original speech in temporal and spectral appearance and perception. This, however, should not be overly surprising since transmission of the error signal requires at least 26,000 bits/s (4-bit log coding at a 6.5 kHz sampling rate). The corresponding synthesizer driving signal based upon explicit pitch, voicing decision, and gain requires 305 bits/s—$6500(5+4)/192 = 305$—a reduction of nearly two orders of magnitude. This system has also been successfully implemented in fixed-point arithmetic [Markel et al., 1973c; Markel and Gray, 1974b]. No noticeable degradation over the floating-point simulation described above is obtained under the following conditions:

1) A maximum sampling frequency of 6.5 kHz was used with 16-bit two's complement truncation arithmetic.

2) The autocorrelation calculation for the coefficient analysis is implemented in double-precision fixed-point arithmetic.

3) The synthesis filter is implemented with a bias reversal modification.

Bias generated with two's complement arithmetic causes an increasingly severe problem as the word length β decreases for voiced speech. For example, with $\beta = 12$, the mean square error (excluding the zero*th* spectral term) may be only 2 dB^2, while the error at zero frequency may be greater than 20 dB^2. In the actual synthesis, this bias is so severe that clipping of the output signal is obtained, resulting in servere distortion. A novel solution to this problem is to compute multiplications, e.g., $c = (a)(b)$ every other sample by $c = -(a)(-b)$. The effect is to shift the negatively biased distribution of numbers introduced by truncation to alterations of plus and minus distributions, and thus, on the average, cancel the bias since adjacent samples will generally be highly correlated for voiced speech. The frequency domain effect of this modification is essentially to eliminate the bias at $f = 0$ and cause a much broader and lower energy distribution of bias near $f_s/2$. Qualitatively, the bias shift can be seen by sampling a constant at the rate f_s with alternating sign.

The analysis portion of this system has also been implemented with fixed-point arithmetic in a high-speed signal processor [Knudsen, 1975]. Fourteen coefficients and a log magnitude spectrum are computed in real time at a frame update rate of 100 Hz and a sampling frequency of 10 kHz.

Makhoul, et al. [1974b] have developed a variable bit rate linear prediction vocoder system based upon the autocorrelation method that results in a low peak bit rate along with a low average bit rate. The system is similar to the previous autocorrelation method vocoder system discussed with the following additional features:

1) The system uses a variable order predictor based upon an information theoretic criterion [Akaike, 1974] for minimizing the number of necessary spectral coefficients for transmission.

2) The log likelihood distance measure defined as the logarithm of (10.10), is applied to determine the significantly changing portions of the spectrum to implement a variable frame rate transmission.

3) A bit allocation scheme for distributing a fixed number of bits among the transmission parameters was applied to maximize the representation accuracy.

4) Based upon the quantized parameter value statistics a variable-length Huffman coding scheme was applied which results in nearly a 20 percent reduction in transmission rate without any effect on the information transmitted (assuming no channel errors).

5) Time-synchronous (as opposed to pitch-synchronous) updating of parameters and interpolation is performed at the synthesizer. This procedure slightly simplifies the computations and has been informally judged to yield better synthesis' quality.

The system performs analysis at a frame rate of $f_r = 100$ Hz with a sampling rate of $f_s = 10$ kHz, $M = 11$ coefficients, and $N = 160$ samples per frame. The coefficient analysis is performed using the autocorrelation method. Pitch extraction is performed using a modified center clipping method. The gain used for transmission is $\sigma = \sqrt{\alpha_M C}$ as in the previously described system. Pitch and gain are quantized based upon experimentally obtained bit allocations. For the fixed order case before further encoding, 41 and 43 bits/frame were required for unvoiced and voiced frames, respectively. The threshold for the variable frame rate system was chosen as 1.4. Based upon these parameters, system bit rates versus conditions were achieved as shown in Table 10.3. All operations were performed with floating-point arithmetic.

The quality of the synthesized speech was judged to differ little between systems 1, 2, and 3, and between 5, 6, and 7. Speech quality of system 5 was only slightly lower than that of system 4, in spite of a reduction in the transmission

Table 10.3. System bit rates versus analysis conditions. [From Makhoul, et al., 1974b].

System No.	Analysis Frame Rate (frames/s)	Order	Transmission Frame Rate	Huffman Coding	Average Transmission Rate (bps)
1	50	Fixed	Fixed	No	2650
2	50	Fixed	Fixed	Yes	2000
3	50	Variable	Fixed	Yes	1750
4	100	Fixed	Fixed	No	5300
5	100	Fixed	Variable	No	2150
6	100	Fixed	Variable	Yes	1650
7	100	Variable	Variable	Yes	1500

rate by a factor of about 2.5. These results illustrate the successful performance of the variable frame rate transmission scheme. Although the bit rate of system 5 is lower than that of system 1 by about 20 percent, speech quality was found to be actually better for system 5 than for system 1. This suggests that starting with a higher analysis rate and transmitting only when necessary produces a better dynamic modeling of speech from the point of view of perception.

10.3.3 Covariance Method Vocoders

Atal [1970a] and Atal and Hanauer [1971b] first presented analysis—synthesis results using the covariance method of linear prediction. A phonograph record was attached to the 1971 paper to demonstrate the synthesis quality obtainable at various bit rates. The input speech was recorded in a high signal-to-noise environment, low-pass filtered to 5 kHz, and then sampled at $f_s = 10$ kHz. The analysis interval was set equal to one pitch period P for voiced samples and 10 ms for unvoiced samples. The predictor coefficients $\{a_i\}$ were obtained using the covariance method with $N = Pf_s$. The gain term σ was computed using the first procedure described in Section 10.2.4 so that the energy of the input speech in each synthesis interval matched the energy of the corresponding analysis interval. Pitch extraction was performed by an autocorrelation analysis of the signal obtained by filtering and cubing the input speech to emphasize the high amplitude portions of the speech waveform [Atal, 1968a].

The step-down procedure described in Chapter 5 was applied to each analysis frame for checking the stability of $1/A(z)$. If $1/A(z)$ was unstable, the roots were obtained using a polynomial root solving program. Let $A(z) = \prod_{m=1}^{M} (1 - z_m z^{-1})$ describe the filter polynomial. If the root z_m lies outside the unit circle, i.e., $|z_m| > 1$, then z_m is replaced by $z_m^*/|z_m|^2$. This replacement guarantees that the spectral shape remains unchanged although the original minimization criterion is no longer satisfied.

The polynomial with all roots inside the unit circle can then be constructed from a recursion such as

$$A'_m(z) = A'_{m-1}(z)(1 - z_m z^{-1}) \tag{10.33}$$

for $m = 1, 2, \ldots, M$, with $A'_M(z)$ replacing $A(z)$. Note that z_m will in general be complex. The modified filter coefficients were coded and quantized in two different ways: 1) the frequencies and bandwidths of the roots z_m were quantized to a total of 60 bits per frame (assuming $M = 12$) and 2) the acoustic tube areas ($\mathscr{A}_m$) (see Chapter 4) were quantized to a total of 60 bits per frame. Both of these schemes ensure stability of the synthesizer filter even if linear interpolation is applied. The other transmission parameters were pitch P, a voiced-unvoiced decision V/UV, and gain coded to 6, 1, and 5 bits, respectively. Transmission bit rate was therefore $B_r = f_r(6 + 1 + 5 + 60) = 72 f_r$. Frame rates of 100, 67, and 33 Hz were used, resulting in bit rates of 7200, 4800, and 2400 bps, respectively.

A direct form filter, driven pitch-synchronously, was used to synthesize the speech. The excitation was based upon a unit sample at initiation of each period or uniformly distributed pseudo-random samples from a zero mean, unit variance generator being multiplied by the gain σ. The voicing parameter V determines which excitation is applied. Since parameters are transmitted at a constant rate f_r, linear interpolation was used to determine pitch-synchronous parameters.

To ensure stability, the sequence $\{a_i\}$ was transformed into the first $M+1$ samples of the autocorrelation sequence $\{r(n)\}$. After interpolation, the sequence $\{r(n)\}$ was transformed back to an interpolated set of parameters $\{a_i\}$ and then applied to a direct form filter for synthesis.

The quality of the synthetic speech was informally judged to be very close to that of the original speech. Several factors relating to the quality of synthetic speech obtained from this system should be considered. The goal of this system was to obtain the highest possible quality for a given bit rate without major concern over computational issues. To perform the analysis with a pitch period dependent window length requires performing the pitch period analysis ahead of the coefficient analysis and, in addition, having very accurate pitch analysis. As Schroeder [1966] has pointed out, a one percent error rate in pitch extraction may be intolerable. The actual pitch period analysis algorithm used requires many decision operations and up to four or five frames of delay (required data buffers) for determination of whether a voiced frame should be changed to unvoiced, etc.

To achieve synthesis quality as illustrated by the record, it is also necessary to have a very high signal-to-noise ratio (45 – 50 dB). Furthermore, results are somewhat dependent upon how well the speech satisfies a complex exponential model during a single pitch period. In addition, all operations were performed with full floating-point precision.

Considerable disappointment will be felt if one implements the system with the assumption that a few short cuts will help computational requirements without affecting the quality. There do not at this time appear to be straightforward procedures (in the sense that the detailed algorithm can be presented with a series of algebraic equations) for implementing systems with high quality and low bit rates. For example, the autocorrelation form of analysis is straightforward in the sense that if sufficient precision is used in the computation, stability of $1/A(z)$ is theoretically assured. Unfortunately, the synthesized quality is often lower than that of the covariance method under ideal conditions (e.g., pitch-synchronous analysis, high signal-to-noise ratio). The covariance method, on the other hand, requires ad hoc operations to ensure against unstable synthesis filters (polynomial root testing and placing roots within the unit circle so that the minimization criterion is no longer satisfied).

A linear prediction vocoder system based upon this analysis-synthesis system has been evaluated by Haskew, et al. [1973]. The objective was to optimize the system in terms of performance and complexity for a wide range of speakers at data rates of 3600 and 7200 bps. In their study, the speech was bandlimited to 4000 Hz and sampled at $f_s = 8000$ Hz. In addition, a fixed analysis interval was used. Experiments were performed using six different sentences, each from a different speaker, to determine a reasonable number of filter coefficients M and analysis interval N. By testing $N = 64$, 128, and 256, an $N = 128$ (16 ms) interval was chosen. The shorter interval was susceptible to instability problems, while the longer interval caused excessive spectral smoothing. The predictor order was chosen as $M=12$, with $f_s=8$ kHz (in contrast to $M=12$ for the $f_s=10$ kHz of the Atal-Hanauer implementation) for providing uniformly good quality synthesis. It was observed that there is essentially no improvement in the synthesis for frame rates above 200 Hz, and that the speech quality degrades gracefully from $f_r=200$ Hz to around $f_r=30$ Hz.

For computational reasons, it was judged to be desirable to eliminate the polynomial root solving. First, the filter $A(z)$ was transformed into an accoustic tube (see Chapter 4). Necessary and sufficient conditions for stability of $1/A(z)$ are that the areas $\{\mathscr{A}_m\}$ be positive since

$$\mathscr{A}_m/\mathscr{A}_{m-1}=(1+k_m)/(1-k_m), \tag{10.34}$$

and $|k_m|<1$ is necessary for stability. If an area was negative, $A(z)$ was replaced by $A(rz)$ where $r>1$ forces all poles towards the origin. A new area function was then computed. The procedure was continued until the modified polynomial had all positive area functions. In addition, it was found to be necessary to ensure that all bandwidths were greater then 30 Hz. The requirement was satisfied by ensuring that a contraction of the unit circle by 1.01 did not result in unstable area functions.

Considerable effort was then spent on determining an efficient coding method for the area functions. The most efficient coding arrangement was found to be logarithmic coding of the area ratios. Based upon a frame rate $f_r=50$ Hz, the best choices for a data rate of 3600 bps were found to be

Area ratios $1-2$	6 bits
Area ratios $3-8$	5 bits
Area ratios $9-12$	4 bits
Pitch and voicing	8 bits
Gain	5 bits.

For a 7200 bps system, simply doubling the frame rate was judged to produce the best results.

The quality of the system at 3600 bps was evaluated using a category judgment method. Thirty sentences (ten talkers speaking three sentences apiece) were run through the simulation program. Thirty listeners rated these sentences (twice each) in a 60-item test using categories of excellent, good, fair, poor, and bad. The listeners were anchored to the extreme categories with speech from a telephone input and a 3600 bps channel vocoder. Results indicate that the overall quality rating of the 3600 bps simulation on this scale is midway between fair and good. There are indications that the ratings are talker and, to a lesser extent, text sensitive. In general, male voices tend to achieve a higher rating than female voices, but exceptions to this observation do exist. For most talkers and texts, the 3600 bps simulation represents a significant improvement in quality over previous channel vocoders.

A paired comparison test was also administered in order to evaluate the quality differences between systems at 3600 and 7200 bps. The 30 sentences employed in the category judgment test were processed by the simulation in a 7200 bps configuration that operated with a 10 ms frame and 72 bits per frame. The sentences for both data rates were combined into a 30-item preference test.

The results for all talkers and sentences showed a 53 percent preference for the higher data rate. A 58 percent figure resulted when only female talkers were considered. This suggests degradation for interpolation of short (female) pitch periods with a 20 ms frame. However, the small preferences involved indicate there is no substantial increase in subjective quality with an increase of 3600 bits in the data rate.

The investigation of implementation considerations led to an estimate of 9200 operations per frame (4200 for transmitter, 5000 for receiver) or at 3600 bps, 461,000 operations per second. In addition, it was assumed that a floating-point processor would be necessary to perform the operations.

Based upon the Atal-Hanauer system and the Haskew, et al. study, adaptations and modifications have been introduced by Welch, et al. [1974] which allow implementation using a high-speed digital processor.

The system, known as LONGBRAKE, has been implemented using fixed-point arithmetic and performs the complete vocoder operation in real time. It is apparently the first operational real time linear prediction vocoder system. An engineering development model of LONGBRAKE is shown in Fig. 10.12 along with a

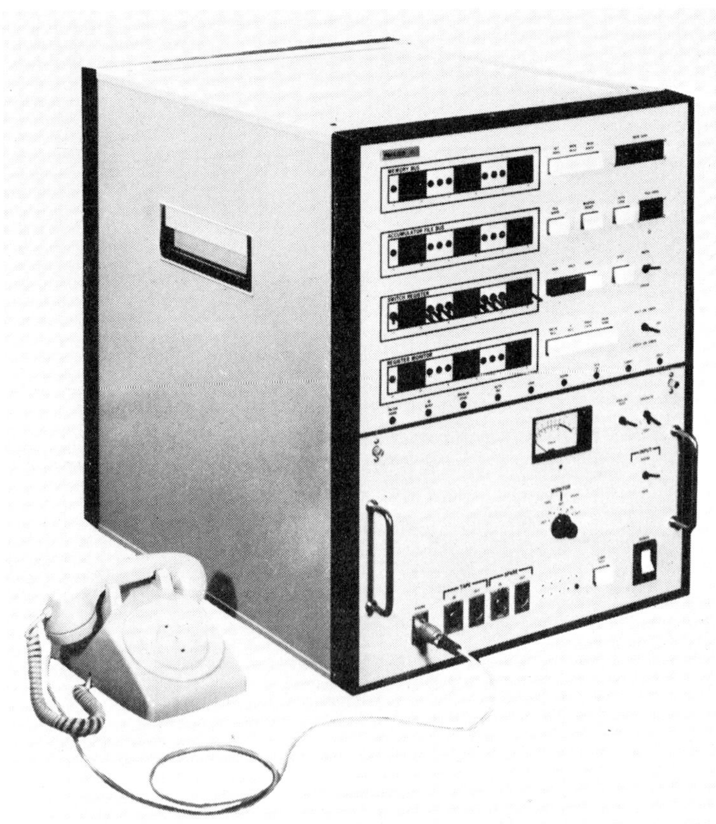

Fig. 10.12 An engineering development model of the LONGBRAKE system. [From Welch, 1974]

telephone-type input for size comparison. The input speech is bandpass filtered with a gradual roll-off below 200 Hz and a sharp cutoff above 3200 Hz. It is then sampled at 6400 Hz. The speech samples are stored into both a pitch analysis buffer and a predictor coefficient analysis buffer at a frame rate of 44.4 Hz. First order digital pre-emphasis with a 700 Hz cutoff is applied to the data to be stored in the predictor analysis buffer.

This buffer is low-pass filtered and then applied to an Average Magnitude Distance Function pitch extraction algorithm [Ross, et al., 1974] and a voicing detector. The voicing detector uses an energy measure, zero crossings analysis, and the AMDF max-to-min ratio to make decisions.

A tenth order analysis is implemented with the capability of performing either the covariance or the autocorrelation method. The window length is $N=102$ samples, but a variable starting point is used to line up successive analysis frames at pitch period multiples. The covariance coefficients are obtained by scaling the input data and applying double-precision accumulation of products. The Choleski decomposition method (see Chapter 3) is used to solve the covariance equations. Block scaling is used to provide high accuracy using only single-precision 16-bit arithmetic. This system uses the generalized reflection coefficients which precisely correspond to the minimizing filter $A(z)$ only in the autocorrelation method. Even though the polynomial obtained by performing the step-up procedure using the generalized reflection coefficients is not the minimizing polynomial $A(z)$, the differences have been experimentally judged to be insignificant. The computational advantage of this approach is the elimination of a step-down operation for obtaining reflection coefficients (which do correspond to the minimizing polynomial) from $A(z)$ obtained in the analysis. Practical considerations dictate repetition of the previous frame whenever any generalized reflection coefficient exceeds 0.97 in magnitude. The coefficients and gain values are delayed by two frames to account for delays introduced in pitch period error detection and correction.

Voicing is coded to one bit while pitch is coded in an approximately logarithmic manner to 6 bits. Gain is coded to 5 bits with linear coding from 0 to 5 and logarithmic coding from 5 to 1024. These parameters are transformed into a serial bit stream with additional error protection for the most significant bits.

The synthesizer decodes the transmitted parameters and linearly interpolates pitch, gain, and the generalized reflection coefficients. The coefficients are interpolated at the mid-point of the frame. Due to timing considerations, this system uses a step-up procedure (see Chapter 5) to transform the generalized reflection coefficients into filter coefficients. A direct form filter is then used for synthesis since only one multiply-add operation per stage is necessary.

The output energy level is matched to that of the input speech energy by scaling the synthesizer output on a pitch period by pitch period basis from (10.28). Unvoiced samples are generated by reading random numbers from a table with a random starting point. The output is then applied to a digital-to-analog converter and sharply filtered to 3200 Hz. Finally, a first order analog de-emphasis filter is applied to the output with a 700 Hz cutoff.

The bit rate of this system is adjustable as 4800, 3600, or 2400 bits per second depending upon the frame rate and bit protection coding. An example output from LONGBRAKE is shown in Fig. 10.13 for the utterance "linear prediction". By

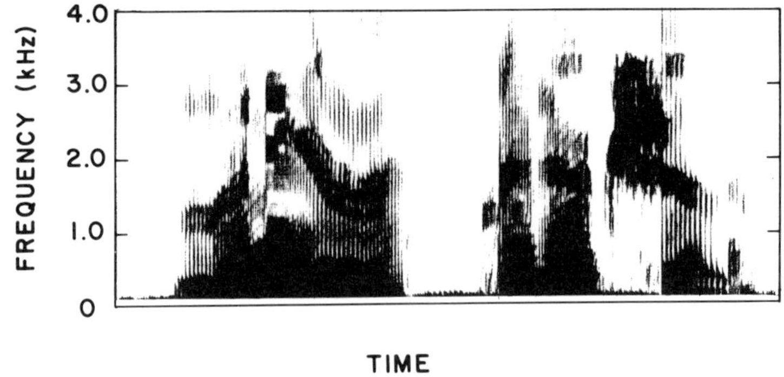

Fig. 10.13 Spectrogram of utterance "linear prediction" from the LONGBRAKE system.

comparing with the original speech spectrogram in Chapter 1, the synthetic speech is seen to closely match in both the time and frequency domain. The cutoff frequency of the analysis was 3.2 kHz. The only obviously gross differences are seen in the segment following the nasal /n/ of li*n*ear, where quantization effects appear in the formant structure.

10.4 Base-Band Excited Vocoders

Two closely related linear prediction vocoder systems have been proposed that eliminate the requirement for explicit pitch extraction (while retaining reasonable low bit rates) by transmitting excitation information over only a narrow baseband. These systems have been referred to as RELP for residual-excited linear prediction vocoder [Magill and Un, 1974] and VELP for voice-excited linear prediction vocoder [Weinstein, 1974, 1975b].

In both systems, linear prediction analysis is used to compute sets of reflection coefficients $\{k_i\}$ on a time-synchronous basis for transmission. The receiver then applies these coefficients to the synthesis filter $1/A(z)$ in a time-synchronous manner. In the pitch-excited vocoder systems just discussed, explicit pitch and gain values are transmitted, and then from these, a synthesizer driving function is reconstructed. In the base-band exited vocoder, a low-frequency portion of the signal or a spectrally flattened version of the signal is transmitted without explicit pitch and gain extraction. From this signal, a synthesizer driving function is then reconstructed. In Fig. 10.14 a simplified block diagram of the

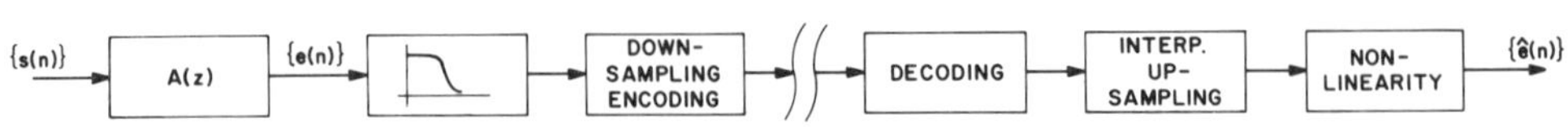

Fig. 10.14 Block diagram of the excitation signal generation in RELP. [After Weinstein, 1974]

excitation signal generator for the RELP system is shown. The digital samples $\{s(n)\}$ obtained by sampling the input speech $s(t)$ at the rate f_s are passed through the inverse filter $A(z)$ (whose coefficients are uniquely related to the transmitted reflection coefficients) to generate the error signal $\{e(n)\}$. This high information rate signal is then low-pass filtered with a bandwidth of f_c, and then down-sampled to $2f_c$. This signal is then coded and transmitted at the receiver, the signal is decoded and applied to an interpolator, and up-sampled to generate a signal at the original sampling frequency f_s. Since the frequency content of the transmitted signal is limited to the range $f_c \ll f_s/2$, a non-linear operation must be employed to generate harmonics out to $f_s/2$. One procedure is to apply an asymmetrical half-wave rectifier and a differencer. What is desired is a reconstructed residue signal $\{\hat{e}(n)\}$ as the synthesizer driving function which retains a flat spectral trend and the same energy as the original residue $\{e(n)\}$.

The only major distinction between the VELP and RELP system is that in the VELP system, the speech signal is directly applied to the base-band filter instead of first passing it through the inverse filter $A(z)$ in Fig. 10.14 [Weinstein, 1974].

An adaptive spectral flattener was introduced at the output of the nonlinearity by Weinstein to assure maximum flatness of the reconstructed excitation signal. The process of spectral flattening was discussed in Chapter 6. Briefly, by applying the autocorrelation method to the output of the non-linearity, a new inverse filter is defined. The output of this filter (the excitation signal) has the property that its spectral flatness has been maximized for the specified filter order.

VELP and RELP systems with and without spectral flattening have been implemented in real time on the Fast Digital Processor (FDP) at Lincoln Laboratories [Weinstein, 1974]. System parameters are as follows: $f_s = 7.69$ kHz, $N = 150$ with a Hamming window, frame rate $f_r = 60$ frames/s, baseband filter from 192 to 577 Hz, and a 10th order linear predictor for both analysis and excitation signal flattening. Several observations were made from the simulations:

1) In both RELP and VELP, spectral flattening produces significant improvements over using only a rectifier and differencer. The synthetic speech is perceived as being smoother and more clear.

2) Both systems could benefit from a somewhat higher order spectral flattening filter.

3) The excitation signal in RELP after spectral flattening is flatter than the one in VELP, apparently due to the fact that the RELP baseband signal is already a flattened signal.

4) The two systems sound somewhat different but it is difficult to judge which is preferable. VELP has occasional energy irregularities due to a large excitation signal harmonic coinciding with a sharp formant. RELP, however, appears to have a generally rougher quality.

5) Finally, techniques for incorporating the original baseband signal into the excitation signal may significantly improve the perceived synthetic speech quality.

Typical characteristics for a baseband excited vocoder are illustrated for the VELP system in Fig. 10.15. Figures 10.15A and B show wideband spectrograms of the respective input and processed speech for the utterance "We've changed the measures". The baseband is coded to 5950 bps, resulting in a total bit rate of 8072 bps when the coding of the filter coefficients is also included. The most obvious effects of the processing are irregularities in the pitch striations and the appearance of noise superimposed on the data. The synthetic speech is perceived to be slightly hoarse. Atal, et al. [1975] have considered a similar technique with the introduction of a pitch predictor in an attempt to reduce the bit rate to 3500 – 4000 bps.

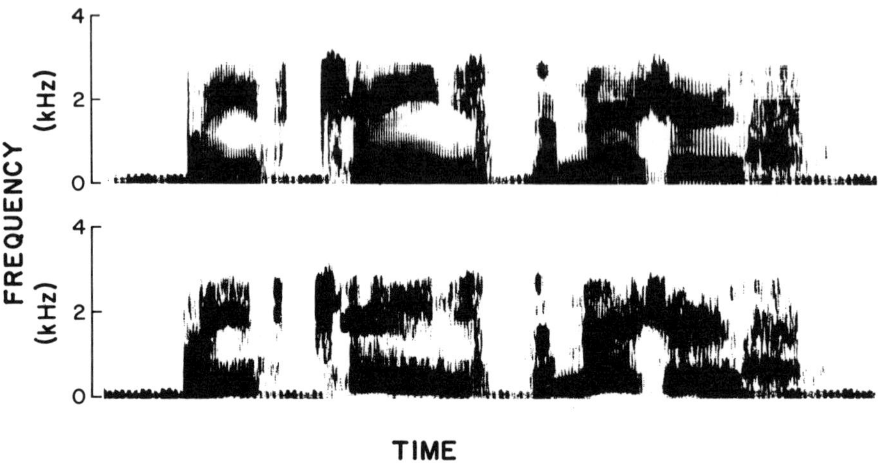

Fig. 10.15 Results from a baseband excited vocoder system. A) input speech. B) processed speech. [From Weinstein, 1975b]

11. Further Topics

The purposes of this final chapter are to briefly describe some applications of linear prediction other than those previously discussed, to summarize a few important properties of linear prediction, and finally to present several thoughts about further research directions.

The area of speech recognition or speech understanding is so vast in terms of the necessary ingredients such as acoustic-phonetics, syntax, and semantics, that the role played by linear prediction for acoustical processing cannot be adequately covered here. The interested reader is referred to the special issue of the *IEEE Transactions on Acoustics, Speech, and Signal Processing*, February 1975, devoted to speech recognition. In six of the nine papers on systems and feature extraction, some discussion of linear prediction is included. In addition, Makhoul (1975c) has discussed the application of linear prediction to speech recognition.

In this chapter, possible applications of linear prediction for speaker identification, verification, word recognition, acoustical detection of laryngeal pathology, the estimation of spectral zeros in addition to poles, and possible future directions for applications of linear prediction in speech processing are discussed.

11.1 Speaker Identification and Verification

Linear prediction parameters are easily and efficiently obtained from speech, and in addition, through synthesis experiments, have been shown to retain a considerable degree of naturalness from the original speech. These factors have motivated investigations into the applicability of linear prediction for speaker identification and verification.

The object of speaker identification is to determine whether or not a speech sample from an unknown talker can be associated with a speech sample from one of several reference talkers. The purpose of speaker verification is to determine whether the speaker is who he says he is. Due to the binary nature of verification, it is an inherently more manageable problem and has met with considerable success for single test utterance comparisons [Doddington, 1970; Lummis, 1973].

Pfeifer [1974] has investigated the applicability of the filter coefficients, reflection coefficients, and the inverse filter spectrum for speaker identification using an unweighted Euclidian distance measure, d_j, where

$$d_j^2 = (\mathbf{t} - \mathbf{r}_j)^t (\mathbf{t} - \mathbf{r}_j). \tag{11.1}$$

The jth reference vector is $\mathbf{r}_j$, and the test vector is $\mathbf{t}$. The dimension of the vectors is the order of the filter, M, for each set of coefficients except for the filter spectrum where the dimension is determined by the size of the FFT used to determine the spectrum.

Analog recordings were low-pass filtered to 3.25 kHz and then sampled. Each desired sound segment was manually located, and analysis was then performed about the center of the sound. The autocorrelation method was used with analysis conditions given by $f_s = 6.5$ kHz, $N = 128$, $M = 10$, with pre-emphasis and a Hamming window applied before analysis. An FFT was applied to the filter coefficients to obtain 256 equally-spaced values of $ln|A[\exp(j\theta)]|$. The reference data consisted of 10 samples of each sound with each sample from a different speaker. Eq. (11.1) was then used to compute the distances. A forced choice was made to determine the speaker. By pooling results from three different sounds — /ɛ/, /n/, and /ɔ/ — a 100 percent correct identification score was achieved over ten male speakers.

Use of the filter coefficients or reflection coefficients as components of the vectors led to significantly poorer results, with the best recognition score reduced to 80 percent, even though the same total information is contained in those parameter sets. The implication is that the distance measure of (11.1) is probably not appropriate for those parameters.

Atal [1974a] investigated the properties of a weighted Euclidian distance measure for both speaker identification and verification tasks. He investigated a number of parameter sets including the filter coefficients, the impulse response of $1/A(z)$, its autocorrelation, the acoustic tube area functions, and the cepstral coefficients of the inverse filter. All these parameter sets are uniquely, but non-linearly, related to each other as shown in Chapter 10. The distance measure used was a weighted quadratic measure of the form

$$d_j^2 = (\mathbf{t} - \mathbf{r}_j)^t \mathbf{W}^{-1} (\mathbf{t} - \mathbf{r}_j). \tag{11.2}$$

The reference vectors are found by averaging all of the separate measurement vectors for each individual speaker. If $\mathbf{x}_{ij}$ represents the ith utterance of speaker j, then

$$\mathbf{r}_j = AVG_i(\mathbf{x}_{ij}), \tag{11.3}$$

where AVG_i denotes an average over i. $\mathbf{W}^{-1}$ is the inverse of $\mathbf{W}$, the pooled intra-speaker, that is, within speaker covariance matrix, where

$$\mathbf{W} = AVG_{i,j}[(\mathbf{x}_{ij} - \mathbf{r}_j)(\mathbf{x}_{ij} - \mathbf{r}_j)^t], \tag{11.4}$$

and $AVG_{i,j}$ denotes any average over both i and j.

The measure of (11.2) has the important property that it is unchanged by nonsingular (invertable) linear transformations on the measurement vectors. It can be derived from maximum likelihood principles if one assumes Gaussian statistics of the observation vectors.

A data base of six repetitions of the sentence "May we all learn a yellow lion roar" from ten speakers was generated for the tests by Atal. Five repetitions from each speaker were used for generating $\mathbf{r}_j$, and the sixth was used as a test. Each utterance was segmented into forty equally long segments (about 50 ms each) to have approximate time alignment of the utterances. Each time segment thus had a different set of reference vectors and a different pooled intraspeaker covariance matrix, $\mathbf{W}$. The covariance method was used with $f_s = 10$ kHz and $M = 12$. Figure 11.1 shows the accuracy of identification for each separate time segment or frame for the various parameter set choices. The average over the frames shows that the cepstral coefficients provided the highest identification accuracy. Since the cepstrum is linearly related to the log spectrum of the inverse filter, this seems consistent with Pfeifer's results.

By averaging over the separate segments, the accuracy for all of the parameter sets is increased. Atal achieved 80 percent accuracy over 0.1 s and 98 percent for intervals exceeding 0.5 s.

11.2 Isolated Word Recognition

Itakura [1975] has investigated a log likelihood ratio, based upon signal and error energy as described in Chapter 10, as it applies to the problem of isolated word recognition. For each speech segment the minimum predictor error energy for a segment is compared to that computed by using a reference inverse filter to generate the error signal.

The isolated word recognition scheme has been implemented on a DDP-516 computer. It allows for 200 isolated words spoken by an individual to be used in generating the training utterances. Each utterance is input to the computer from a conventional telephone set. The sampling frequency is 6.667 kHz and each word is contained within the fixed interval of 1.2 s. A 200-sample Hamming window is advanced in steps of 100 samples, and autocorrelation analysis is applied.

To correct for possible spectral deviations caused by physical factors such as transducer and line response, and human factors such as stress and physical condition, a second-order inverse filter is first applied to a large interval of the speech to perform spectral flattening. The first six autocorrelation coefficients of the preprocessed speech are then used to define the reference inverse filters. On the DDP-516 this system operates in about 22 times real time and has a recognition rate of 97.3 percent. These results were obtained over a three-week period with a designated talker recording 2000 test utterances.

Sambur and Rabiner [1975] developed a speaker independent digit recognition system, where each word was a spoken digit. A block diagram of the system is

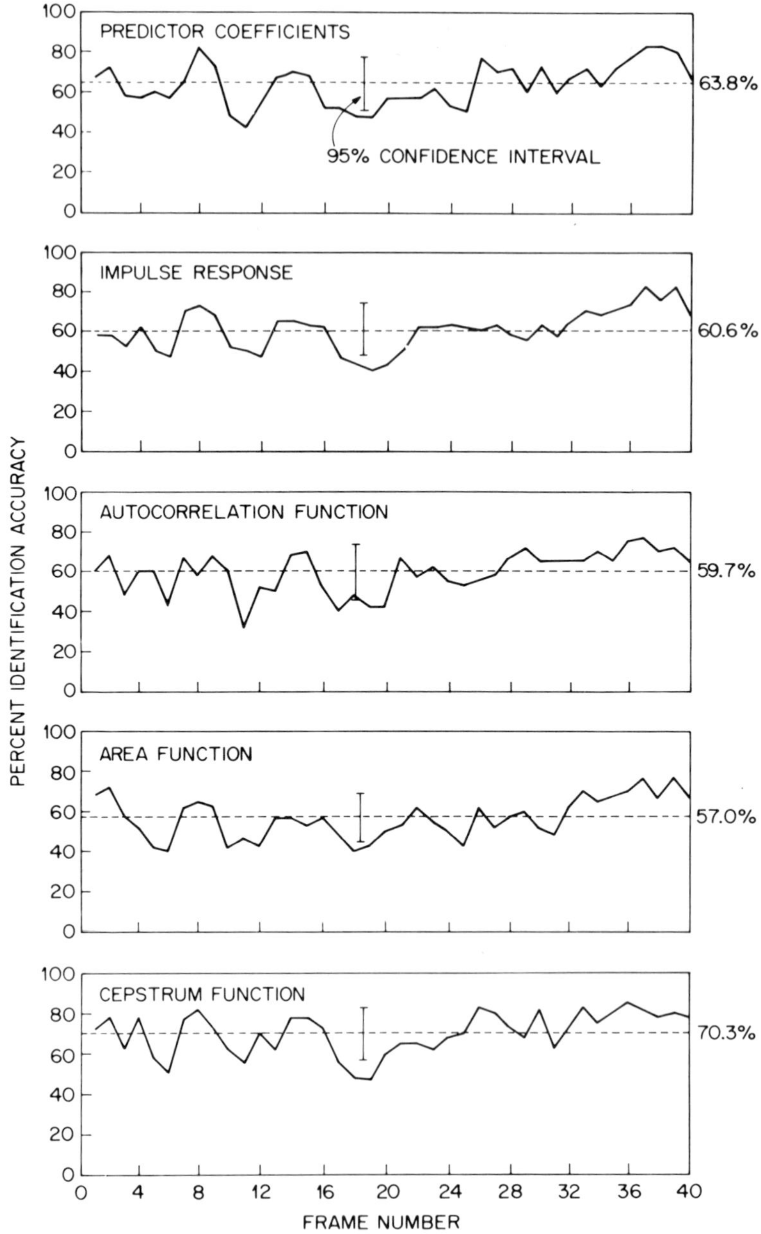

Fig. 11.1 Percent identification accuracy for different parametric representations of speech based on a 50-ms-long speech segment. [From Atal, 1974a]

shown in Fig. 11.2. In this system a second-order linear prediction analysis is utilized [Makhoul and Wolf, 1973b], with other factors used in the recognition being zero crossing rate, energy, and total squared error α.

After performing endpoint alignment to isolate the interval containing each word (digit), the speech is analyzed every 10 ms to obtain the parameters. The second-order linear predictor generally results in a resonance situated between the first and second formants of the speech. As the first formant moves over a smaller range than the second formant, the computed pole from the linear predictor tends to follow the motion of the second formant. The normalized error α_2/α_0 generally increases from sonorants to vowels and then to fricatives. Within the three vowel types, the back vowels have the lowest normalized error and the front vowels the highest. Figure 11.3 shows the normalized error, zero crossing rate, pole frequency, and signal energy for the word "two". After the region of frication in /t/, which is marked by high error and low energy, the normalized error uniformly decreases. The basic recognition approach is not that of a simple distance measure, but rather a combination of measurements. Each interval is segmented and a preliminary decision algorithm places the word into a digit class. Final decisions are made on the basis of the presence or absence of key features in utterances. An experimental test of this system was made using five men and five women over a five-week period with each recording ten repetitions of the ten digits resulting in an average error rate of 2.7 percent.

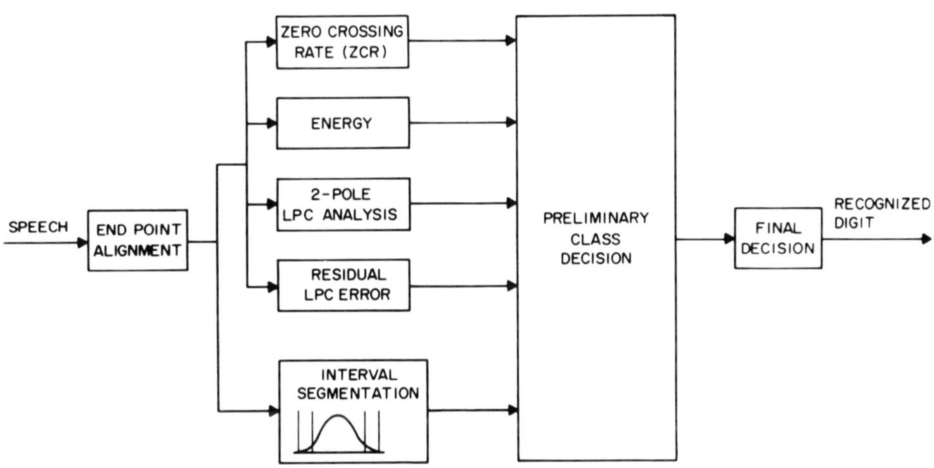

Fig. 11.2 Block diagram of the overall digital recognition system. [From Sambur and Rabiner, 1975]

11.3 Acoustical Detection of Laryngeal Pathology

Koike and Markel [1975] have investigated the application of linear prediction for the detection of laryngeal pathology. Although previous spectral analysis studies have shown wide variability [Winckel, 1952; Yanagihara, 1967], they also suggest the feasibility of utilizing the acoustic signal to reveal the existence of pathology in the larynx. One of the problems associated with this type of study has

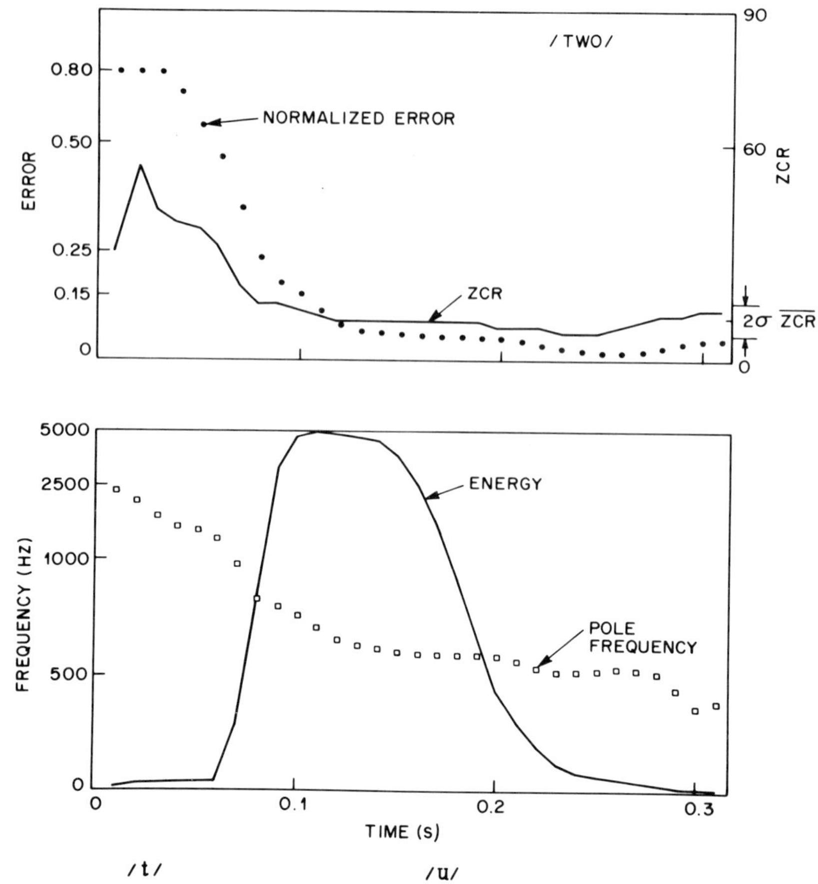

Fig. 11.3 Complete set of measurements for one example of the word "two". [From Sambur and Rabiner, 1975]

been the effect of the vocal tract which modifies the acoustic wave traveling from the glottis to the lips. It is not a trivial task to explicate the changes actually attributable to the laryngeal pathology from the speech spectrum.

An indirect technique for obtaining information about laryngeal vibratory behavior is inverse glottal wave filtering [Miller, 1959]. There are two serious disadvantages in using this approach to study pathological speech. First, it is difficult to extract the necessary parameters of the model, and second, the resultant waveform contains only low frequency information. By inspection of sonograms, pathological voices often show deviant behavior from normal voices in the higher frequency ranges above 1 kHz.

The error signal or residue at the output of the inverse filter obtained by linear prediction, on the other hand, contains an equal weighting of all frequency components, in the sense that the spectrum is flattened. Based upon the linear speech production model, it is expected that for normal voiced speech the error signal will show clear spikes at the initiation of each period. For pathological cases where an incomplete glottal closure exists, the assumptions of vocal tract and source separability in the linear speech production model would be invalidated. Therefore, a less distinct pattern of periodic spikes would be expected, depending upon the severity of the pathology. For a single small nodule on the surface of the vocal fold, little difference between a normal voice and a pathological voice would be expected, but for the severe case of vocal fold fixation, where glottal closure would be negligible, no periodic behavior would be expected in the residue.

A population of ten normal and ten pathological voices was analyzed using the autocorrelation method to determine characteristics of the residue signal. One hundred ms portions of the vowel /ɑ/ ("ah") were analyzed based upon 40 ms windows. The speech signal and residue for a normal male subject with fundamental frequency of 130 Hz are shown in Figs. 11.4A und B, respectively. The acoustic waveform for this subject is seen to be very regular with approximately eight pitch periods. There is only slight amplitude modulation seen by looking at the peaks of the speech waveform.

The residue signal consists of sharp spikes at the initiations of the pitch periods, with relatively low amplitude irregular behavior between the spikes. The cyclic oscillations within the pitch periods of the speech signal have been effectively removed by the inverse filtering. The noise or irregular behavior in the residue signal decays such that during the last half of each pitch period it is much

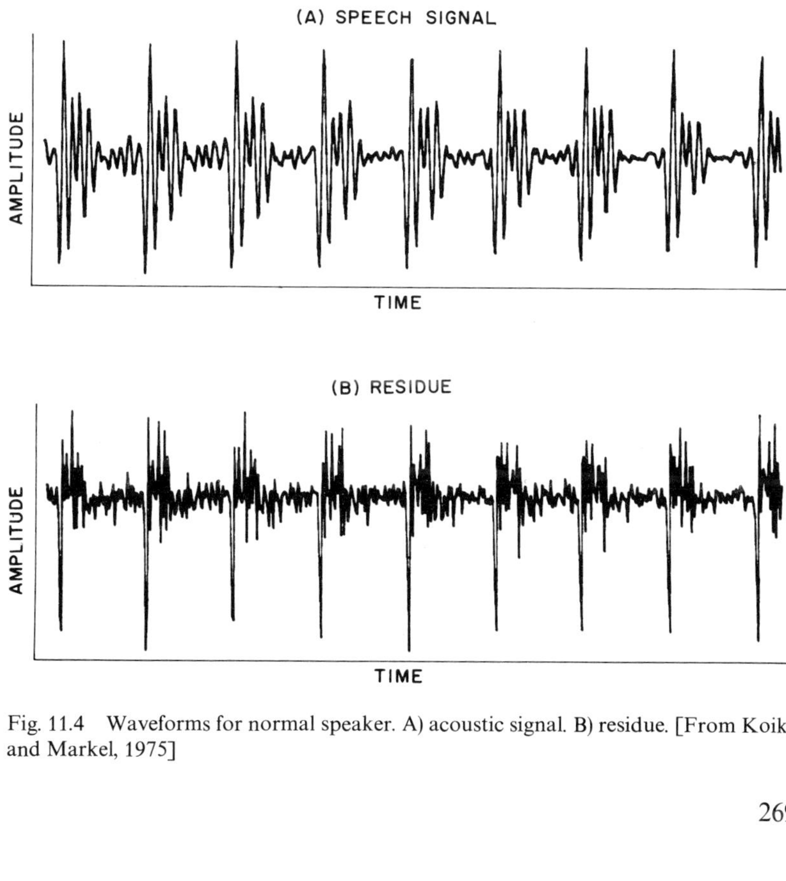

Fig. 11.4 Waveforms for normal speaker. A) acoustic signal. B) residue. [From Koike and Markel, 1975]

smaller than the spikes at the pitch period initiations. The most important feature of the residue signal for the normal speaker is the high peak signal (spike amplitude) to noise (during the last half of the pitch period) ratio.

The speech signal and residue for a patient with an advanced case of laryngeal cancer are shown in Figs. 11.5A and B, respectively. The recording was made several days before a total laryngectomy. The acoustic waveform is seen to

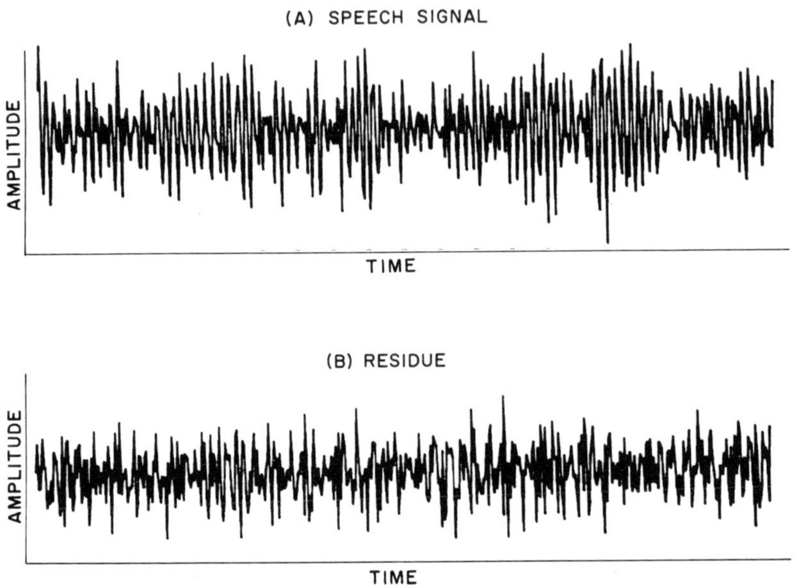

Fig. 11.5 Waveform for speaker with advanced laryngeal cancer. A) acoustic signal. B) residue. [From Koike and Markel, 1975]

contain little, if any, periodic component, even though some cyclic behavior due to vocal tract resonances can be seen. The voice sounds extremely harsh and rough. The residue signal shows the cyclic behavior eliminated, leaving a very noisy non-periodic signal. Comparison of Fig. 11.4B with Fig. 11.5B illustrates the extremes one might expect to see between a normal subject and an advanced pathological subject. For these extremes the differences are obvious from either the acoustic signals or the residues. Of major interest is the detection of the early stages of pathological development.

An example of a pathological voice that would probably pass undetected on the basis of the acoustic signal alone is shown in Fig. 11.6A. Its residue is in Fig. 11.6B. The speaker had a small nodule on the left vocal fold and moderate hyperemia at the margins of the vocal folds, as shown by indirect laryngoscopy. Perceptually, the voice was judged to have only slight hoarseness by an experienced listener. The acoustic waveform has strong periodicity and similarity between pitch periods, with a slight decrease in amplitude near the end of the analysis window shown. Due to the fact that the pathology is in an early stage, it would be expected that the waveform would lie somewhere between those for a normal speaker and a pathological speaker. Indeed this is the case as seen from the residue, Fig. 11.6B. Moderate sharpness of spikes at the initiation of each pitch period is observable. There are rather large variations in several peaks, however. The noise level appears to be somewhat higher than that of the normal voices. In a recent study by Davis [1976], the same ten normal and ten pathological voices were correctly separated by using computer measurements of residue signal features in a pattern classification scheme. Although these observations and results are of a preliminary nature, it is believed that the error signal or residue from linear prediction analysis does contain valuable information about the state of the vocal folds.

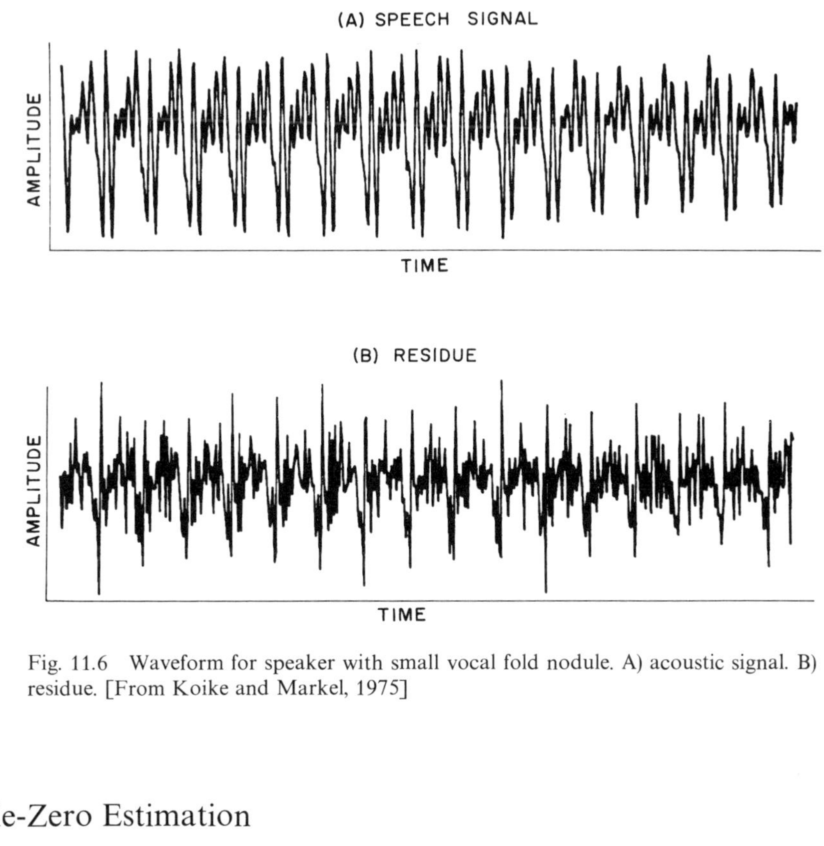

Fig. 11.6 Waveform for speaker with small vocal fold nodule. A) acoustic signal. B) residue. [From Koike and Markel, 1975]

11.4 Pole-Zero Estimation

It has been demonstrated in previous chapters that linear prediction is a powerful tool in speech processing, even though the analysis has been restricted to an all-pole model. Atal [1974b] has suggested that the perceived differences between real speech and the best synthetic speech obtainable using linear prediction are at least partially due to the all-pole model restriction. It is most certainly a factor in the analysis of nasals such as /n/ in li*n*ear where transfer function zeros theoretically occur in a network model with a nasal branch [Flanagan, 1972, pp. 77–80]. In addition, coarticulation in an utterance such as "linear" will cause adjacent vowels to also be nasalized. It therefore seems logical to consider the extension of the all-pole linear prediction model to also include zeros. Unfortunately, a direct least squares estimation immediately leads to non-linear equations for the numerator terms, even in the simplest cases. Solutions to such equations are necessarily iterative and cannot be guaranteed to converge to give a global minimum.

There are approaches that theoretically work, and indeed can be applied to pure synthetic speech, generated from the linear model. One such approach was alluded to in the discussion of Prony's method in Chapter 2. There it was assumed that within a pitch period, the data sequence had approximately a z-transform given by (2.25), rewritten here as

$$X(z)=P(z)/A(z)=\sum_{i=0}^{M-1} p_i z^{-i} / \sum_{i=0}^{M} a_i z^{-i}. \tag{11.5}$$

The polynomial $A(z)$ was obtained approximately following Prony's method which was shown to be equivalent to the covariance method. Given sufficient numerical accuracy, $A(z)$ is found exactly if the data sequence used actually has a z-transform given by (11.5). If the roots of $A(z)$ are denoted by $z_1, z_2, \ldots, z_M$, then the sequence $x(n)$ of (11.5) would be given by (2.34), rewritten here as

$$x(n)=\sum_{i=1}^{M} u_i(z_i)^n+\eta(n) \tag{11.6}$$

where $\eta(n)$ represents an error that might result from the fact that (11.5) is only approximate. But, from (11.5), this can also be expressed in terms of the unit sample response $h(n)$, from a filter $1/A(z)$, as

$$x(n)=\sum_{i=0}^{M-1} p_i h(n-i)+\eta(n). \tag{11.7}$$

A least squares solution can be applied to either (11.6) or (11.7) to find the set of coefficients $\{u_i\}$ or $\{p_i\}$ which minimizes the sum of the squares of $\eta(n)$, giving exact results for the model if there is no error. That is, (11.5) is exact so that there will be a set of coefficients which make $\{\eta(n)\}$ equal to zero.

This approach was tested quite successfully on many frames of pure synthetic speech. However, when applied to real speech the results appeared to be meaningless. Spectra that were obtained contained sharp cusps and were entirely unsatisfactory. This same procedure was used by Shanks [1967] for the design of digital filters based upon their unit sample response. If the data sequence, speech, or unit sample response does indeed have a z-transform given by (11.5), then both numerator and denominator can be precisely found by using the covariance method and at least $2M$ adjacent samples taken anywhere from the sequence.

It was noted above that Shank's method assumes analysis of a single time-synchronized pitch period. Actually, this assumption is common to all methods of pole-zero analysis. It is a statement of the need to know both the input and output of a system being identified. One way of satisfying this requirement is based on the technique of homomorphic deconvolution [Oppenheim, et al., 1968]. By retaining only the low-time portion of the cepstrum of a windowed speech segment, a minimum-phase counterpart of the vocal tract impulse response (including glottal source) can be estimated. A minimum phase signal by definition has no linear phase component and thus is properly synchronized for analysis by Shank's or some other method.

This approach underlies a new technique for pole-zero modeling of speech, called *homomorphic prediction* [Oppenheim, et al., 1976], which capitalizes on the robustness of linear prediction by using it to directly estimate both poles and zeros. The basic strategy is to transform the zeros of the original signal into poles and then use linear prediction to locate them.

One way of doing this [Kopec, 1975] relies on the following observation. Let $x(n)$ denote a signal and $X(z)$ its z-transform. By inverting the spectrum of $x(n)$ a new signal $x^{-1}(n)$ can be created whose z-transform is $1/X(z)$. The poles of $1/X(z)$ are the zeros of $X(z)$ and vice versa. The operation of inverting the spectrum of $x(n)$ can be accomplished by using the cepstrum. When linear prediction is applied to $x^{-1}(n)$, the roots of the resulting prediction polynomial are estimates of the zeros of $X(z)$, the z-transform of $x(n)$.

In speech analysis, $x(n)$ corresponds to an estimate of the vocal tract impulse response obtained from the low-time portion of the cepstrum of the windowed speech signal. The effectiveness of the method is enhanced by first applying linear prediction to $x(n)$ and then filtering $x^{-1}(n)$ through the resulting all pole filter. This procedure has the effect of approximately removing the zeros of $x^{-1}(n)$. Sample spectra illustrating this technique are presented in Fig. 11.7. The speech segment is 54 ms of intervocalic /m/ from the sentence "Say mo*m*o again". Figure 11.7A shows the pre-emphasized speech spectrum with each horizontal line corresponding to 10 dB. Each vertical line corresponds to 1 kHz. Figure 11.7B shows the cepstrally smoothed spectrum while Fig. 11.7C shows the all-pole linear prediction estimate of the spectrum using 12 coefficients. Figure 11.7D shows the log difference between Figs. 11.7B and C as the smoothed spectral representation of the zero behavior. Note the two major resonances away from the origin corresponding to the zeros or valleys of the speech spectrum. By applying linear prediction to the signal whose spectrum is Fig. 11.7D, Fig. 11.7E is obtained, showing very clearly two major resonances corresponding to the zeros of the speech spectrum. Finally, Fig. 11.7F is the overall pole-zero estimate of the speech, obtained by subtracting Fig. 11.7E from Fig. 11.7C. The first anti-resonance of Fig. 11.7E at around 600 Hz compares with Fujimura's [1962] results showing major /m/ first anti-resonances in the interval between 750 and 1250 Hz.

Preliminary evaluation of the above procedure confirms that it is considerably more reliable than Shank's method for estimating zeros from real speech [Kopec, 1975]. However, as with any non-iterative two-pass scheme which identifies poles and zeros separately, there is reason to expect difficulty when trying to resolve nearly coincident pole-zero pairs.

A second approach to estimating zeros by first transforming them into poles was proposed by Oppenheim and Tribolet [1973]. If $x(n)$ has a z-transform of the form (11.5), the z-transform of its complex cepstrum will be

$$C(z) = ln[P(z)/A(z)] = ln[P(z)] - ln[A(z)]. \tag{11.8}$$

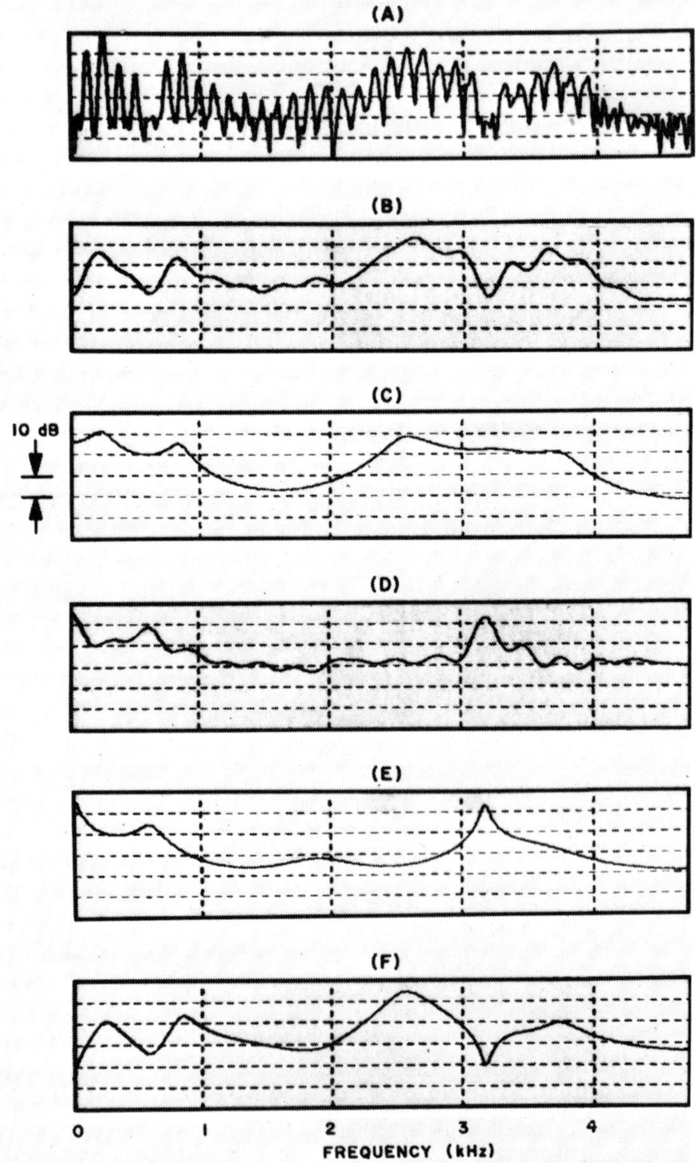

Fig. 11.7 Illustration of homomorphic prediction for pole-zero analysis. [After Kopec, 1975]

By definition, $C(z)$ is given by

$$C(z) = \sum_{n=-\infty}^{\infty} c(n) z^{-n}. \tag{11.9}$$

Differentiating (11.9) with respect to z and using primes to denote derivatives, the following expression is obtained:

$$-zC'(z) = \sum_{n=-\infty}^{\infty} n c(n) z^{-n} = \frac{zA'(z)P(z) - zP'(z)A(z)}{A(z)P(z)}. \tag{11.10}$$

Thus, the sequence $\{n c(n)\}$ has a z-transform whose poles are the poles and zeros of $X(z)$. The denominator $A(z)P(z)$ can be estimated by applying linear prediction to $\{n c(n)\}$. In using this method it is necessary to classify each root of $A(z)P(z)$ as either a pole or zero of the original signal. One way of doing this is to estimate $A(z)$ by linear prediction analysis of $\{x(n)\}$. The remaining roots of $A(z)P(z)$ are then identified as zeros. Although this approach works well with synthetic speech, it has not yet been very successful for real speech. Makhoul [1975a] has discussed both iterative and non-iterative approaches to pole-zero estimation. What he refers to as the *inverse linear predictor method* is quite similar (in its end results) to Kopec's method if cepstral smoothing is applied before inverting the spectrum.

11.5 Summary and Future Directions

We have attempted to cover the topic of linear prediction of speech in some depth, ranging from theoretical considerations through practical implementation and applications. It was shown that linear prediction has been applied as a tool in almost all facets of acoustical speech processing.

A number of properties and extensions of the basic linear prediction mathematics have been discussed, including parameter transformations, recursive stability tests, speech synthesis structures, fundamental and formant frequency estimation, and spectral analysis. Hopefully, this book will lay a foundation for further research into the myriad challenges of understanding and manipulating features of human speech. Numerous areas remain to be studied. We shall discuss here some of the future directions and problem areas.

Along with linear prediction, it is certainly possible to use similar approaches to perform linear interpolation. Rather than estimate a data value from past values, one can use least squares minimization techniques to estimate unknown data values in terms of past and future values (with reference to the unknown value). Dibbern [1974] has studied this problem and shown that the total squared error for interpolation in the center of a data sequence is nearly an order of magnitude below the prediction result. Such approaches may lead to more accurate pitch extraction in interpolating between data samples, and may have application in understanding non-minimum phase characteristics of speech where an all-pole model is no longer assumed.

In recent years there has been considerable interest in applying Kalman filtering [Sage, 1968] and related sequential estimation methods to speech processing [Boll, 1973; Matsui, et al., 1972; Gibson, et al., 1975]. The linear prediction equations can be viewed as extremely simplified cases of the general Kalman filter theory. It would appear that if one were willing to pay a price in complexity, that some benefit should be received. Unfortunately, at the present in any case, the value of Kalman filter theory for the processing of real speech has not been demonstrated. There are at least two serious problems. First, the computational effort is not merely slightly greater, it is actually enormous for general cases. Second, the use of a priori estimates implies a considerable knowledge of the speech signal. Kalman filter theory has been successfully used in rocket trajectory estimation and correction, for example, because the mathematics of the motion (while not the random disturbances) from start to finish are known. Direct applications to speech modeling implies, loosely speaking, that one knows a priori what the person is about to say! In spite of these problems, Kalman filter theory and sequential estimation techniques have potential for improving upon the linear prediction methods presented. Before such improvements are realized, however, it will be necessary to understand more fully the inherent properties of the speech waveforms and how they relate to the mathematics of Kalman filter theory.

New extensions and applications for estimating the vocal tract shapes and glottal waveforms, such as described in Chapter 4, are being considered by various researchers.

Crichton and Fallside [1974a, b] have investigated the applicability of the acoustic tube model as an aid for deaf speech training. An important issue is the meaningfulness of the display. Figure 11.8 A shows how the device might be used by a deaf child, while Fig. 11.8 B shows an enlarged display of the screen seen by the child. The screen shows a target trace (preferably made by a normal child so that vocal tract normalization is not a significant problem) and the result from the deaf child who attempts to match the reference pattern. The discrete area functions are smoothed by parabolic interpolation before display. In this application, the vocal tract shape needs to be normalized lengthwise so that comparisons from different speakers can be made. This normalization is also considered to play an important role in speech recognition problems [Wakita, 1975a]. Attempts have been made to estimate the vocal tract length directly from acoustic data [Paige and Zue, 1970; Wakita, 1974a], but significant difficulties remain to be solved. One important unresolved problem in computing vocal tract area functions is the validity of the model whereby the vocal tract is assumed to be a lossless acoustic tube, with all losses lumped into a source impedance at the glottis. Distributed losses due to heat conduction and viscosity [Flanagan, 1972, pp. 23–35] are not represented by the model of Chapter 4. Correction of the frequency shift due to losses in the vocal tract should give more accurate estimation of the area function. Similarly, more accurate estimation of bandwidths may also be needed, requiring more precise knowledge about the relationships between physiological data and effects on acoustic data. In this regard, studies of speech synthesis models based on physiological data [Ishizaka and Flanagan, 1972] may provide more useful information in acoustic tube modeling of the vocal tract.

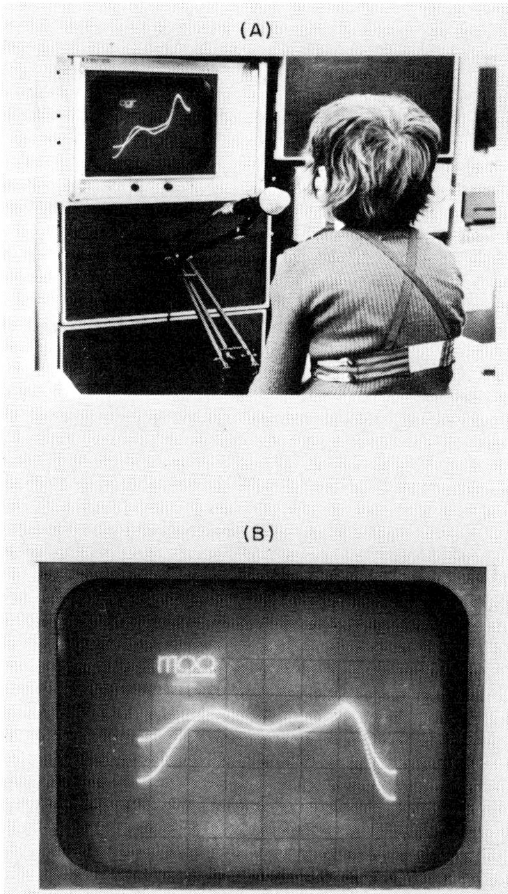

Fig. 11.8 System for deaf speech training. A) illustration of how device might be used by child. B) enlarged display of screen. [From Crichton and Fallside, 1974a]

The areas of formant and fundamental frequency estimation will always be areas for fruitful research due to the fact that no direct mathematical transformation exists for obtaining these basic speech parameters. As greater accuracy is needed, more and more ad hoc decisions are necessary. Algorithms were presented for both formant and fundamental frequency estimation in Chapters 7 and 8. By applying these to real speech, it will become evident that if more decisions are introduced, better results can be obtained. Vocoding is also an open-ended area in terms of potential improvements. The conflicting goals are natural speech quality and a low transmission bit rate.

The linear prediction techniques discussed have brought about major advances. To obtain further advances, we believe that along with the mathematics it will be necessary to have a better understanding of the underlying physiological mechanisms so that the complex nature of the acoustical speech waveform can be more accurately and efficiently represented.

References

Akaike, H.: A New Look at the Statistical Model of Identification. IEEE Trans. AC-**19**, 716–723 (1974).

Åström, K. J., Jury, E. I., Agniel, R. G.: A Numerical Method for the Evaluation of Complex Integral. IEEE Trans. AC-**15**, 468–471 (1970).

Atal, B. S., Schroeder, M. R.: Predictive Coding of Speech Signals. Proc. 1967 Conf. Commun. and Process., 360–361 (1967).

Atal, B. S.: *Automatic Speaker Recognition Based on Pitch Contours*. Ph. D. Thesis (Polytechnic Institute of Brooklyn, 1968a).

Atal, B. S., Schroeder, M. R.: Predictive Coding of Speech Signals. Reports of 6th Int. Cong. Acoust., ed. by Y. Kohasi (Tokyo) C-5-5 (1968b).

Atal, B. S., Schroeder, M. R.: Predictive Coding of Speech Signals. 1968 WESCON Technical Papers, Paper 8/2 (1968c).

Atal, B. S.: Speech Analysis and Synthesis by Linear Prediction of the Speech Wave. J. Acoust. Soc. Am. **47**, 65(A) (1970a).

Atal, B. S.: Determination of the Vocal Tract Shape Directly from the Speech Wave. J. Acoust. Soc. Am. **47**, 65(A) (1970b).

Atal, B.S., Schroeder, M. R.: Adaptive Predictive Coding for Speech Signals. Bell System Tech. J. **49**, 1973–1986 (1970c).

Atal, B. S.: Characterization of Speech Signals by Linear Prediction of the Speech Wave. Proc. IEEE Symposium on Feature Extraction and Selection in Pattern Recognition, Argonne, Illinois, 202–209 (1970d).

Atal, B. S.: Sound Transmission in the Vocal Tract with Applications to Speech Analysis and Synthesis. Proc. 7th Int. Cong. Acoust., Budapest, Hungary, 169 (1971a).

Atal, B. S., Hanauer, S. L.: Speech Analysis and Synthcsis by Linear Prediction of the Speech Wave. J. Acoust. Soc. Am. **50**, 637–655 (1971b).

Atal, B. S.: Effectiveness of Linear Prediction Characteristics of the Speech Wave for Automatic Speaker Identification and Verification. J. Acoust. Soc. Am. **55**, 1304–1312 (1974a).

Atal, B. S.: Recent Advances in Predictive Coding-Applications to Speech Synthesis, Proceedings of the 1974 Stockholm Speech Communications Seminar, C.G.M. Fant, Ed., John Wiley and Sons, N.Y., N.Y. (To be published) (1974b).

Atal, B. S., Schroeder, M. R., Stover, V.: Voice-Excited Predictive Coding System for Low Bit-rate Transmission of Speech. Proc. Int. Conf. Commun. (1975).

Bell, C. G., Fujisaka, H., Heinz, J. M., Stevens, K. N., House, A. S.: Reduction of Speech Spectra by Analysis-by-Synthesis Techniques. J. Acoust. Soc. Am. **33**, 1725–1736 (1961).

Bergland, G. D.: A Radix-Eight Fast Fourier Transform Subroutine for Real Valued Series. IEEE Trans. AU-**17**, 138–144 (1969).

Blackman, R. B., Tukey, J. W.: *The Measurement of Power Spectra* (Dover Publications, Inc., New York, 1958).

Blankinship, W. A.: Note on Computing Autocorrelations. IEEE Trans. ASSP-**22**, 76–77 (1974).

Boll, S. F.: *A Priori Digital Speech Analysis*. Ph. D. Dissertation (Computer Science Department, University of Utah, 1973).

Boll, S. F.: *Selected Methods for Improving Synthesis Speech Quality Using Linear Predictive Coding; System Description, Coefficient Smoothing, and STREAK*. UTEC-CSc-74-151, Computer Science Department, University of Utah (1974).

Brigham, E. O.: *The Fast Fourier Transform* (Prentice-Hall, Englewood Cliffs, New Jersey, 1974).

Broad, D. J.: Formants in Automatic Speech Recognition. Int. J. Man-Mach. Studies **4**, 411–424 (1972).

Chandra, S., Lin, W. C.: Experimental Comparison Between Stationary and Non-stationary Formulations of Linear Prediction Applied to Voiced Speech. IEEE Trans. ASSP-**22**, 403–415 (1974).

Cheney, E. W.: *Introduction to Approximation Theory* (McGraw-Hill Book Co., New York, 1966).

Chiba, T., Kajiyama, M.: *The Vowel, Its Nature and Structure* (Tokyo Kaiseikan Pub. Co., Tokyo, 1941).

Clasen, R. J.: Numerical Methods for Inverting Positive Definite Matrices (The Rand Corporation. Santa Monica, California. AD637–930 (1966).

Cooley, J. W., Lewis, P. A. W., Welch, P. D.: The Fast Fourier Transform Algorithm: Programming Considerations in the Calculation of Sine, Cosine and Laplace Transforms. J. Sound Vib. **12**, 315–337 (1970). (Also see Rabiner and Rader, 271–293, 1972.)

Crichton, R. G., Fallside, F.: Linear Prediction Model of Speech Production with Applications to Deaf Speech Training. Proc. IEE **121**, 865–873 (1974a).

Crichton, R. G., Fallside, F.: A Real-Time Articulatory Display for Deaf Speech Training. (1974b).

Davis, S. B.: Computer Evaluation of Laryngeal Pathology Based on Inverse Filtering of Speech, Ph. D. Dissertation (University of California at Santa Barbara, 1976).

Dibbern, U.: Speech Analysis by Wiener Filtering with Interpolation. Unpublished memorandum, Philips Forschungslaboratorium, Hamburg (1974).

Doddington, G. R.: *A Method of Speaker Verification.* Ph. D. Dissertation (University of Wisconsin, 1970).

Dudley, H.: Remaking Speech. J. Acoust. Soc. Am. **11**, 169–177 (1939).

Dunn, H. K.: The Calculation of Vowel Resonances, and an Electrical Vocal Tract. J. Acoust. Soc. Am. **22**, 740–753 (1950).

Dunn, J. G., Cowan, J. R., Russo, A. J.: Progress in the Development of a Digital Vocoder Employing an Itakura Adaptive Predictor. Telecommunications Conference Record, IEEE Publ. No. 73, CHO 805–2, 29B 1–6 (1973).

Ekstrom, M. P.: A Spectral Characterization of the Ill-Conditioning in Numerical Deconvolution. IEEE Trans. AU-**21**, 344–348 (1973).

Faddeeva, D. K., Faddeeva, V. N.: *Computational Methods of Linear Algebra* (W. H. Freeman and Company, San Francisco, 1963) pp. 144–147.

Fant, G. C. M.: *Acoustic Theory of Speech Production* (Mouton and Co., 's-Gravenhage, The Netherlands, 1960).

Flanagan, J. L.: Source-System Interactions in the Vocal Tract. Ann. New York Acad. Sci. **155**, 9–15 (1968).

Flanagan, J. L.: *Speech Analysis Synthesis and Perception*, 2nd ed. (Springer-Verlag, Berlin, Heidelberg, New York, 1972).

Flanagan, J. L., Rabiner, L. R., ed.: *Speech Synthesis* (Dowden, Hutchinson and Ross, Stroudsburg, Pennsylvania, 1973).

Flinn, E. A., Robinson, E. A., Treitel, S., eds.: Special Issue on MIT Geophysical Analysis Group Reports. Geophys. **32**, 411–582 (1967).

Fujimura, O.: Analysis of Nasal Consonants. J. Acoust. Soc. Am. **34**, 1866–1875 (1962).

Fujimura, O.: An Approximation to Voice Aperiodicity. IEEE Trans. AU-**16**, 68–72 (1968).

Gentleman, W. M., Sande, G.: Fast Fourier Transforms for Fun and Profit. 1966 Fall Joint Computer Conference, AFIPS Proc. **20** (Spartan Books, Washington, D. C., 1966) pp. 563–578.

Gibson, J. D., Jones, S. K., Melsa, J. L.: Digital Speech Analysis Using Sequential Estimation Techniques, IEEE Trans. ASSP-**23** 362–369, 1975.

Gold, B., Rader, C. M.: *Digital Processing of Signals*. (McGraw-Hill, New York, 1969).

Gray, A. H., Jr., Markel, J. D.: Digital Lattice and Ladder Filter Synthesis. IEEE Trans. AU-**21**, 491–500 (1973).

Gray, A. H., Jr.: Log Spectra of Gaussian Signals. J. Acoust. Soc. Am. **55**, 1028–1033 (1974a).

Gray, A. H., Jr., Markel, J. D.: A Spectral Flatness Measure for Studying the Autocorrelation Method of Linear Prediction of Speech Analysis. IEEE Trans. ASSP-**22**, 207–217 (1974b).

Gray, A. H., Jr., Markel, J. D.: A Normalized Digital Filter Structure. IEEE Trans. ASSP-**23**, 268–277 (1975a).

Gray, A. H., Jr., Markel, J. D.: A Digital Elliptic Filter Program. Submitted for publication (1975b).

Grenander, U., Szego, G.: *Toeplitz Forms and Their Applications* (University of California Press, Berkeley, California, 1958).
Haskew, J. R., Kelly, J. M., McKinney, T. H.: Results of a Study of the Linear Prediction Vocoder. IEEE Trans. COM-**21**, 1008–1014 (1973).
Heinz, J. M.: Perturbation Functions for the Determination of Vocal Tract Area Functions from Vocal Tract Eigenvalues. R. Inst. Tech., Stockholm Speech Trans. Lab. Q. Prog. and Status Rep., 1–14 (1967).
Hildebrand, F. B.: *Introduction to Numerical Analysis* (McGraw-Hill, New York, 1956).
Ishizaka, K., Flanagan, J. L.: Synthesis of Voiced Sounds from a Two-Mass Model of the Vocal Cords. Bell System Tech. J. **51**, 1233–1268 (1972).
Itakura, F., Saito, S.: Analysis Synthesis Telephony Based Upon the Maximum Likelihood Method. Reports of 6th Int. Cong. Acoust., ed. by Y. Kohasi, Tokyo, C-5-5, C17–20 (1968). (Also see Flanagan and Rabiner, 289, 292, 1973.)
Itakura, F., Saito, S.: Speech Analysis-Synthesis System Based on the Partial Autocorrelation Coefficient. Acoust. Soc. of Japan Meeting (1969).
Itakura, F., Saito, S.: A Statistical Method for Estimation of Speech Spectral Density and Formant Frequencies. Electron. Commun. Japan **53-A**, 36–43 (1970). (Also see Flanagan and Rabiner, 293–304, 1973.)
Itakura, F., Saito, S.: Extraction of Speech Parameters Based Upon the Statistical Method. Proc. Speech Info. Process., Tohoku University, Sendai, Japan, 5.1, 5.12 (1971a). (In Japanese)
Itakura, F., Saito, S.: Digital Filtering Techniques for Speech Analysis and Synthesis. 7th Int. Cong. Acoust., Budapest, 25 C 1 (1971b).
Itakura, F., Saito, S.: Speech Information Compression Based on the Maximum Likelihood Spectral Estimation. J. Acoust. Soc. Japan **27**, 463–472 (1971c).
Itakura, F.: *Speech Analysis and Synthesis Systems Based on Statistical Method.* Doctor of Engineering Dissertation (Department of Engineering, Nagoya University, Japan, 1972). (In Japanese).
Itakura, F., Saito, S.: On the Optimum Quantization of Feature Parameters in the PARCOR Speech Synthesizer. Conf. Record IEEE 1972 Conf. Speech Commun. and Process., New York, paper L4, 434–437 (1972b). (Also see Flanagan and Rabiner, 289–292, 1973.)
Itakura, F., Saito, S., Koike, Y., Sawabe, H., Nishikawa, M.: An Audio Response Unit Based on Partial Correlation. IEEE Trans. COM-**20**, 792–797 (1972c).
Itakura, F.: Minimum Prediction Residual Principle Applied to Speech Recognition. IEEE Trans. ASSP-**23**, 67–72 (1975).
Jury, E. I.: *Theory and Application of the Z-Transform Method* (John Wiley and Sons, New York, 1964).
Kailath, T.: A View of Three Decades of Linear Filtering Theory. IEEE Trans. IT-**20**, 146–181 (1974).
Kelly, J. L., Jr., Lochbaum, C.: Speech Synthesis. Proc. 4th Int. Cong. Acoust., G42, 1–4 (1962). (Also see Flanagan and Rabiner, 1973.)
Kendall, W. B.: A New Algorithm for Computing Correlations. IEEE Trans. C-**23**, 88–90 (1974).
Klayman, C., Kaufman, G., Snyder, A., Teacher, C., Welch, J., Brandenstein, W., Tremain, T.: Real-Time Implementation of a Linear Predictive Coding System. Nat. Telecommun. Conf. Record, IEEE publ. 73 CHO 805–2, 29E 1–7 (1973).
Koike, Y., Markel, J. D.: Application of Inverse Filtering for Detecting Laryngeal Pathology. Ann. Otology, Rhinology, and Laryngology, 117–124 (1975).
Kopec, G. E.: *Speech Analysis by Homomorphic Prediction*, S. M. Thesis (Dept. of Electrical Engineering, Massachusetts Institute of Technology, 1975).
Knudsen, M. J.: Real Time Linear Predictive Coding of Speech on the SPS-41 Micro-programmed Triple-Processor Machine. IEEE Trans. ASSP-**23**, 140–145 (1975).
Levinson, N.: The Wiener RMS (Root Mean Square) Error Criterion in Filter Design and Prediction. J. Math. Phys. **25**, 261–278 (1947). (Also see Wiener, Appendix B, 1966.)
Lummis, R. C.: Speaker Verification by Computer Using Speech Intensity for Temporal Registration. IEEE Trans. AU-**21**, 80–89 (1973).
McCandless, S. S.: An Algorithm for Automatic Formant Extraction Using Linear Prediction Spectra. IEEE Trans. ASSP-**22**, 135–141 (1974a).
McCandless, S. S.: A New Encoding Technique for the K-Parameters: A Statistical Approach. Unpublished memorandum, MIT Lincoln Laboratory, Lexington, Massachusetts (1974b).

McCandless, S. S.: Modifications to Formant Tracking Algorithm of April 1974. Submitted for publication (1975).

McDonough, R. N.: *Matched Exponents for the Representation of Signals*. Ph. D. Dissertation (Dept. of Electrical Engineering, The John Hopkins University, 1963).

Magill, D. T.: Adaptive Speech Compression for Packet Communication Systems. Telecommun. Conf. Record, IEEE Publ. # 73 CHO 805-2, 29D 1–5 (1973).

Magill, D. T., Un, C. K.: Residual Excited Linear Predictive Vocoder. Presented at the 87th Meeting of The Acoust. Soc. Am., New York (1974).

Makhoul, J., Wolf, J.: Linear Prediction and the Spectral Analysis of Speech. NTIS No. AD-749066, BBN Report No. 2304 Bolt Beranek and Newman, Inc., Cambridge, Massachusetts, (1972).

Makhoul, J.: Spectral Analysis of Speech by Linear Prediction. IEEE Trans. AU-**21**, 140–148 (1973a).

Makhoul, J., Wolf, J.: The Use of a Two-Pole Spectral Model in Speech Recognition. BBN Report No. 2537 Bolt Beranek and Newman, Inc., Cambridge, Massachusetts (1973b).

Makhoul, J., Viswanathan, R.: Adaptive Preprocessing for Linear Predictive Speech Compression Systems. J. Acoust. Soc. Am. **55**, 475(A) (1974a).

Makhoul, J. Viswanathan, R., Cosell, L., Russell, W.: Natural Communication with Computers: Speech Compression Research at BBN. BBN Report No. 2976 **II** Bolt Beranek and Newman, Inc., Cambridge, Massachusetts (1974b).

Makhoul, J.: Linear Prediction vs. Analysis-by-Synthesis. Proceedings of the 1974 Stockholm Speech Communications Seminar, C. G. M. Fant, Ed., John Wiley and Sons N. Y., N. Y. (to be published). (1974c).

Makhoul, J.: Linear Prediction: A Tutorial Review. Proc. IEEE **63**, 561–580 (1975a).

Makhoul, J.: Spectral Linear Prediction: Properties and Applications. IEEE Trans. ASSP-**23**, 283–296 (1975b).

Makhoul, J.: Linear Prediction in Automatic Speech Recognition. Speech Recognition: Invited Papers presented at the 1974 IEEE Symposium, D. R. Reddy, ed. (Academic Press, New York, 183–220, 1975c).

Maksym, J. N.: Real Time Pitch Extraction by Adaptive Prediction of the Speech Waveform. IEEE Trans. AU-**21**, 149–154 (1973).

Markel, J. D.: The Prony Method and Its Application to Speech Analysis. J. Acoust. Soc. Am. **49**, 105(A) (1971a).

Markel, J. D.: Formant Trajectory Estimation from a Linear Least-Squares Inverse Filter Formulation. SCRL Monograph No. 7, Speech Communications Research Laboratory, Santa Barbara, California (1971b).

Markel, J. D.: FFT Pruning. IEEE Trans. AU-**19**, 305–311 (1971c).

Markel, J. D.: Automatic Formant and Fundamental Frequency from a Digital Inverse Filter Formulation. Conference Reprints, 1972 Int. Conf. Speech Commun. Process., Boston, Massachusetts, paper 89, 81–84 (1972a).

Markel, J. D.: Digital Inverse Filtering, A New Tool for Formant Trajectory Estimation. IEEE Trans. AU-**20**, 129–137 (1972b).

Markel, J. D.: The SIFT Algorithm for Fundamental Frequency Estimation. IEEE Trans. AU-**20**, 367–377 (1972c).

Markel, J. D., Gray, A. H., Jr.: On Autocorrelation with Application to Speech Analysis. IEEE Trans. AU-**21**, 69–79 (1973a).

Markel, J. D.: Basic Formant and F_0 Parameter Extraction from a Digital Inverse Filter Formulation. IEEE Trans. AU-**21**, 154–160 (1973b).

Markel, J. D., Gray, A. H., Jr., Wakita, H.: Linear Prediction of Speech—Theory and Practice. SCRL Monograph No. 10, Speech Communications Research Laboratory, Santa Barbara, California (1973c).

Markel, J. D., Gray, A. H., Jr.: A Linear Prediction Vocoder Simulation Based Upon the Autocorrelation Method. IEEE Trans. ASSP-**22**, 124–134 (1974a).

Markel, J. D., Gray, A. H., Jr.: Fixed-Point Truncation Arithmetic Implementation of a Linear Prediction Autocorrelation Vocoder. IEEE Trans. ASSP-**22**, 273–281 (1974b).

Markel, J. D., Gray, A. H., Jr.: Fixed-Point Implementation Algorithms for a Class of Orthogonal Polynomial Filter Structures. IEEE Trans. ASSP-**23**, 486–494 (1975a).

Markel, J. D., Gray, A. H., Jr.: Roundoff Noise Characteristics of a Class of Orthogonal Polynomial Structures. IEEE Trans. ASSP-**23**, 473–486 (1975b).

Markel, J. D., Gray, A. H., Jr.: An Optimal Linear Prediction Synthesizer Structure for Array Processor Implementation. Speech Recognition: Invited Papers presented at the 1975 IEEE Symposium, D. R. Reddy, ed. (Academic Press, New York, 231–242, 1975c).

Matsui, E., Nakajima, T., Suzuki, T., Omura, H.: An Adaptive Method for Speech Analysis Based Upon Kalman Filter Theory. Bull. Electro. Tech. Lab. **36**, Tokyo 42–51 (1972). (In Japanese).

Mermelstein, P.: Determination of the Vocal-Tract Shape from Measured Formant Frequencies. J. Acoust. Soc. Am. **41**, 1283–1294 (1967).

Miller, R. L.: Nature of the Vocal Cord Wave. J. Acoust. Soc. Am. **31**, 667–677 (1959).

Morf, M.: *Fast Algorithms for Multivariable Systems*. Ph. D. Dissertation (Stanford University, 1974).

Morris, L. R.: A Simple Real-Time Error Waveform Extraction Using Analog Adaptive Prediction of the Speech Waveform. Submitted for publication (1975).

Morse, P. M., Ingard, K. V.: *Theoretical Acoustics* (McGraw-Hill Book Co., New York, 1968).

Nakajima, T., Omura, H., Ishizaki, S.: Estimation of Vocal Tract Area Functions by Adaptive Inverse Filtering Methods and Identification of Articulatory Model. Proceedings of the 1974 Stockholm Speech Communications Seminar, C.G.M. Fant, Ed., John Wiley and Sons N.Y., N.Y. (to be published) (1974).

Narasimha, M. J., Shenoi, K., Peterson, A. M.: A Hilbert Space Approach to Linear Predictive Analysis of Speech Signals. Tech. Report 3606-10, Radioscience Lab., Stanford Electronics Lab., Stanford University, California (1974).

Oppenheim, A. V., Schafer, R. W., Stockham, T. C.: Non-linear Filtering of Multipled and Convolved Signals. Proc. IEEE **56**, 1264–1291 (1968).

Oppenheim, A. V., Tribolet, J. M.: Pole-Zero Modeling Using Cepstral Prediction. QPR No. 111, Res. Lab., Electronics, M.I.T., Cambridge, Massachusetts, 157–159 (1973).

Oppenheim, A. V., Schafer, R. W.: *Digital Signal Processing* (Prentice-Hall, Englewood Cliffs, New Jersey, 1975).

Oppenheim, A. V., Kopec, G. E., Tribolet, J. M.: Signal Analysis by Homomorphic Prediction, to appear in IEEE Trans. ASSP, 1976.

Paige, A., Zue, V. W.: Computation of Vocal Tract Area Functions. IEEE Trans. AU-**18**, 7–18 (1970).

Peacock, K. L., Treitel, S.: Predictive Deconvolution Theory and Practice. Geophys. **34**, 155–169 (1969).

Peterson, G. E., Shoup, J. E.: A Physiological Theory of Phonetics. J. Speech Hear. Res. **9**, 5–67 (1966a).

Peterson, G. E., Shoup, J. E.: The Elements of an Acoustic Phonetic Theory. J. Speech Hear. Res. **9**, 68–100 (1966b).

Pfeifer, L. L.: Multiplication Reduction in Short-Term Autocorrelation. IEEE Trans. AU-**21**, 556–558 (1973).

Pfeifer, L. L.: Inverse Filter for Speaker Identification. RADC-TR-74-214, Final Report, Speech Communications Research Laboratory, Santa Barbara, California (1974).

Pinson, E. N.: Pitch Synchronous Time Domain Estimation of Formant Frequencies and Bandwidths. J. Acoust. Soc. Am. **35**, 1264–1273 (1963).

Prony, R.: Essai experimental et Analytique sur les lois de la dilatabilité des fluides élastiques et sur celles de la force expansive de la vapeur de l'eau et de la vapeur de l'alcool, à différentes temperatures. J. l'Ecole Polytech. (Paris) **1**, 24–76 (1795).

Rabiner, L. R.: Digital-Formant Synthesizer for Speech-Synthesis. J. Acoust. Soc. Am. **43**, 822–828 (1968).

Rabiner, L. R.: The Chirp-z Transform Algorithm. IEEE Trans. AU-**17**, 86–92 (1969).

Rabiner, L. R., Rader, C. M., eds.: *Digital Signal Processing* (IEEE Press, New York, 1972).

Rabiner, L. R., Gold, B.: *Theory and Application of Digital Signal Processing* (Prentice-Hall, Inc., Englewood Cliffs, New Jersey, 1975).

Roberts, L., Wessler, B.: The ARPA Computer Network. *Computer Communications Networks*, ed. by Abramson and Kuo (Prentice-Hall, Englewood Cliffs, New Jersey (1972).

Robinson, E. A.: *Statistical Communication and Detection with Special Reference to Digital Data Processing of Radar and Seismic Signals* (Hafner Publishing Co., New York, 1967).

Robinson, E. A., Treitel, S.: Principles of Digital Wiener Filtering. Geophys. Prosp. **XV**, 311–333 (1967).

Rosenberg, A. E.: Effect of Glottal Pulse Shape on the Quality of Natural Vowels. J. Acoust. Soc. Am. **49**, 583–590 (1971).

Ross, M. J., Shaffer, H. L., Cohen, A., Freudberg, R., Manley, H. J.: Average Magnitude Difference Function Pitch Extractor. IEEE Trans. ASSP-**22**, 353–362, (1974).

Runge, C., Konig, H.: *Vorlesungen über Numerisches Rechnen*, vol. 11 of series *Die Grundlehren der Mathematischen Wissenschaften* (Springer-Verlag, Berlin, 1924), p. 231.

Sage, A. D.: *Optimum Systems Control* (Prentice-Hall, Englewood Cliffs, New Jersey, 1968).

Saito, S., Itakura, F.: The Theoretical Consideration of Statistically Optimum Methods for Speech Spectral Density. Report No. 3107, Electrical Communication Laboratory, N.T.T., Tokyo (1966). (In Japanese).

Sambur, M. R., Rabiner, L. R.: A Speaker-Independent Digit-Recognition System. BSTJ **54**, 81–102 (1975).

Schafer, R. W., Rabiner, L. R.: System for Automatic Formant Analysis of Voiced Speech. J. Acoust. Soc. Am. **47**, 634–678 (1970).

Schroeder, M. R., Mermelstein, P.: Determination of Smoothed Cross-Sectional Area Function of the Vocal Tract from Formant Frequencies. Rep. 5th Int. Cong. Acoust. 1a, A-24, D. E. Commins, ed. (1965).

Schroeder, M. R.: Vocoders: Analysis and Synthesis of Speech. Proc. IEEE **54**, 720–734 (1966).

Schroeder, M. R.: Determination of the Geometry of the Human Vocal Tract by Acoustic Measurements. J. Acoust. Soc. Am. **41**, 1002–1010 (1967).

Shanks, J. L.: Recursion Filters for Digital Processing. Geophys. **32**, 33–51 (1967).

Skinner, T. E.: Unpublished memoranda on Fundamental Frequency Analysis, Sperry Univac, Defense Systems Division, St. Paul, Minnesota (1973).

Sondhi, M. M.: New Methods of Pitch Extraction. IEEE Trans. AU-**16**, 262–266 (1968).

Sondhi, M. M., Gopinath, B.: Determination of Vocal Tract Shape from Impulse Response at Lips. J. Acoust. Soc. Am. **49**, 1867–1873 (1971).

Sorenson, H. W.: Least-Squares Estimation: From Gauss to Kalman. IEEE Spect. **7**, 63–68 (1970).

Stevens, K. N., Kasowski, S., Fant, C. G. M.: An Electrical Analog of the Vocal Tract. J. Acoust. Soc. Am. **25**, 734–742 (1953).

Strube, H. W.: Determination of the Instant of Glottal Closure from the Speech Wave. J. Acoust. Soc. Am. **56**, 1625–1629 (1974).

Timothy, L. K.: A Formulation of Linear Predictive Coding. Unpublished notes, University of Utah, Salt Lake City (1973).

Todd, J.: The Condition of Certain Matrices, I. Quart. J. Mech. and Appl. Math. **2**, 469–472 (1949).

Turing, A. M.: Rounding-off Errors in Matrix Processes. Quart. J. Mech. and Appl. Math. **1**, 287–308 (1948).

Viswanathan, R., Makhoul, J.: Quantization Properties of Transmission Parameters in Linear Predictive Systems. IEEE Trans. ASSP-**23**, 309–321 (1975).

Wakita, H.: Estimation of the Vocal Tract Shape by Optimal Inverse Filtering and Acoustic/Articulatory Conversion Methods. SCRL Monograph No. 9, Speech Communications Research Laboratory, Santa Barbara, California (1972).

Wakita, H.: Direct Determination of Input Impedance Singularities from Speech for Obtaining the Vocal Tract Area Function. J. Acoust. Soc. Am. **53**, 294–295 (A) (1973a).

Wakita, H.: Direct Estimation of the Vocal Tract Shape by Inverse Filtering of Acoustic Speech Waveforms. IEEE Trans. AU-**21**, 417–427 (1973b).

Wakita, H.: Estimation of the Vocal-Tract Length from Acoustic Data. J. Acoust. Soc. Am. **55**, Suppl. J7 (1974a).

Wakita, H., Gray, A. H., Jr.: Some Theoretical Considerations for Linear Prediction of Speech and Applications. Proceedings of the 1974 Stockholm Speech Communications Seminar, C.G.M. Fant, Ed., John Wiley and Sons N.Y., N.Y. (to be published) (1974b).

Wakita, H.: A Theory of Linear Prediction Acoustic Tube Model for Estimating Vocal-Tract Area Functions. Unpublished report, Speech Communications Research Laboratory, Santa Barbara, California (1974c).

Wakita, H.: An Approach to Vowel Normalization. J. Acoust. Soc. Am. **57**, Suppl. A12 (1975a).

Wakita, H., Gray, A. H., Jr.: Numerical Determination of the Lip Impedance and Vocal Tract Area Functions. IEEE Trans. ASSP-**23**, 574–580 (1975b).

Weinstein, C. J.: Comments on VELP and RELP. Unpublished memorandum, M.I.T. Lincoln Lab., Lexington, Massachusetts (1974).
Weinstein, C. J., McCandless, S. S., Mondshein, L. F., Zue, V. W.: A System for Acoustic-Phonetic Analysis of Continuous Speech. IEEE Trans. ASSP-**23**, 54–67 (1975a).
Weinstein, C. J.: A Linear Prediction Vocoder with Voice Excitation. EASCON Proc. (1975b).
Welch, P. D.: A Fixed-Point Fast Fourier Transform Error Analysis. IEEE Trans. AU-**17**, 151–157 (1969).
Welch, J.: LONGBRAKE II. Final Report, Contract No. DAAB03-74-C-0098, Philco-Ford Corporation, Willow Grove, Pennsylvania (1974).
Wiener, N.: *Extrapolation Interpolation and Smoothing of Stationary Time Series* (M.I.T. Press, Cambridge, Massachusetts, 1966).
Winckel, F.: Electroakustische Untersuchungen an der menschlichen Stimme. Folia Phoniat. **4**, 93–113 (1952).
Yanagihara, N.: Significance of Harmonic Changes and Noise Components in Hoarseness. J. Speech Hear. Res. **10**, 531–541 (1967).

Subject Index

acoustic phonetics 164
acoustic pressure wave 1
acoustic tube
  relationship to linear prediction 71–77
  losses 88–89
  stability 90–91
  equation derivation 61–71
adaptive pre-emphasis 82
all-pole synthesis 92, 94
analysis conditions 151–159
analysis model 8
analysis by synthesis 129
  uniform weighting property of 136
anchor point 180
  flow chart 181
approximate maximum likelihood 22
area function 61, 229, 230
  estimation 78–79
  examples 80–82
  evaluation 82–83
area ratios 75, 229–235
Arpanet 235
articulators 2
artificial section 69, 71
AUTO subroutine 51, 219
autocorrelation method 14, 15, 42–43
autocorrelation analysis 200
autocorrelation coefficients
  efficient calculation of 217
  finite word length properties of 223
autocorrelation equations
  solution of 15, 50–58
  finite word length properties of 224
automatic formant estimation
  general procedure of 166
  algorithm 1 176
  algorithm 2 180
automatic speech recognition 164
average magnitude difference function (AMDF) 259

backward prediction error 33, 198
bandwidths 7, 168
base-band excited vocoders 260
boundary conditions 68–74

Cauchy-Schwartz inequality 47, 48
cepstral analysis 164
cepstral coefficients 229–233, 238–239
cepstral smoothing 172
cepstrum 133, 230
chirp z-transform 162
Cholesky decomposition 44, 59
circuit-switched system 235
circular correlation 201
coefficient transformations 229
condition number 214
conditional maximum likelihood 22
contained correlation 201
continuity conditions 65
correlation matching 18, 31–32, 131
correlation matrix 51
covariance coefficients, efficient calculations of 220
covariance equations solution 14, 50–58
COVAR subroutine 52, 221
covariance method 14, 15, 19, 43, 219, 229

decoding 233
DELCO algorithm 237
driving function models 144
difference equations 213
DIRECT subroutine 119
discrete fourier transform 159
discrete signal representation 6
discrete spectral computation 212
down-sampling 207

eigenvalues 214
elliptic filters 208
encoding 233
equal area coding 235
error signal 192–194
EVAL subroutine 112

FACTAC factored autocorrelation subroutine 218
fast digital processor (FDP) 261
FFT subroutine 160
filter coefficients 233
filter energy evaluation 112
filter order choice 154
FINDPK subroutine 168
finite word length considerations 222–226

FORMNT subroutine 177
formant 164
  estimation 165–185
  smoothing 183
  accuracy 185–187
forward prediction error 33, 198
forward-traveling wave 64
frame rate f_r 170
frequency f 9
fundamental frequency F_0 2, 9, 190
  preprocessing by spectral flattening 191
  error signal filtering 195

gain 17, 31, 129–132, 239–242
Gauss-Siedel method 214
Gaussian statistics 19
generalized reflection coefficients 43, 57, 102, 259
glottal
  shaping model 7, 77
  termination 71
  volume velocity waveform 2, 5
  impedance 69
  effects of opening and closure 152
glottis 2
Gram-Schmidt orthogonalization 44, 54, 55, 222

higher pole correction factor 6
Hilbert space formulation 18, 44
homomorphic prediction 273
Huffman coding 254

ill-conditioning 212
  measure of 214
  condition number 214
  relation to spectral flatness 215
impedance 69
  lip radiation 69
  glottal 69
  characteristic 70
inner product 35–38, 44–50, 103
  signal processing 35, 103
  properties 47–48, 103–111
Interpolation
  pitch synchronous 244
  time synchronous 254
inverse filter 8, 18, 20, 142
  relation to predictor filter 12
  detailed structure of 40
  orthogonality solution 45
  transfer function 76
inverse glottal wave filtering 268
inverse sine coding 234

Kalman filter 103, 276
Kelly-Lochbaum acoustic tube structure 67
KLOCH subroutine 123

laryngeal pathology 267
Levinson's method 44, 55
linear interpolation 275
linear prediction
  analysis considerations 151
  error 10
  coefficients 10
  model 10
  relation to inverse filter 12
  relation to speech production model 1, 6, 12
  matrix forms 58–59
  weighting property 136
  minimum predictor total squared error 133
  equiripple spectral behavior of 138
  comparison with cepstral smoothing 172
  synthesizer program 242
  vocoder systems 245–262
linear speech production model 1, 6
lip impedance 84, 85
  poles and zeros 86–87
lip radiation model 7
log area ratios 229–235
log likelihood distance 237
log magnitude spectrum $LM(\cdot)$ 9, 131
log spectral distance measures 238
LONGBRAKE vocoder system 258
LPTRAN transformation subroutine 232

matrix solutions 58, 212
matrix loading
  computational considerations in 212
  factored methods for 217
maximum likelihood 18–23, 134, 246
minimax spectral matching 136
minimum predictor total squared error 133
minimum variance formulation 23–25
modified autocorrelation method 199–201
modified tap parameters 114, 125

norm 2, 13, 36, 46
NORMAL subroutine 127
normalized filter 123
normalized frequency θ 9
normalized log spectrum 140
normalized squared error 133

off-axis spectral evaluation 161
ONEMUL subroutine 124
orthogonality principal 35, 37, 48–50, 106, 111

packet switched systems 235
parabolic interpolation 167
PARCOR formulation 18, 32–41, 107, 247, 249
partial correlation coefficients 32, 40
peak picking 167

pi-parameter 114
pitch 2
pitch period 3, 190
pitch synchronous analysis 157, 166, 180
pitch synchronous synthesis 8
pole enhancement 161
pole-zero estimation 271
  Shanks method for 272
  homomorphic prediction for 273
  inverse linear prediction for 275
pre-emphasis 80, 81–158, 166, 171
  optimum choice of 216
pre-filtering 216
prediction error 10
predictive convolution 15
predictor coefficients 10
predictor error variance 24
Prony's method 18, 25–31

quantizers 227

raw data 166, 167, 169
  bandwidth estimates from 173
reflection coefficients 16, 43, 56
  linear quantization of 234
  spectral sensitivity 234
  sample-by-sample estimation of 249
  as a product of inverse filter roots 98
  stability bound 98–101
  alternate expressions for 99
  of acoustic tube 66
RELP 260–261
residual error 30
reverse-traveling wave 64
Robinson's method 44
root solving 163

sampling rate f_s 153
segmentation 185
selective linear prediction 146–151
sequential estimation methods 276
SIFT algorithm 206
  subroutines STEP1, STEP2 for 209–210
sign parameters 121
SLP selective linear prediction subroutine 150
sonograms 3
speaker identification 263
speaker verification 263
spectra 238
spectra decomposition 142
spectral flatness 18, 139–146
spectral flatness measure $\Xi(\cdot)$ 140
STREAK subroutine 198
spectral irregularities in voiced speech 193
spectral matching 18
spectral model 129
  zero mean property 130
  non-uniform weighting property 134–136
  minimax property 136–139
spectral resonances 129
spectrogram 3
speech production model 1, 6
speech recognition 263
speech spectrum, dynamic range of 214
speech synthesis structures 117
  stability of 93, 101
  numerator representation 108
  energy evaluation 109
  summary of relationship for 110–111
stability 111
  of acoustic tube 90–91
  of synthesis structures 93–102
steady state vowel 7
STEPDN subroutine 96
step-down procedure 93, 95–98
STEPUP subroutine 95
step-up procedure 93, 94, 110, 111
STREAK subroutine 197
synthesis filter structures
  direct form 118
  two-multiplier form 118
  Kelly-Lochbaum Model 121
  one-multiplier model 123
  normalized filter model 123

tap parameters 114
total squared error 13, 222, 133
transfer function-parameter evaluation 103
TWOMUL subroutine 120

unvoiced sounds 3
up-sampling 261

variable frame rate transmission 230, 235
vector spaces 44
VELP 261
vocal folds 1
vocal mechanism 2
vocal tract 2
  representations 62
  area function estimation algorithm 78–82
  losses 88–89
  area functions 79
  length estimation 276
  transfer function model 7, 9, 75
vocoder 17, 227
  pitch excited 227
  base-band excited 260
  real-time 227
  maximum likelihood 229, 246
  PARCOR 246
  real-time implementations 249, 253, 258
  autocorrelation method 249
  covariance method 255

quality issues 256, 258
voice excited linear prediction (VELP) 260
voice periodicity 188
voicing parameter 179
voiced sounds 1
voiced-unvoiced decision 255
voiceprint 3
volume velocity
glottal 2, 77
acoustic tube 63
at the lips 75

wave equation (one dimensional) 63
Webster horn equation 63
weighting matrix 236
Wiener filtering 18
windowing 157
word recognition 265

z-transform 6

Communication and Cybernetics

Editors: K. S. Fu, W. D. Keidel, H. Wolter

Vol. 1 W. Meyer-Eppler
Grundlagen und Anwendungen der Informationstheorie
2. Auflage, neu bearbeitet und erweitert von G. Heike und K. Löhn

Vol. 2 B. Malmberg
Structural Linguistics and Human Communication
An Introduction into the Mechanism of Language and the Methodology of Linguistics. 2nd revised edition

Vol. 3 J. L. Flanagan
Speech Analysis / Synthesis and Perception
2nd edition

Vol. 4 G. Herdan
The Advanced Theory of Language as Choice and Chance

Vol. 5 G. Hammarström
Linguistische Einheiten im Rahmen der modernen Sprachwissenschaft

Vol. 6 J. Peters
Einführung in die allgemeine Informationstheorie

Vol. 7 K. Weltner
The Measurement of Verbal Information in Psychology and Education
Translated from the German by B. M. Crook

Vol. 8 **Facts and Models in Hearing**
Proceedings of the Symposium on Psychophysical Models and Physiological Facts in Hearing, held at Tutzing, Oberbayern, Federal Republic of Germany, April 22–26, 1974. Edited by E. Zwicker, E. Terhardt

Vol. 9 G. Hammarström
Linguistic Units and Items

Springer-Verlag Berlin Heidelberg New York

Communication and Cybernetics

(continued)

Vol. 10 **Digital Pattern Recognition**
Edited by K. S. Fu. With contributions by T. M. Cover, E. Diday, K. S. Fu, A. Rosenfeld, J. C. Simon, T. J. Wagner, J. S. Weszka, J. J. Wolf

Vol. 11 **Structure and Process in Speech Perception**
Proceedings of the Symposium on Dynamic Aspects of Speech Perception, held at I.P.O., Eindhoven, Netherlands, August 4–6, 1975. Edited by A. Cohen und S. G. Nooteboom

Vol. 13 R. G. Busnel A. Classe
Whistled Languages

Topics in Applied Physics

Founded by H. K. V. Lotsch

Vol. 6 **Picture Processing and Digital Filtering**
Edited by T. S. Huang. With contributions by H. C. Andrews, F. C. Billingsley, J. G. Fiasconaro, B. R. Frieden, T. S. Huang, R. R. Read, J. L. Shanks, S. Treitel

Vol. 11 **Digital Picture Analysis**
Edited by A. Rosenfeld. With contributions by S.J. Dwyer, R.M. Haralick, C.A. Harlow, G. Lodwick, R.L. McIlwain, K. Preston, A. Rosenfeld, J.R. Ullmann

Springer-Verlag Berlin Heidelberg New York